LIBRARY

ASTRONOMY AND ASTROPHYSICS LIBRARY

ASTRONOMY AND ASTROPHYSICS LIBRARY

The Stars By E. L. Schatzman and F. Praderie

Modern Astrometry 2nd Edition
By J. Kovalevsky

The Physics and Dynamics of Planetary Nebulae By G. A. Gurzadyan

Galaxies and Cosmology By F. Combes, P. Boissé, A. Mazure and A. Blanchard

Observational Astrophysics 2nd Edition
By P. Léna, F. Lebrun and F. Mignard

Physics of Planetary Rings Celestial Mechanics of Continuous Media
By A. M. Fridman and N. N. Gorkavyi

Tools of Radio Astronomy 4th Edition, Corr. 2nd printing
By K. Rohlfs and T. L. Wilson

Tools of Radio Astronomy Problems and Solutions 1st Edition, Corr. 2nd printing
By T. L. Wilson and S. Hüttemeister

Astrophysical Formulae 3rd Edition (2 volumes)
Volume I: Radiation, Gas Processes and High Energy Astrophysics
Volume II: Space, Time, Matter and Cosmology
By K. R. Lang

Galaxy Formation By M. S. Longair

Astrophysical Concepts 4th Edition
By M. Harwit forthcoming

Astrometry of Fundamental Catalogues The Evolution from Optical to Radio Reference Frames
By H. G. Walter and O. J. Sovers

Compact Stars. Nuclear Physics, Particle Physics and General Relativity 2nd Edition
By N. K. Glendenning

The Sun from Space By K. R. Lang

Stellar Physics (2 volumes)
Volume 1: Fundamental Concepts and Stellar Equilibrium
By G. S. Bisnovatyi-Kogan

Stellar Physics (2 volumes)
Volume 2: Stellar Evolution and Stability
By G. S. Bisnovatyi-Kogan

Theory of Orbits (2 volumes)
Volume 1: Integrable Systems and Non-perturbative Methods
Volume 2: Perturbative and Geometrical Methods
By D. Boccaletti and G. Pucacco

Black Hole Gravitohydromagnetics
By B. Punsly

Stellar Structure and Evolution
By R. Kippenhahn and A. Weigert

Gravitational Lenses By P. Schneider, J. Ehlers and E. E. Falco

Reflecting Telescope Optics (2 volumes)
Volume I: Basic Design Theory and its Historical Development. 2nd Edition
Volume II: Manufacture, Testing, Alignment, Modern Techniques
By R. N. Wilson

Interplanetary Dust
By E. Grün, B. Å. S. Gustafson, S. Dermott and H. Fechtig (Eds.)

The Universe in Gamma Rays
By V. Schönfelder

Astrophysics. A New Approach 2nd Edition
By W. Kundt

Cosmic Ray Astrophysics
By R. Schlickeiser

Astrophysics of the Diffuse Universe
By M. A. Dopita and R. S. Sutherland

The Sun An Introduction. 2nd Edition
By M. Stix

Order and Chaos in Dynamical Astronomy
By G. J. Contopoulos

Astronomical Image and Data Analysis 2nd Edition
By J.-L. Starck and F. Murtagh

The Early Universe Facts and Fiction 4th Edition By G. Börner

Continued after Index

Paul Lorrain
François Lorrain
Stéphane Houle

Magneto-Fluid Dynamics

Fundamentals and Case Studies of Natural Phenomena

Springer

Paul Lorrain died peacefully on June 29, 2006. After his retirement from the Département de physique of the Université de Montréal, he worked many years in the Department of Earth and Planetary Sciences at McGill University (also in Montréal), where he did most of the research that led to this book.

François Lorrain
Département de mathématiques et d'informatique
Collège Jean-de-Brébeuf
3200, chemin de la Côte-Sainte-Catherine
Montréal (Québec), Canada H3T 1C1
florrain@brebeuf.qc.ca

Stéphane Houle
Département de physique
Collège Jean-de-Brébeuf
3200, chemin de la Côte-Sainte-Catherine
Montréal (Québec)
Canada H3T 1C1
Stephane.Houle@brebeuf.qc.ca

Cover illustration: Solar spicules near the limb of the Sun – the luminous jets of plasma can be as much as 10 000 kilometers long. See Chapter 14.

Library of Congress Control Number: 2006923636

ISBN 10: 0-387-33542-0
ISBN 13: 978-0387-33542-1

Printed on acid-free paper.

Printed in the United States of America. (TB/MVY)

9 8 7 6 5 4 3 2 1

springer.com

Contents

Part IV Natural Dynamos

List of Figures

Preface

Magnetohydrodynamics (MHD) concerns the interaction between magnetic fields and conducting fluids. We are concerned here with macroscopic interactions and, when the conducting fluid is a plasma, time scales are very much longer than the plasma period. Plasma periods vary widely, but are short, say 10^{-10} second.

We prefer the term Magneto-*Fluid*-Dynamics (MFD) because the discipline concerns mostly plasmas, various liquid conductors, and the liquid part of the Earth's core. It seems that the only applications of MFD to water are the induction of electric currents in the oceans by the Earth's magnetic field, and ship propulsion. But even MFD is not quite appropriate because that term also includes solid conductors that move in magnetic fields.

This book is meant for graduate and upper-division undergraduate students in Physics, Geophysics, and Astrophysics, as well as for practicing scientists in these fields.

This book is no more than a brief *introduction* to MFD because this vast subject is closely related to many others, namely Astrophysics, Electrodynamics, Fluid Dynamics, Geophysics, Oceanography, Plasma Physics, Thermonuclear Fusion, etc. We sketch the fundamentals, and provide many Examples, as well as Case Studies related to natural phenomena.

MFD sorely needs a rethink: it must of course be totally compatible with Physics. On the contrary, it is the custom to discuss the shapes of imaginary magnetic field lines, without ever referring to the required electric currents. This fundamental error leads to much nonsense. Of course, a given magnetic field $\boldsymbol{B}$ and its specific current distribution, of current density $\boldsymbol{J}$, are not distinct quantities: the magnetic field is just one property of the current distribution. Moreover, because of Ohm's law for moving conductors (Chapter 6), $\boldsymbol{J}$ is a function of $\boldsymbol{B}$! Indeed, the $\boldsymbol{J}$ that generates the $\boldsymbol{B}$ field depends on $\boldsymbol{B}$ in three different ways. Even the electric potential V is a function of $\boldsymbol{v} \times \boldsymbol{B}$, as in Chapter 7.

Here, we focus our attention on the current density $\boldsymbol{J}$. Our approach leads to new results, and disproves a number of misconceptions.

Hannes Alfvén, who was possibly the most prominent figure of Magneto-Fluid-Dynamics in the twentieth century, often stressed the importance of considering electric currents in MFD phenomena. Geophysicists and Astrophysicists should re-read Alfvén:

> *[In pseudo-MHD,] the electric current is traditionally eliminated, so that it does not appear explicitly, [so that] the formalism is not suited for describing phenomena associated with currents [...] Most serious is that the transfer of energy cannot be described in an adequate way [...] The "field line reconnection" concept is misleading [...] Frozen-in picture is often completely misleading [...] The transfer of energy [... is] lost. This has disastrous consequences [...] a better understanding [...] is possible if we derive the current system [...] Forgetting the "frozen-in" picture [...] We have no use for "magnetic field line reconnection" [...] this monstrous concept is a product of the frozen-in picture* in absurdum *[... etc.]*
> (Alfvén, 1943, 1975, 1981; Alfvén and Fälthammar, 1963)

Unfortunately, Alfvén's admonitions have gone unheeded.

Also, in our approach to MFD we take into account the electric field $-\nabla V$ of the space, or volume, charges *inside* conductors that move in magnetic fields. Although this electric field is hugely important, it is invariably ignored. That field is important because it opposes, and even cancels, the more prominent $\boldsymbol{v} \times \boldsymbol{B}$ field. Disregarding this space charge leads to absurd situations.

Finally, our approach is different in that we calculate numerical values, whenever possible.

This book comprises 16 chapters, each one preceded by its Table of Contents and a short introductory text, and followed by a Summary. Chapters 2 to 8 provide many Examples. Chapters 9 to 16 are Case Studies. Some of these Case Studies concern thought experiments, while the others concern natural magnetic fields. Two Appendices follow.

The book comprises five parts. We sketch here the overall structure; for further details, see "Looking Ahead", after this Preface.

There is a certain amount of repetition in the Case Studies because they are not meant to be read in precisely the same order as they are in the book.

Part I, The Early History (Chapter 1), is a brief historical introduction.

Part II, Fundamentals (Chapters 2 to 5), sketches the fundamentals of electromagnetic fields. These chapters are meant as a review, but *it is essential to master this material thoroughly before proceeding further.* The Examples relate largely to the material in the chapters that follow.

Readers who wish to delve more deeply into the basic principles of electromagnetic fields can refer to three other books by the undersigned and colleagues, all published by W.H. Freeman, New York: *Electromagnetic Fields and Waves* (third edition 1988), *Electromagnetism: Principles and*

Applications (second edition 1990), and *Fundamentals of Electromagnetic Phenomena* (2000). All three books provide many numerical examples and problems. The first one is also published in Arabic, Chinese, French, German, Portuguese, and Spanish; the second is also published in Chinese and in Korean, while the third is also published in French. Those foreign editions could be useful for readers who are not too familiar with English.

Part III, Moving Conductors (Chapters 6 to 9), explores the electric and magnetic phenomena associated with conductors that move in magnetic fields. *These chapters are the most important in the book.*

Chapter 8 provides examples related to Chapters 6 and 7.

Chapter 9 is a Case Study of the azimuthal magnetic field induced by the rotation of the Earth's core.

In **Part IV, Natural dynamos** (Chapters 10 to 14), we explore the problem of the *generation* of natural magnetic fields through self-excited dynamos. The disk dynamo of Chapter 10 is the classical model.

Chapter 11 discusses the generation of local electric currents in natural magnetic flux tubes and flux ropes, including an interesting hypothesis: that these tubes and ropes might act as light guides, like gigantic optical fibers.

In Chapters 12 and 13 we investigate solar magnetic elements and sunspots as magnetic flux tubes. Then, in Chapter 14, we propose a self-excited dynamo that possibly generates proton beams that we observe as solar spicules. Chapters 15 and 16 concern solar coronal loops. Chapter 16 discusses at length the channeling of broad proton beams in the solar atmosphere.

Essentially all the material that follows the first five introductory chapters has been published in various journals and is now rewritten, coordinated, and much augmented. References to the original papers are given. These papers were often written with collaborators, enumerated below.

Warning! In our Case Studies we propose models for the phenomena investigated. These models are at best plausible; they are in no way definitive! Hopefully, readers will find alternative, superior, models. Our aim is to spur further investigation.

In **Part V, Appendices**, Appendix A discusses characteristic lengths and times, a technique that is commonly used for evaluating the order of magnitude of variables. We show that these concepts are quite subtle, even though they seem to be no more than common sense.

The book ends with an extensive list of references and an index.

I am grateful to the many people who made this book possible. My thanks go first to my wife who tolerates my addiction to physical phenomena and to MFD. It is thanks to David Crosley that I first became interested in the Earth's dynamo, many years ago. That led me into MFD and into some of its many ramifications.

My thanks go first to Anthony Peratt of the Los Alamos National Laboratory for his encouragement and to Professors A.E. Williams-Jones and Al Mucci, successively chairpersons of the Department of Earth and Planetary Sciences of McGill University, where most of this book was written.

Many thanks to Professors Ernesto Martín and José Margineda of the Departamento de Física of the Universidad de Murcia, Spain, to Jean-Louis Lemouël, head of the Institut de physique du Globe de Paris, and to Professor Antonio Castellanos of the Departamento de Física of the Universidad de Sevilla, Spain, for their hospitality.

Many thanks also to the heads of the Physics or Electrical Engineering Departments of the following Universities in the People's Republic of China, again for their hospitality: Nankai University in Tianjin (Tientsin), Qing Hua University in Beijing (Peking), Anhui University in Hefei, the University of the Science Academy, also in Hefei, and Zongshan University in Guangzhou (Canton).

I am particularly grateful to Serge Koutchmy of the Institut d'astrophysique de Paris, with whom I have had innumerable discussions over many years, both in Paris and through e-mail. The chapters related to the Sun mostly concern papers written in close collaboration with him. He is the one who kindled my interest in the Sun, through one of his papers on magnetic elements (Chapter 12). All the images of the Sun that will be found here, save one (Fig. 13.1), were supplied by him. Serge Koutchmy observed the Sun over many years at the Sacramento Peak Solar Observatory, and then through the Web. He has also organized innumerable expeditions all over the world to observe the solar corona during eclipses.

Many thanks to the other coauthors of the papers on which the Case Studies are based:

Olga Koutchmy of the Université Marie et Pierre Curie in Paris,
James McTavish, of the Liverpool John Moores University,
Dragan Redžić of the Physics Faculty of the University of Belgrade, and
Nikos Salingaros of the University of Texas at San Antonio.

Jean-Pierre Drucbert of the Office national d'études et de recherches aérospatiales of Toulouse, France, provided software and much advice for generating the chapter tables of contents.

I am indebted to Paul LaFrance, who checked the first few chapters of the manuscript. Many thanks to Joseph Miskin, who supplied the .eps files for all the figures, and to François Girard who also worked many hours on the figures.

Acknowledgment

This work was subsidized in part by the NATO Cooperative Research Grant 940291.

Paul Lorrain
Département de physique, Université de Montréal,
and Department of Earth and Planetary Sciences, McGill University,
Montréal (Québec), Canada

Looking Ahead

Most of this book consists of Examples and Case Studies. There is no need to run through all of them, but it is important to first study carefully Chapters 2 to 5.

Part I. The Early History

Chapter 1, page 3

The story unfolds over two and one half millennia, from −600 to the 1920's. The pace was accelerating all the while. In 1908, Hale observed magnetic fields in sunspots. Then Gouy, and then Larmor, realized that self-excited dynamos can exist in convecting conducting fluids such as the solar plasma: they had founded Magneto-Fluid-Dynamics. Focus on Heaviside, who played a major role in developing the so-called Maxwellian theory of electromagnetism.

Part II. Fundamentals

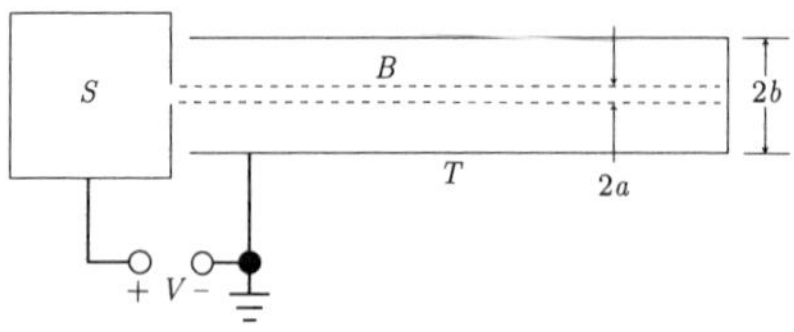

Chapters 2 to 5, pages 19 to 63

The Maxwell equations are the four keys to all electromagnetic phenomena. All about $\boldsymbol{E}$ and $\boldsymbol{B}$, and the many ways in which they are related.[1]

Part III. Moving Conductors

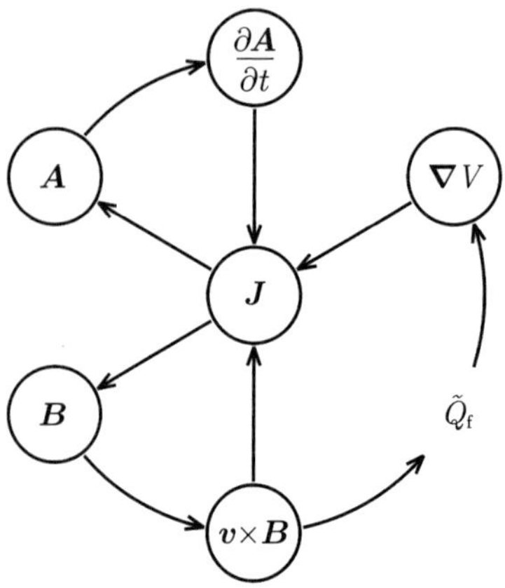

Chapter 6. Ohm's Law for Moving Conductors, page 73

This is where we enter Magneto-Fluid-Dynamics proper. We need a bit of Relativity here because we occasionally have to think in terms of two reference frames, one that is fixed, and another one that moves with the conductor. An observer in one frame does not see the same things as an observer in the other frame, so we need a few simple transformation equations. Simple, because fluid speeds are so much smaller than the speed of light. Ohm's law now needs a new term, namely $\boldsymbol{v} \times \boldsymbol{B}$, that adds to $\boldsymbol{E}$.

[1] Don't just say "Oh I know all this." For three reasons. First, Nature is infinitely complex and there is no end to its subtleties; you keep stumbling on new questions, and you are never through understanding. Second, the Magneto-Fluid-Dynamics literature contains several fundamental errors related to the Maxwell equations: there is a lot of pseudo-Physics out there! See the quotations from Alfvén in the Preface. Third, there is a world of difference between knowing an equation or a principle, and applying it correctly in a specific case. At the very least, study the Examples given in these four chapters.

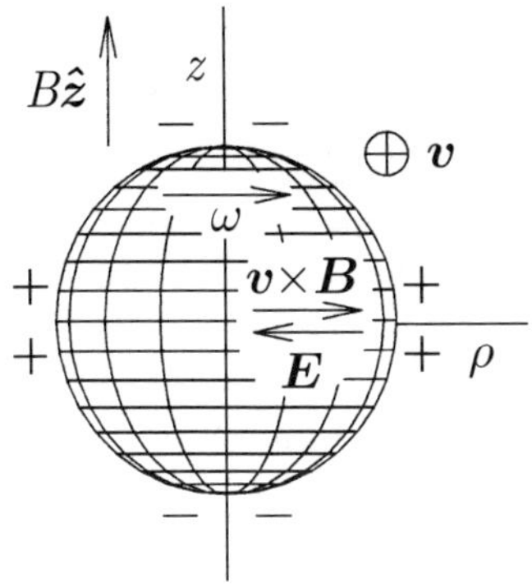

Chapter 7. Charges Inside Moving Conductors, page 87

When a conductor moves at a velocity $\boldsymbol{v}$ in a magnetic field $\boldsymbol{B}$, it usually carries a space, or volume, charge *inside* it that is a simple function of the product $\boldsymbol{v}\times\boldsymbol{B}$. The electric field of this space charge is always ignored in the literature, despite the fact that it opposes, and even cancels, $\boldsymbol{v}\times\boldsymbol{B}$. The Faraday disk and the rotating sphere serve as Examples.

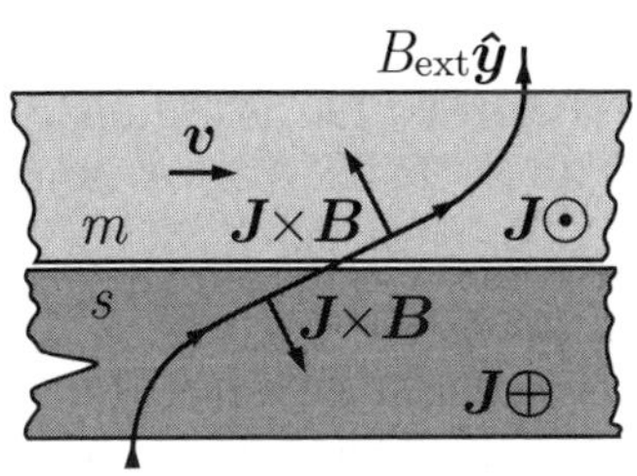

Chapter 8. Magnetic Fields in Moving Conductors, page 103

What happens to a magnetic field if you move a conducting body in it? As a rule, currents are induced in the conductor because of the $\boldsymbol{E}+(\boldsymbol{v}\times\boldsymbol{B})$ field, and the net magnetic field is the sum of the applied and induced fields. However, in some cases, the moving conductor has no effect at all on the magnetic field. There are nine Examples, here.

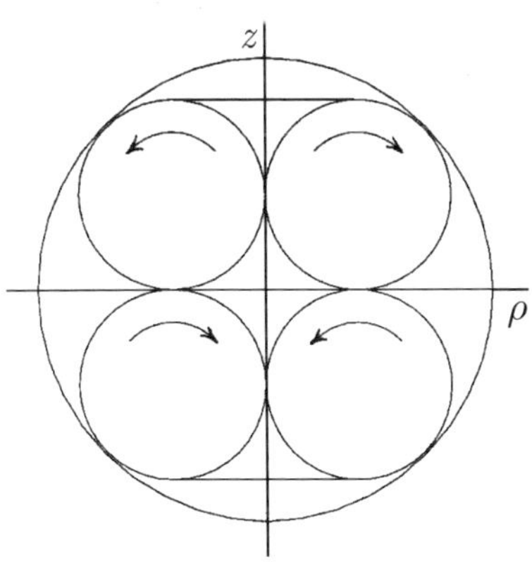

Chapter 9. Case Study: The Azimuthal Magnetic Field in the Earth's Core, page 125

Does the Earth's magnetic field rotate? Good question! The answer is both Yes, and No: part of the field rotates, and the other part does not. It turns out that the Earth's rotation generates an azimuthal magnetic field in the core. We investigate the case where the core rotates as a solid, and then the case where there is differential rotation.

Part IV. Natural Dynamos

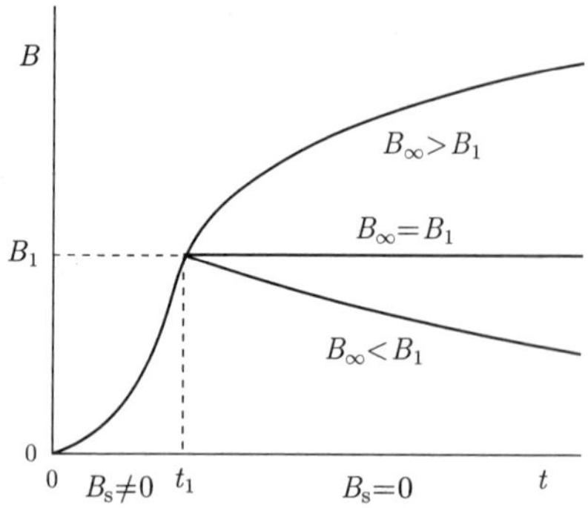

Chapter 10. The Disk Dynamo Model, page 147

The self-excited disk dynamo is the classical model for natural dynamos. The model is instructive, despite the fact that it bears little relation to natural dynamos.

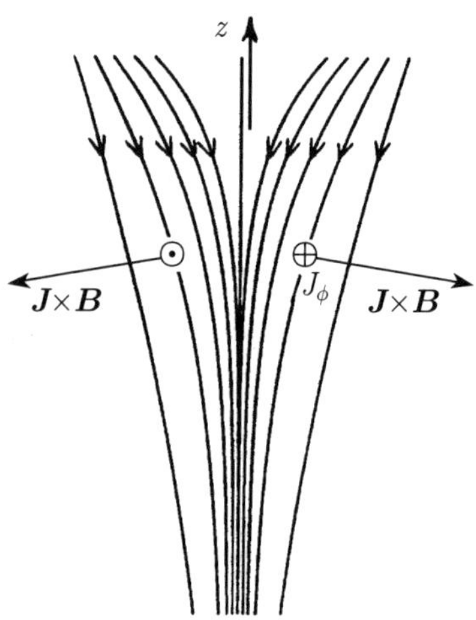

Chapter 11. Magnetic Flux Tubes, Flux Ropes, and Flux Coils, page 159

A magnetic flux *tube* is a bundle of magnetic field lines, as inside a solenoid. The configuration is common in convecting, conducting fluids, like the solar plasma, where they are self-excited dynamos. In a flux *rope* the magnetic field lines are helical. We investigate the hypothesis that both tubes and ropes might act as light guides, like optical fibers. In a flux *coil*, the lines are azimuthal.

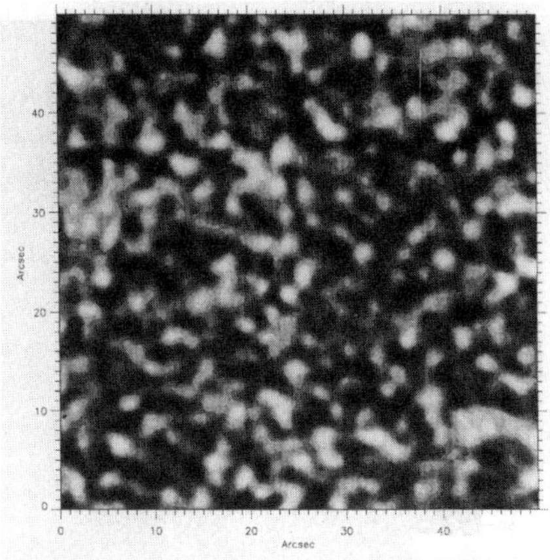

Chapter 12. Case Study: Solar Magnetic Elements, page 189

Solar magnetic elements are small bright spots at the surface of the Sun. They are roughly vertical magnetic flux tubes, presumably the result of downwelling in the solar plasma. In downwelling, the plasma moves both downward and inward, and the inward motion of the plasma in the magnetic field of the flux tube generates an azimuthal electric current that adds to any existing axial magnetic field: the flux tube is a self-excited dynamo. Solar magnetic elements are bright, we hypothesize, because the flux tube acts as a light guide and permits us to see deep into the photosphere, where the temperature is higher.

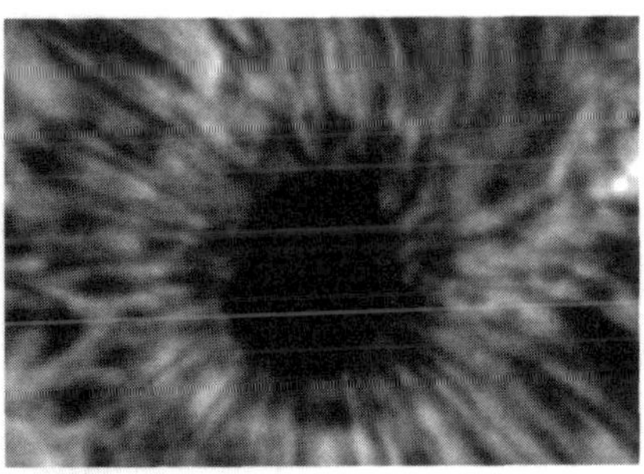

Chapter 13. Case Study: Sunspots, page 205

Sunspots are again more or less vertical magnetic flux tubes but, contrary to magnetic elements, they are dark, and not bright. We investigate the current density, the magnetic pressure, and the gas density. We discuss briefly the Wilson depression, where sunspots are situated.

Chapter 14 Case Study: Solar Spicules, page 223

Solar spicules are luminous jets that emerge from the surface of the Sun. They are somewhat similar to coronal loops (Chapters 15 and 16), in that they are presumably proton beams, but shorter by orders of magnitude. Spicules are presumably generated in particle accelerators at their base, in the solar plasma.

Chapters 15 and 16. Case Study: Solar Coronal Loops as Proton Beams, I and II, pages 239 and 253

In these two chapters we attempt to explain three incredible phenomena related to coronal loops. 1. Loops have a uniform diameter over hundreds of megameters, despite orders of magnitude changes in the ambient magnetic flux density and gas density. 2. Loops follow magnetic field lines, but only up to a point: they neither fan out nor pinch in a non-uniform field. 3. Loops do not interact. At times one sees dozens of loops that emerge from a single sunspot, and that do not interact at all.

Part V. Appendices

Appendix A. Characteristic Lengths and Times, a Justification, page 273
It is often useful, in Magneto-Fluid-Dynamics, to evaluate the relative orders of magnitude of two related quantities. This involves *characteristic* lengths $\mathcal{L}$ and times $\mathcal{T}$. For example, in discussing the magnetic field of the Earth, one might use the radius of the Earth as a characteristic length. If one has to deal with the divergence or the curl of a certain field $\boldsymbol{F}$, and if one is only concerned with orders of magnitude, it is the custom to substitute $F/\mathcal{L}$ for $|\boldsymbol{\nabla} \cdot \boldsymbol{F}|$ or for $|\boldsymbol{\nabla} \times \boldsymbol{F}|$. Although this seems to make sense, things are much more complex than that. Indeed, applying this procedure to $\boldsymbol{\nabla} \cdot \boldsymbol{B} = 0$ yields $B = 0$! We discuss the proper use of characteristic lengths and times.

List of Symbols

Space, Time, Mechanics

Length	l, L, s, r
Characteristic length	$\mathcal{L}$
Area	$\mathcal{A}$
Volume	v
Solid angle	Ω
Unit vector along $\boldsymbol{r}$	$\hat{\boldsymbol{r}}$
Unit vectors in Cartesian coordinates	$\hat{\boldsymbol{x}}$, $\hat{\boldsymbol{y}}$, $\hat{\boldsymbol{z}}$
Unit vectors in cylindrical coordinates	$\hat{\boldsymbol{\rho}}$, $\hat{\boldsymbol{\phi}}$, $\hat{\boldsymbol{z}}$
Unit vectors in spherical coordinates	$\hat{\boldsymbol{r}}$, $\hat{\boldsymbol{\theta}}$, $\hat{\boldsymbol{\phi}}$
Unit vector normal to a surface	$\hat{\boldsymbol{n}}$
Wavelength	λ
Time	t
Characteristic time	$\mathcal{T}$
Period	$T = 1/f$
Frequency	$f = 1/T$
Angular frequency	$\omega = 2\pi f$
Angular velocity	ω
Velocity, speed	$\boldsymbol{v}$, v
Mass	m
Mass per unit volume	$\widetilde{m}$
Force	$\boldsymbol{F}$
Force per unit volume	$\widetilde{\boldsymbol{F}}$
Pressure	p
Energy	$\mathcal{E}$
Energy per unit volume	$\widetilde{\mathcal{E}}$
Power	P
Power per unit volume	$\widetilde{P}$

Electricity and Magnetism

Speed of light	c
Electronic charge	e
Boltzmann constant	k
Planck's constant	h
Electric charge	Q
Electric charge per unit volume	$\widetilde{Q}$
Electric charge per unit area	$\widetilde{S}$
Electric potential	V
Electric field strength	$\boldsymbol{E}$
Permittivity of vacuum	ϵ_0
Relative permittivity	ϵ_{r}
Electric current	I
Electric current per unit area	$\boldsymbol{J}$
Conductivity	σ
Mobility	$\mathcal{M}$
Resistance	R
Capacitance	C
Vector potential	$\boldsymbol{A}$
Magnetic flux density	$\boldsymbol{B}$
Magnetic flux	Φ
Permeability of vacuum	μ_0
Relative permeability	μ_{r}
Self-inductance	L

Mathematical Symbols

Approximately equal to	$\approx$
Of the same order of magnitude as	$\sim$
Of the same or a smaller order of magnitude	$\lesssim$
Proportional to	$\propto$
Vector	$\boldsymbol{F}$
Magnitude of $\boldsymbol{F}$	F
Gradient of V	$\boldsymbol{\nabla} V$
Divergence of $\boldsymbol{E}$	$\boldsymbol{\nabla} \cdot \boldsymbol{E}$
Curl of $\boldsymbol{E}$	$\boldsymbol{\nabla} \times \boldsymbol{E}$
Laplacian of V	$\boldsymbol{\nabla}^2 V \equiv \boldsymbol{\nabla} \cdot (\boldsymbol{\nabla} V)$

Part I

The Early History

1 The Early History

Contents

Our story covers two and one half millennia. It starts with Thales of Miletus who wrote in −600 about the mineral magnetite. It ends with Gouy and Larmor who proposed at the beginning of the 20th century the concept of self-excited dynamo to explain the existence of magnetic fields on the Sun. All this while, the pace was accelerating.

Mankind has labored over two and one half millennia to arrive at our present knowledge of natural magnetic fields. A striking aspect of the process, for an early 21st century scientist, is its accelerating pace. For example, the compass was discovered in China during the 3rd century B.C., but it came into use for navigation only fourteen centuries later, both in Europe and in China, more or less simultaneously. The science of natural magnetic fields provides a good illustration of the acceleration of history.

Magnetic fields in nature originate either in certain solids, or in convecting, conducting fluids. The story of magnetism starts with the magnetism of solids, and it is only at the beginning of the 20th century that physicists, astrophysicists, and geophysicists became aware of the fact that convecting, conducting, fluids can spontaneously generate magnetic fields, given a seed field. In this book we are concerned with the second type of magnetic field, but this first chapter concerns mostly the first type because that is where it all started.

The first step was the discovery of magnetite, Fe_3O_4, which is a fairly common mineral, and which is magnetic. At first a subject of curiosity because of its properties of attraction and repulsion, it later permitted the fabrication of compasses. Compasses were first used for divining purposes in China, and the orientation of compasses in the Earth's field remained a mystery until Pierre de Maricourt, in 1269, and then Gilbert, in 1600, found plausible, but wrong, explanations.

The Gouy and Larmor papers to which we refer below mark the beginning of Magneto-Fluid-Dynamics, over two millennia after the discovery of magnetism.

1.1 Magnetism in China

Magnetite was discovered in China in the course of the 3rd century B.C., about three centuries after it was discovered in Greece.[1] The first reference in Chinese literature to the orientation of a magnetic stone in the Earth's field is in the *Book of the Devil Valley Master* dated −295. The stone was called a *South pointer.* Of course the idea that North is "up" is arbitrary, and the Chinese considered instead that South is "up".

About the year −1000, the Chinese had invented a special kind of chariot, called a *chariot pointing South.* It was equipped with a set of gears that served to maintain its orientation, despite clouds of smoke emitted by an enemy; the mechanism had nothing to do with magnetism.

By the first century B.C., but possibly as early as the second century B.C., Chinese compasses had the shape of magnetite spoons that could rotate on a polished bronze plate and that were called *South-controlling spoons.* They were used for divination.

Magnetized needles were apparently first used in China in the sixth century A.D. In 1088 Shen Kua describes the needle compass in his *Essays from the Torrent of Dreams.* He describes how a needle can be magnetized, either by rubbing it on a magnetic stone, or by heating and then cooling it, while oriented in the North–South direction.

Magnetic declination is the angle formed by a compass needle with respect to the true North. It appears that magnetic declination was discovered in China between the 7th and the 10th centuries A.D. Shen Kua is familiar with magnetic declination, but he does not mention the use of the compass in navigation.

[1] References for this section about China: Haudricourt and Needham (1957, pp. 483–485); Hellemans and Bunch (1988, pp. 19, 37, 39, 75); Macdonald (1957, p. 1508); Needham and Ling (1954–1956, Vol. 1, p. 128, and Vol. 2, p. 576); Needham, Ling, and Robinson (1962, pp. 229–230, 232, 249–250, 279–280, 334); Sarton (1927–1948, Vol. 1, pp. 755–756, 764).

The first mention of the use of compasses for navigation in Chinese literature is to be found in *P'ingchow Table Talk* by Chu Yuk, dating from 1111–1117. The compass was used by Chinese sailors between Canton and Sumatra. The compass had the form of a small magnetized dish shaped like a fish that floated on water. The Chinese continued to use floating compasses until the 16th century.

It is probable that, by the end of the 11th century, the Chinese used compasses for drawing maps.

1.2 Magnetism in Europe

The word *magnetism* possibly comes from *Magnesia*, a region in the North–East of Greece, where magnetite was discovered. However, two cities of Asia Minor were also called Magnesia, one of which is now Manisa, in Lydia, close to the Western extremity of Turkey, North–East of Izmir, and the term *Lydian stone* that was used at the time possibly refers to that region.

It seems that Thales of Miletus (ca. −625 to ca. −547) was the first Greek author to mention magnets.[2] He is considered to be the founder of Natural Philosophy in the Western World; none of his texts have survived, but some of his work was reported by later authors. Aristotle (−384 to −322) refers to Thales's statements about magnetite. At that time there was no clear distinction between the electrostatic and the magnetostatic phenomena of attraction and repulsion.[3]

Several other Greek and Latin authors also discussed magnetic phenomena.[4] For example, the Elder Pliny (23–79), in his *Natural History*, relates that the architect Dinochares had undertaken the construction of the vault of a temple of the goddess Arsinoe in Alexandria with magnetite so that the statue of the goddess could be suspended from it.[5]

1.2.1 Navigation by Compass in Europe

The use of compasses for navigation is mentioned by several European authors before the end of the 12th century.[6] Alexander Neckham (1157–1217), in his *De Naturis Rerum* published around 1190–1199, mentions the use of the compass in navigation, without stressing its novelty. In fact the use of the

[2] Needham, Ling, and Robinson (1962, p. 231); Sarton (1927–1948, Vol. 1, pp. 65, 71, 72).

[3] Lenoble (1958, p. 325).

[4] See Brothers Arnold and Potamian (1904). The authors provide many references to Latin and medieval writings, with quotations.

[5] Pliny the Elder (1855, Chap. XXXIV, 137).

[6] Beaujouan (1957, p. 573); Hellemans and Bunch (1988, p. 79); Needham, Ling, and Robinson (1962, pp. 230, 279–280); Sarton (1927–1948, Vol. 2, pp. 24, 385, 589, 592–593, 671).

compass for navigation was perhaps secret, so that its use could have preceded its divulging by centuries.

As we saw above, the first reference to the use of the compass for navigation in Chinese literature is dated 1111–1117, while the first reference in European literature is dated 1190–1199. Does this imply that compasses were first used for navigation in Europe at about the same time as in China? One plausible, but unproven, explanation is that Muslim sailors, who traveled extensively between the Far East, India, Persia, Arabia, and Africa, spread the technology. The first reference in Muslim literature is dated about 1230, later than in Europe, but that is probably not meaningful.[7]

One interesting fact is that European *surveyors and astronomers* used compasses that showed the South direction, like all Chinese compasses, while European *sailors* used compasses that showed the North direction.[8] So the use of the compass for surveying was possibly transmitted overland from China to Europe by surveyors.

1.2.2 Pierre de Maricourt (Petrus Peregrinus)

Pierre de Maricourt was presumably born in Maricourt, in Picardy.[9] The dates of his birth and of his death are unknown. His surname *Peregrinus* (*The Pilgrim*) recalls the fact that he had visited the Holy Land in the course of the seventh crusade, led by Louis IX. His *Epistola de Magnete* (Letter on the Magnet) to a friend in Picardy, is probably the first treatise of experimental physical science.

Pierre de Maricourt was probably a military engineer. He wrote *Epistola* in 1269 during the siege of Lucera, in South–East Italy, by Charles I of Anjou, brother of Louis IX. The Letter was first published three centuries later, in 1558, about one century after the invention of the printing press.

He notes that a magnet has both a North and a South pole, that similar poles repel, while contrary poles attract, that a magnet broken into two parts gives two new magnets, each with its North and South poles, and that the original magnet can be reconstituted to form a single one. Moreover, he states that a compass moves because it is submitted to forces that act on its poles. He fabricated a spherical magnet and observed its field lines, noting that the magnetic force is largest near the poles. He describes in detail floating and pivoted compasses, as well as an instrument that could serve both as a compass and as an astrolabe.

Pierre de Maricourt does not mention magnetic declination but states, rather, that a compass shows, not the North star, but the true North of the

[7] Macdonald (1957, p. 1504); Needham, Ling, and Robinson (1962, pp. 229–230); Sarton (1927–1948, Vol. 2, p. 630).

[8] Needham, Ling, and Robinson (1962, p. 330).

[9] Arnold and Potamian (1904); Beaujouan (1957, p. 542); Hellemans and Bunch (1988, pp. 61, 81); Lenoble (1958, pp. 326–327); Merrill and McElhinny (1983, pp. v, 5); Sarton (1927–1948, Vol. 2, pp. 24, 630).

celestial sphere. He rejects the idea that there might be large iron deposits at the poles.

He was much admired by his contemporary and student Roger Bacon (1214–1280), but his work was unfortunately ignored until William Gilbert arrived on the scene over three centuries later.

1.2.3 From Gilbert on

The year 1600 saw the publication of *De Magnete* by William Gilbert (1540–1603), President of the Royal College of Physicians, and physician of Elizabeth the First. Gilbert attributes the orientation of a compass to the Earth itself.[10] He assumes the existence of a huge permanent magnet near the center, and that the planets are maintained on their orbits by magnetic forces. Gilbert was wrong in many ways, but the statement that the orientation of the compass was to be attributed to the Earth was a major step in our understanding of natural magnetic fields.

At the time, it was thought that the odor of garlic neutralized the magnetism of compasses, and sailors were strictly forbidden to eat it, but Gilbert showed experimentally that garlic had no such effect.

Magnetic *declination* was discovered in China between the 7th and the 10th centuries, and centuries later in Europe, at the beginning of the 15th century.[11] Christopher Columbus was probably the first to notice that magnetic declination varied from one point to another.[12] According to legend, his sailors were on the point of mutiny when the compass showed a declination of 10 degrees west but, during the night, Columbus turned the compass card.

Magnetic *inclination* is the angle between the Earth's magnetic field and the horizontal direction. It is said to have been discovered in Rome in 1544 by George Hartmann, a parish priest in Nuremberg.[13] Like Pierre de Maricourt, Hartmann described his discovery in a letter that was lost for three centuries; it was found in 1831.

The first laboratory dedicated to the observation of the Earth's magnetic field was that of Sir Nicholas Millet in London, between 1652 and 1670.[14] But it was George Graham (1674–1751) who discovered in 1722, in London, rapid fluctuations in the Earth's magnetic field by observing a compass under a microscope.[15] We know today that these fluctuations result from random

---

10 Fleming (1939, pp. 2–3); Malin (1989, pp. 7–8); Merrill and McElhinny (1983, pp. 6–7, 12). See also *Physics Today* (June 2000), p. 52. There exists an extensive site on Gilbert's life and work at www.spof.gsfc.nasa.gov/earthmag/demagint.htm

11 Beaujouan (1957, p. 573).

12 Fleming (1939, p. 2); Hellemans and Bunch (1988, p. 99).

13 Merrill and McElhinny (1983, p. 6).

14 Malin (1989, p. 35).

15 Malin (1989, p. 21–22); Merrill and McElhinny (1983, p. 11).

electric currents induced in the ionosphere by winds in the Earth's magnetic field.

1.3 Faraday

Fig. 1.1. Despite the fact that Michael Faraday (1791–1867) had no education, he became one of the greatest scientists of his time. He is the father of field theory, and he provided the experimental, and much of the theoretical, ideas that led Maxwell to develop electromagnetic theory. Faraday was forty years younger than Maxwell

Michael Faraday (1791–1867) was born in what is now South London.[16] His life was strikingly similar to that of Heaviside (Sect. 1.6). Both were born in the same district in very poor families. Neither had a University education: Heaviside had to leave school at sixteen, while Faraday learned to read, write, and cipher in his church's Sunday School! They had no other education. Both became among the greatest scientists of their time.

Faraday's father was a "Northerner" who migrated to the South of England in search for work. He was a blacksmith, but he was often ill and unable to work, with the result that the family just barely had enough to eat. At one

[16] Faraday (1852a,b,c, 1965); Gillispie (1970–1978).

time he was given a loaf of bread that had to last him for one week. Luckily, when he was fourteen, he was apprenticed to a book-binder, and started to read books that were brought in for re-binding.

Later, he attended a lecture at the Royal Institution given by Humphry Davy, one of the greatest chemists of his generation. He took notes and sent a copy to Davy, in the hope obtaining a job in Davy's laboratory. Some time later, Davy offered him employment and, in 1812, Faraday began a long series of experiments that revolutionized Chemistry. Then, in 1821, he settled at the Royal Institution, and undertook another long series of experiments that then revolutionized Physics.

In 1831 he started to work with Charles Wheatstone, one of Heaviside's uncles (Sec. 1.6), and invented the first electric motor, although in a crude and impractical form. Possibly his most important discovery was magnetic induction, which is now expressed as the Maxwell equation for the curl of $\boldsymbol{B}$. He discovered that a magnetic field can rotate the plane of polarization of light in glass. He then discovered that some materials tend to move closer to a magnet, while others tend to move away: he had discovered paramagnetism and diamagnetism. Faraday was obsessed by what we now call fields, a concept that is fundamental to the work of Maxwell and to all of present-day Physics. Faraday was thus the founder of field theory.

Later in life, Queen Victoria offered him a knighthood, but he refused, insisting that he would always be called Mr. Faraday. He died in 1867.

1.4 The Story of the "Maxwell" Equations

The story of the "Maxwell" equations is captivating.[17]

First, as we shall see, the "Maxwell" equations should be called, more appropriately, the *Heaviside* equations.

Second, the story covers less than the fourth quarter of the nineteenth century, despite the many fruitless attempts by *Maxwellians* and others at understanding electromagnetic phenomena.

Third, there were few Maxwellians, the main ones being James Clerk Maxwell (1831–1879), Oliver Heaviside (1850–1925), George Francis FitzGerald (1851–1901), Oliver Lodge (1851–1940), and Heinrich Hertz (1857–1894).

Fourth, these were all British, except for Hertz, who was German. Maxwell, who was Scottish, worked at Cambridge but, after his death, there were no Maxwellians either there or at Oxford.

Fifth, Heaviside, FitzGerald, and Lodge were all born in 1850 and 1851.

[17] Bork (1963); Domb (1979); Gillispie (1970–1978); Harman (1998); Heaviside (1950); Hunt (1991).

Fig. 1.2. James Clerk Maxwell (1831–1879) is best known for the four equations that bear his name, but his contributions to other aspects of Physics, and even to engineering, is highly impressive. He was Scottish, and much of his scientific writing was done in his large estate in southwest Scotland. He was the first professor of experimental Physics at Cambridge, and he planned and developed the Cavendish Laboratory

1.5 Maxwell

James Clerk Maxwell was born in Edinburgh, Scotland, in 1831.[18] His formulation of electromagnetic theory was published in 1873,[19] six years before his premature death at the age of 48. The Maxwell theory was later reformulated extensively by several people,[20] in particular by Oliver Heaviside. See below.

Maxwell's contribution to Physics, and particularly to Electromagnetic Theory, was immense. He distinguished between $\boldsymbol{B}$ and $\boldsymbol{H}$, between $\boldsymbol{J}$ and $\boldsymbol{E}$, and between "intensities" and "fluxes". He introduced the terms "gradient", "divergence", and "curl" in Cartesian coordinates, and wrote that $\boldsymbol{E} = -\partial \boldsymbol{A}/\partial t$ in a changing magnetic field.

[18] For the origin of the name Maxwell, see Gillispie (1970–1978).
[19] Maxwell (1873).
[20] Hunt (1991).

But Maxwell's *Treatise on Electricity and Magnetism*, first published in 1873, is so disorganized that it is unreadable. He explained later that "the aim of the *Treatise* was not to expound his theory finally to the world but to educate himself by presenting a view of the stage he had reached." He did not live long enough to reformulate it properly. It is a wonder that the Maxwellians managed to master it. Maxwell had great admiration for Faraday, and he was attempting to put Faraday's field concept into mathematical form.

The *Treatise* does *not* state the "Maxwell" equations, because Maxwell reasoned mostly in terms of the potentials V (then called ψ) and $\boldsymbol{A}$. The shift from V and $\boldsymbol{A}$ to $\boldsymbol{E}$ and $\boldsymbol{B}$, initiated by Heaviside, caused much debate about "the murder of ψ".

The Maxwellians faced enormous problems. The concept of field used by Faraday was thought to be particularly abstruse. Instantaneous action at a distance, and the ether, were primary preoccupations. The Maxwellians devised various *mechanical* models for the ether, with vortices, wheels, rubber bands, cogwheels, and racks, which are all laughable today, but which were desperate attempts at understanding what were deeply mysterious phenomena. It is this context that Heaviside worked.

1.6 Heaviside

Since Heaviside (1850–1925) is little known, compared to Maxwell, it is appropriate to recall his work here.[21] He was born in 1850 "in a very mean street in London", where he lived thirteen years "in miserable poverty", as he later wrote. He left school when he was sixteen and, two years later, in 1868, started work as telegraph operator of the Anglo-Danish submarine cable in Newcastle. One of his uncles was Sir Charles Wheatstone, the inventor of the Wheatstone bridge and the father of telegraphy. Recall that Faraday worked with Wheatstone, starting in 1831. So Heaviside possibly knew Faraday, despite the fact that the latter was 59 years older. Heaviside remained in Newcastle until 1874, when he returned to live with his parents. He had no real job for the rest of his life.

Heaviside was a prolific and colorful writer, endowed with a biting wit. He battled all his life against the scientific and mathematical establishments, despite the fact that he had a poor health and lived as a recluse, partly because of his deafness, and partly because of his poverty. His texts are crystal clear, contrary to those of Maxwell, for whom he had immense admiration. For many years Heaviside published his theories in *The Electrician*, which was a trade journal. For this he was paid 40£ per year, which was less than the salary of a laborer.

---

[21] Gillispie (1970–1978).

Fig. 1.3. Oliver Heaviside (1850–1925) reformulated the Maxwell theory in the form of the present-day vector equations that bear the name of Maxwell. An outstanding physicist and applied mathematician, he battled all his life against the scientific and mathematical establishments of his time. Like Faraday, it is probably because he was self-taught that he was so much ahead of his contemporaries

Although he was self-taught and worked on the fringes of the scientific community, he became possibly the best physicist–mathematician of his generation. It is interesting to reflect upon the fact that two of the greatest British physicists of all time, Faraday and Heaviside, had no University education at all. Both were geniuses who were a long way ahead of their contemporaries, and a university education would have spoilt their creativity.

Heaviside focused on the electromagnetic *field*, like Faraday and Maxwell. He finally recast the Maxwell theory in the form of the four "Maxwell" vector equations in 1884, the same year that Hertz stated them in Cartesian form.[22] Hertz admitted that Heaviside had the priority, in view of the fact that his own "proof" was "very shaky". Clearly, the "Maxwell" equations should rather be called the *Heaviside* equations.

Modern Physics owes much to Heaviside. It has been said that "The scope and amount of his work within the field of electromagnetism is *staggering.*"

[22] Gillispie (1970–1978) attributes the discovery of the vector formulation to Lorentz (1853–1928) in 1895, eleven years after Heaviside.

He introduced the use of rationalized units[23] in electromagnetic theory. Heaviside also discovered the "Poynting" theorem[24] in 1884, in the same year as Poynting (1852–1914).

The Kennelly–Heaviside layer, which reflects radio waves in the ionosphere, was one of his ideas. He also discovered the "Lorentz" force (Sect. 6.3) in 1880–1881, fifteen years before Lorentz. He foresaw superconductivity, discussed the skin effect, and founded transmission-line theory. He coined the word "waveguide" and realized that an electromagnetic wave travels *outside* wires. He also reasoned that cathode rays are streams of electrons.

As early as 1888 he calculated the $\boldsymbol{E}$ and $\boldsymbol{B}$ fields of a moving charge, and the longitudinal contraction that involves the factor γ of relativity! This phenomenon was later called the "FitzGerald" contraction.

Heaviside's numerous papers on Electromagnetic Theory were later edited by Ernst Weber.[25]

Heaviside's contributions to Mathematics are equally impressive. He introduced in 1882 the use of vector analysis as we know it today, independently of J.W. Gibbs (1839–1903) in the United States. Sixty years later, electromagnetism was still routinely taught without vectors. Heaviside also invented the impulse function and Operational Calculus, which both bear his name. Operational Calculus created such an uproar in the mathematical establishment that the Royal Society stopped publishing his papers on the subject in its Proceedings.

With time, Heaviside achieved recognition and became a Fellow of the Royal Society in 1891, at the age of 41. From about that time on, he never strayed more than a few miles away from his home in Paington. Starting in 1896, thanks to some friends, he was paid a Civil List Pension of 120£ per year. He received many honors in the early 1900's, and eventually died miserably poor in a nursing home in 1925 at the age of 75.

1.7 Zeeman, Hale, Gouy, and Larmor

During an eclipse in 1889, Bigelow (1889) observed rays diverging from opposite ends of the Sun, as if it were a spherical magnet. He concluded correctly that the Sun has a magnetic field.

---

[23] With unrationalized units, Coulomb's law is written $F = Q_1 Q_2 / r^2$, without the factor of $1/(4\pi\epsilon_0)$. Then the coulomb is a *mechanical* quantity whose dimensions are $\mathrm{M}^{1/2}\mathrm{L}^{3/2}/\mathrm{T}$! That is bad enough; indeed, it is going back to early 19th-century Physics. But also, the fact of "simplifying" Coulomb's law, which is *never* used, complicates most of the useful equations by adding factors of either 4π or $1/(4\pi)$. Some authors still use unrationalized units (esu, emu, Gaussian, modified Gaussian, Heaviside–Lorentz, etc.), which were abandoned officially back in 1939.

[24] Heaviside (1950), Sect. 22.6.

[25] See Heaviside (1950).

A few years later, Peter Zeeman (1865–1943) reported that the D-line in the spectrum of sodium vapor becomes distinctly wider when exposed to the field of an electromagnet.[26] He also reported that the radiation at the edges of the line is polarized. He observed the same phenomenon with the red lines of lithium. Michael Faraday had attempted the same experiment earlier, in 1862, but without success. That was Faraday's last experiment.

The Zeeman paper was first published in the Proceedings of the Academy of Amsterdam, then in the Philosophical Magazine and, two months later, in the Astrophysical Journal. The paper attracted much attention among spectroscopists and, eleven years later, Hale (1868–1938) observed magnetic fields in sunspots.[27] These fields are as intense as those of a powerful permanent magnet, a few tenths of a tesla. Thanks to the Zeeman effect, it had become possible to measure magnetic fields in celestial bodies.

Ever since Bigelow's observations it was suspected that there were magnetic fields on the Sun.

Magnetic fields were well-known at the time, since Maxwell's treatise had been published 35 years earlier. But natural magnetic fields were associated with magnetic ores. How could an incandescent gas like that of the Sun's photosphere generate a magnetic field?

Four years later, in 1912, Gouy (1854–1926) stated that whirling motions in the Sun generate intense magnetic fields, starting with a seed field.[28] He did not elaborate, but wrote that he was planning experiments.

Then, seven more years later, Larmor (1857–1942) elaborated on Gouy's proposal.[29] In his paper, Larmor first rejects magnetic materials as the source of the solar magnetic field, because of the high temperature. He then rejects both electric polarization and the convection of electric charge, because the electric fields would be too large. He then spells out the fundamental principle of Magneto-Fluid-Dynamics, all in about 700 words. He refers neither to Bigelow, nor to Gouy, nor to Hale.

He writes:

> In the case of the Sun [...] internal motion induces an electric field [...] and an electric current will flow [...], which may in turn increase the inducing magnetic field. In this way, it is possible for the internal cyclic motion to act after the manner of the cycle of a *self-exciting dynamo*, and maintain a permanent magnetic field from insignificant beginnings, at the expense of some of the energy of the internal circulation. [Our emphasis.]

That is the principle of operation of natural dynamos.

Larmor was also interested in the Earth's field: he wrote that the above self-generation of magnetic fields

---

[26] Zeeman (1897).
[27] Hale (1908).
[28] Gouy (1912).
[29] Larmor (1919).

Fig. 1.4. Joseph Larmor (1857–1942) was Irish. He is the founder of Magneto-Fluid-Dynamics, apparently following an idea first proposed by Gouy. He struggled with the concept of ether and stressed phenomena, rather than the associated mathematics

> would account for magnetic change, sudden or gradual, on the Earth merely by change of internal conducting channels though, on the other hand, it would require fluidity and residual circulation in deep-seated regions.

We saw above that the rapid changes in the Earth's magnetic field have their origin, not in the Earth's core, but in the ionosphere.

Larmor elaborated his concept of self-excited dynamo in two later papers.[30]

He was Fellow and Secretary of the Royal Society, as well as Fellow and Lecturer at St John's College, Cambridge. "He considered himself as the last follower of the Scotto–Irish school of Physics which dominated the world in the nineteenth century, particularly Hamilton, MacCullagh, Maxwell, Kelvin, and FitzGerald."[31]

We start our study of Magneto-Fluid-Dynamics with a brief discussion of the so-called "Maxwell" equations, as formulated by Heaviside.

[30] Larmor (1929, 1934).
[31] Gillispie (1970–1978).

Part II

Fundamentals

2 The Maxwell Equations

Contents

All electromagnetic phenomena satisfy the Maxwell equations. These are the four magical keys that give access to the Physics of a huge class of natural phenomena, and to the design of electromagnetic devices and processes.

2.1 Introduction

We assume that the reader is familiar with vector analysis and with phasors; otherwise he or she must study Chapters 1 and 2 of FEP,[1] or of EFW,[2] or Chapter 1 and Section 15.7 of EPA,[3] or equivalent texts, before going on.

All macroscopic electric, magnetic, and electromagnetic phenomena satisfy the four Maxwell equations.

It is possible to express these equations in various forms, but we need only two forms, one involving partial derivatives, which we shall use most of the time, and the other involving integrals.

We disregard magnetic materials. Chapter 9 concerns the liquid part of the Earth's core, which consists largely of liquid iron, but the iron is non-magnetic because of its high temperature. The solid part of the core is also non-magnetic for the same reason.

We disregard moving media until Chapter 6.

2.2 The Equations in Differential Form

Excluding magnetic media, the Maxwell equations reduce to

$$\boldsymbol{\nabla} \cdot \boldsymbol{E} = \frac{\widetilde{Q}}{\epsilon_0} , \tag{2.1}$$

$$\boldsymbol{\nabla} \cdot \boldsymbol{B} = 0 , \tag{2.2}$$

$$\boldsymbol{\nabla} \times \boldsymbol{E} + \frac{\partial \boldsymbol{B}}{\partial t} = \boldsymbol{0} , \tag{2.3}$$

$$\boldsymbol{\nabla} \times \boldsymbol{B} = \mu_0 \boldsymbol{J} . \tag{2.4}$$

Here,

ϵ_0 is the *permittivity of free space*,

$$\epsilon_0 = 8.8541878 \times 10^{-12} \text{ farad/meter} ; \tag{2.5}$$

μ_0 is the *permeability of free space*,

$$\mu_0 \equiv 4\pi \times 10^{-7} \text{ henry/meter} \tag{2.6}$$

(we set the relative permeability μ_r equal to unity because we disregard magnetic media);

---

[1] *Fundamentals of Electromagnetic Phenomena*, by Paul Lorrain, Dale R. Corson, and François Lorrain (W.H. Freeman, New York 2000).

[2] *Electromagnetic Fields and Waves* by the same authors and the same publisher, third edition (1988).

[3] *Electromagnetism, Principles and Applications*, by Paul Lorrain and Dale R. Corson, by the same publisher, second edition (1990).

ϵ_r is the *relative permittivity* of the medium, which is a pure number, usually of the order of 3; we assume homogeneous, isotropic, linear, and stationary (HILS) media;
$\boldsymbol{E}$ is the *electric field strength* in volts/meter;
$\boldsymbol{B}$ is the *magnetic flux density* in teslas;
$\widetilde{Q}$ (pronounced Q tilde) is the *net electric space charge density* in coulombs per meter^3:

$$\widetilde{Q} = \widetilde{Q}_f + \widetilde{Q}_b = \widetilde{Q}_f - \boldsymbol{\nabla} \cdot \boldsymbol{P} , \tag{2.7}$$

where $\widetilde{Q}_f$ is the *free* electric charge density, while $\widetilde{Q}_b$ is the *bound* electric charge density that results when a dielectric is subjected to a non-uniform electric field, and where the *polarization*

$$\boldsymbol{P} = \epsilon_0(\epsilon_r - 1)\boldsymbol{E} . \tag{2.8}$$

The quantity ϵ_r is the *relative permittivity* of the medium, and $\epsilon = \epsilon_r\epsilon_0$ is its *permittivity*.

We can also rewrite Eq. 2.1 as follows:

$$\boldsymbol{\nabla} \cdot \boldsymbol{E} = \frac{\widetilde{Q}_f}{\epsilon_r\epsilon_0} . \tag{2.9}$$

In isotropic, linear, and stationary conductors,

$$\boldsymbol{J} = \sigma \boldsymbol{E} , \tag{2.10}$$

where σ is the *electrical conductivity* of the medium. This is *Ohm's law*, Eq. 3.10.

All this applies if the medium, whether a conductor or a non-conductor, is stationary. If the medium is *not* stationary, then some of the above equations must be modified, as we shall see in Chapters 6 and 7.

In Eq. 2.10 we have omitted the *displacement current density*,[4] $\epsilon_r\epsilon_0\partial\boldsymbol{E}/\partial t$. In the case of moving media, we also omit from the $\boldsymbol{J}$ of Eq. 2.4 the *convection current density* $\widetilde{Q}\boldsymbol{v}$, where $\boldsymbol{v}$ is the velocity of the medium at the point considered. Both currents add to $\boldsymbol{J}$, but prove to be negligible here. See Sects. 2.4 and 2.5 below.

The *speed of light* in free space is

$$c = (\epsilon_0\mu_0)^{-1/2} = 2.99792458 \times 10^8 \text{ meters/second} . \tag{2.11}$$

It is usually impossible to measure *all* the variables $\boldsymbol{E}$, $\boldsymbol{B}$, $\widetilde{Q}$, and $\boldsymbol{J}$ in a given situation. For example, the Earth's magnetic field results mostly from electric currents in the liquid part of the core; one can measure $\boldsymbol{B}$ at the

[4] The antique term "displacement" dates back to the nineteenth century; it is seldom used.

surface, and geophysicists try to calculate the field at the surface of the core, but it is impossible to measure the current distribution in the core directly.

A medium is said to be *homogeneous* if its properties are the same at all points; it is *isotropic* if its properties at a given point are the same in all directions. A quantity is *constant* if it is independent of the time. If a property, such as the density, is the same at all points in a certain region, we say that the property is *uniform.*

Note carefully.

1) In the above differential form, Maxwell's equations 2.1 to 2.4 concern the *rates of change* of $\boldsymbol{E}$ and of $\boldsymbol{B}$ with respect to the coordinates and with respect to time, *at a given point and at a given time.* They are *local* relations.

2) We have written Maxwell's equations with $\boldsymbol{E}$ and $\boldsymbol{B}$ on the left, and with $\widetilde{Q}$ and $\boldsymbol{J}$ on the right, as if the latter were the sources of the field $\boldsymbol{E}$, $\boldsymbol{B}$. They are indeed the sources of the field, but they are closely related to both $\boldsymbol{E}$ and $\boldsymbol{B}$.

3) Maxwell's equations are *linear*: all variables and their derivatives appear to the first power, and there are no products or ratios of the variables or of their derivatives.

This linearity is important. It means that, in homogeneous, isotropic, linear, and stationary (HILS) media, sources act independently of each other. Say the field $\boldsymbol{E}_1$, $\boldsymbol{B}_1$ corresponds to $\widetilde{Q}_1$ and $\boldsymbol{J}_1$, and the field $\boldsymbol{E}_2$, $\boldsymbol{B}_2$ corresponds to $\widetilde{Q}_2$, $\boldsymbol{J}_2$. Then, if the sources are $\widetilde{Q}_1 + \widetilde{Q}_2$ and $\boldsymbol{J}_1 + \boldsymbol{J}_2$, the field is $\boldsymbol{E}_1 + \boldsymbol{E}_2$, $\boldsymbol{B}_1 + \boldsymbol{B}_2$. This is the *principle of superposition.*

4) In Eq. 2.3 the *space* rates of change of $\boldsymbol{E}$ depend on the *time* rate of change of $\boldsymbol{B}$, while Eq. 2.4 relates the *space* rates of change of $\boldsymbol{B}$ to the local value of $\boldsymbol{J}$, and hence of $\boldsymbol{E}$.

5) Equation 2.1 is *Gauss's law for* electric *fields* in differential form.

6) Equation 2.2 is *Gauss's law for* magnetic *fields* in differential form; it assumes that magnetic monopoles, or "magnetic charges", do not exist, which is an experimental fact, to date (2005).

7) Equation 2.3 is the differential form of the *Law of Faraday.* If $\boldsymbol{B}$ is either zero or constant, then

$$\boldsymbol{\nabla} \times \boldsymbol{E} = \boldsymbol{0}\,. \tag{2.12}$$

8) Equation 2.4 is the differential form of *Ampère's circuital law.* In regions where $\boldsymbol{J} = \boldsymbol{0}$,

$$\boldsymbol{\nabla} \times \boldsymbol{B} = \boldsymbol{0}\,. \tag{2.13}$$

2.2.1 Example: The Differential Form of Gauss's Law for *Electric* Fields

In the field of a point charge in a vacuum, the charge density is zero. Then we must have that $\boldsymbol{\nabla} \cdot \boldsymbol{E} = 0$. Let us check. Here $\boldsymbol{E}$ is radial, it points away from Q. Setting $\hat{\boldsymbol{r}}$ the unit vector pointing away from Q,

$$\boldsymbol{E} = \frac{Q}{4\pi\epsilon_0 r^2}\hat{\boldsymbol{r}} \quad \text{and} \quad \boldsymbol{\nabla} \cdot \boldsymbol{E} = \frac{Q}{4\pi\epsilon_0}\boldsymbol{\nabla} \cdot \left(\frac{\hat{\boldsymbol{r}}}{r^2}\right) . \tag{2.14}$$

Because of the symmetry, we use spherical coordinates. From the Table on the back of the front cover,

$$\boldsymbol{\nabla} \cdot \left(\frac{\hat{\boldsymbol{r}}}{r^2}\right) = \frac{1}{r^2}\frac{\partial}{\partial r}1 = 0 , \tag{2.15}$$

and $\boldsymbol{\nabla} \cdot \boldsymbol{E} = 0$, except of course at the position of the charge Q.

2.2.2 Example: The Differential Form of Gauss's Law for *Magnetic* Fields

A current I flows in a long straight wire, in the positive direction of the $z-$axis. Its magnetic field is azimuthal and, at a distance ρ,

$$\boldsymbol{B} = \frac{\mu_0 I}{2\pi\rho}\hat{\boldsymbol{\phi}} . \tag{2.16}$$

The magnetic field lines are circles centered on the wire.

In this field we should have that $\boldsymbol{\nabla} \cdot \boldsymbol{B} = 0$. Let us see. We use cylindrical coordinates. The field has only a ϕ-component, and it is independent of both ϕ and z. So $\boldsymbol{\nabla} \cdot \boldsymbol{B} = 0$.

2.2.3 Example: The Differential Form of the Law of Faraday

No simple illustration of this equation comes to mind, but see Sects. 2.3.3 and 5.2.3.

2.2.4 Example: The Differential Form of Ampère's Circuital Law

The magnetic field of the above wire is given above. So we must calculate

$$\boldsymbol{\nabla} \times \left(\frac{\mu_0 I}{2\pi\rho}\hat{\boldsymbol{\phi}}\right)$$

in cylindrical coordinates. Outside the wire, $\boldsymbol{J} = \boldsymbol{0}$ and we should find that $\boldsymbol{\nabla} \times \boldsymbol{B} = \boldsymbol{0}$. The $\boldsymbol{B}$ has only a ϕ-component, and it is independent of z. From the back of the front cover, we do find that, outside the wire, $\boldsymbol{\nabla} \times \boldsymbol{B} = \boldsymbol{0}$.

2.3 The Equations in Integral Form

The integral form of Maxwell's equations is generally less useful than the differential form, but it is intuitively more instructive; one form complements the other. Integrating Eqs. 2.1 to 2.4, assuming a uniform medium, and applying the divergence and Stokes's theorems (see the page facing the front cover), we find that

$$\int_{\mathcal{A}} \boldsymbol{E} \cdot \boldsymbol{d\mathcal{A}} = \frac{1}{\epsilon_r \epsilon_0} \int_v \widetilde{Q}_f \, dv \,, \tag{2.17}$$

$$\int_{\mathcal{A}} \boldsymbol{B} \cdot \boldsymbol{d\mathcal{A}} = 0 \,, \tag{2.18}$$

$$\oint_C \boldsymbol{E} \cdot \boldsymbol{dl} = -\frac{d}{dt} \int_{\mathcal{A}} \boldsymbol{B} \cdot \boldsymbol{d\mathcal{A}} \,, \tag{2.19}$$

$$\oint_C \boldsymbol{B} \cdot \boldsymbol{dl} = \mu_0 \int_{\mathcal{A}} \boldsymbol{J} \cdot \boldsymbol{d\mathcal{A}} \,. \tag{2.20}$$

In Eqs. 2.17 and 2.18, $\mathcal{A}$ is a *closed* surface bounding a volume v. The element of area $\boldsymbol{d\mathcal{A}}$ is a vector that is normal to $\mathcal{A}$ and that points *outward.*

In Eqs. 2.19 and 2.20, $\mathcal{A}$ is an *open* surface bounded by the *closed* curve C. The positive directions for $\boldsymbol{d\mathcal{A}}$, $\boldsymbol{B}$, $\boldsymbol{J}$, and the positive direction for integration around C satisfy the right-hand-screw rule.

Stated otherwise,

$$\int_{\mathcal{A}} \boldsymbol{E} \cdot \boldsymbol{d\mathcal{A}} = \frac{Q_f}{\epsilon_r \epsilon_0} \,, \tag{2.21}$$

$$\int_{\mathcal{A}} \boldsymbol{B} \cdot \boldsymbol{d\mathcal{A}} = 0 \,, \tag{2.22}$$

$$\oint_C \boldsymbol{E} \cdot \boldsymbol{dl} = -\frac{\partial \Phi}{\partial t} \,, \tag{2.23}$$

$$\oint_C \boldsymbol{B} \cdot \boldsymbol{dl} = \mu_0 I_C \,, \tag{2.24}$$

where

Q_f is the net free charge inside the closed surface $\mathcal{A}$,

Φ is the net magnetic flux linking the closed curve C that bounds the open surface $\mathcal{A}$,

$$\Phi = \int_{\mathcal{A}} \boldsymbol{B} \cdot \boldsymbol{d\mathcal{A}} \,, \tag{2.25}$$

and

I_C is the net electric current linking C.

Note the following.

1) These equations concern *closed* volumes and *closed* curves *of finite size*, and not space rates of change *at a point*: they are *non-local* relations.

2) They apply *at a given time.*

3) The way in which the variables change from point to point is irrelevant here; only the integrals matter.

4) Equation 2.17 states that the total efflux of $\boldsymbol{E}$ through a closed surface is proportional to the enclosed electric charge. This is the *integral form of Gauss's law for* electric *fields.*

5) Equation 2.18 states that there is zero net magnetic efflux or influx through a closed surface. This is the *integral form of Gauss's law for* magnetic *fields.*

6) Equation 2.19 states that the line integral of $\boldsymbol{E}$ around a closed curve C is equal to *minus* the time derivative of the magnetic flux linking C. This line integral is the *electromotance* around C. The surface $\mathcal{A}$ is *any* surface bounded by C. This is the *integral form of the Law of Faraday.*

7) According to Eq. 2.20, the line integral of the magnetic flux density $\boldsymbol{B}$ around a closed curve C is proportional to the electric current I_C linking C. This is the *integral form of Ampère's circuital law.*

We now illustrate Maxwell's integral equations with a few simple examples.

2.3.1 Example: The Integral Form of Gauss's Law for *Electric* Fields

Consider a sphere of radius r with a point charge Q at its center. On the surface of the sphere, $\boldsymbol{E}$ is normal, it points away from Q if Q is positive and, according to the integral form of Gauss's law,

$$4\pi r^2 E = Q/\epsilon_0 \, . \tag{2.26}$$

Then, at a distance r from a point charge Q, in a vacuum, we have the familiar relation

$$E = \frac{Q}{4\pi\epsilon_0 r^2} \, . \tag{2.27}$$

If, instead, there is an arbitrary charge distribution inside the sphere, then Eq. 2.21 again applies, ϵ_r being the relative permittivity of the medium. Note that, if there are also charges outside the sphere, then the net electric field is affected, but Eq. 2.21 still applies.

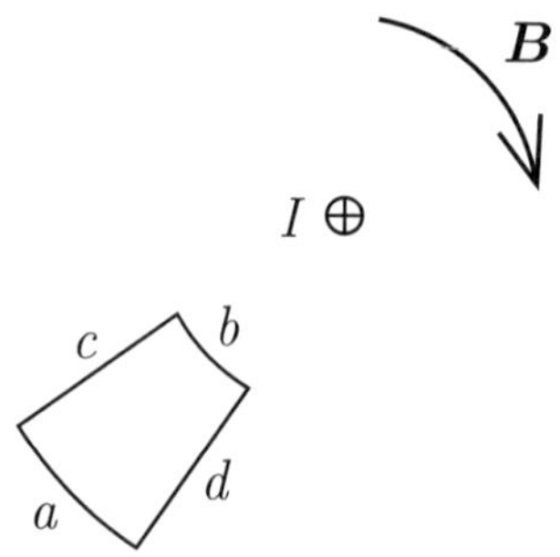

Fig. 2.1. Wire, perpendicular to the paper, carrying an electric current I. The magnetic field lines are azimuthal. a, b, c, d are the sides of an imaginary volume of height h that stands perpendicular to the paper. We show that there is no net magnetic flux leaving that volume

2.3.2 Example: The Integral Form of Gauss's Law for *Magnetic* Fields

We return to the field of Sect. 2.2.2. Consider a volume having the cross-section shown in Fig. 2.1 and of height h. The vector $\boldsymbol{B}$ is tangent to faces a and b, and thus normal to the vector element of area $\boldsymbol{d\mathcal{A}}$. So, on those two surfaces, $\boldsymbol{B} \cdot \boldsymbol{d\mathcal{A}} = 0$. Similarly, on the top and bottom faces, which are parallel to the plane of the paper, $\boldsymbol{B} \cdot \boldsymbol{d\mathcal{A}} = 0$. But, on faces c and d, $\boldsymbol{B}$ is either parallel or antiparallel to $\boldsymbol{d\mathcal{A}}$. The surface integrals of $\boldsymbol{B}$ over the two faces c and d cancel and, for the complete volume,

$$\int_{\mathcal{A}} \boldsymbol{B} \cdot \boldsymbol{d\mathcal{A}} = 0 \,. \tag{2.28}$$

2.3.3 Example: The Integral Form of the Law of Faraday

Figure 2.2 shows a simple-minded transformer. The primary is a solenoid S that carries a current I, and the secondary s feeds the resistance R. If I changes, then the magnetic flux inside the solenoid changes, and a voltage is induced in s.

According to Eq. 2.23, the voltage induced in the secondary is equal to the rate of change of the magnetic flux Φ linking it. If the secondary has N turns, then the secondary voltage is N times larger.[5]

[5] This simple circuit has baffled many an author because, if the solenoid is long, then $B \sim 0$ outside. So how can the secondary be affected by a changing magnetic flux, if it is in a region where there is no magnetic field? We return to this case in Sect. 5.2.3.

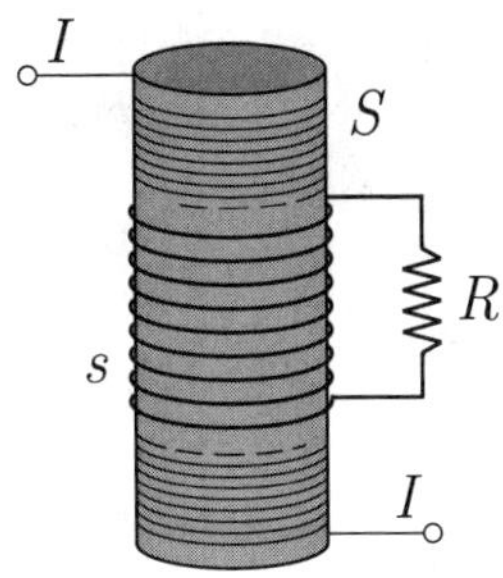

Fig. 2.2. A simple-minded transformer. The solenoid S carries a current I, and the secondary s feeds the load resistance R

2.3.4 Example: The Integral Form of Ampère's Circuital Law

Imagine a circle C of radius ρ that follows a magnetic field line around a straight wire carrying a current I. The current that links C is I. Then we should have that

$$2\pi\rho B = \mu_0 I\ , \quad \text{or} \quad B = \frac{\mu_0 I}{2\pi\rho}\ , \tag{2.29}$$

which is correct.

2.4 The Displacement Current Density

We referred to the displacement current density $\epsilon_r\epsilon_0\partial\boldsymbol{E}/\partial t$ in Eqs. 2.4 and 2.20, and we stated that, for the phenomena investigated in this book, the displacement current density is negligible compared to the conduction current density $\sigma\boldsymbol{E}$ in a stationary medium.

We can justify this approximation as follows. Assume a stationary medium in which $\boldsymbol{E}$ is a sinusoidal function of the time. Then the conduction current density is

$$\sigma\boldsymbol{E} = \sigma\boldsymbol{E}_0 \exp j\omega t\ , \tag{2.30}$$

the displacement current density is

$$\epsilon_r\epsilon_0\frac{\partial\boldsymbol{E}}{\partial t} = \epsilon_r\epsilon_0 j\omega\boldsymbol{E}_0 \exp j\omega t\ , \tag{2.31}$$

and

$$\left|\frac{\epsilon_r\epsilon_0 j\omega\boldsymbol{E}_0 \exp j\omega t}{\sigma\boldsymbol{E}_0 \exp j\omega t}\right| = \frac{\epsilon_r\epsilon_0\omega}{\sigma}\ . \tag{2.32}$$

The displacement current density is therefore negligible, compared to the conduction current density, *if* $\epsilon_r\epsilon_0\omega/\sigma \ll 1$.

For terrestrial and solar media, relative permittivities ϵ_r vary between 1 (solar plasma) and 80 (sea water), and conductivities σ vary between about 5 siemens/meter (sea water) and 3×10^5 siemens/meter (the liquid part of the Earth's core). The worst case is that of sea water, with $\epsilon_r = 80$ and $\sigma = 5$. In that case, after substituting $\epsilon_0 = 8.85 \times 10^{-12}$ farad/meter, we find that

$$\left| \frac{\epsilon_r \epsilon_0 \partial \boldsymbol{E}/\partial t}{\sigma \boldsymbol{E}} \right| = \frac{80 \times 8.85 \times 10^{-12} \omega}{5} = 1.42 \times 10^{-10} \omega \, . \tag{2.33}$$

So, for terrestrial and solar media, the displacement current density is negligible compared to the conduction current density, if the frequency is much less than about one gigahertz. That condition is amply satisfied, here.

The same result follows from the method of Appendix A. From Eq. A.78, we obtain, for typical values of the fields,

$$\frac{\epsilon_r \epsilon_0 |\partial \boldsymbol{E}/\partial t|_{\text{typical}}}{\sigma E_{\text{typical}}} \sim \frac{\epsilon_r \epsilon_0 E_{\text{typical}}/\mathcal{T}}{\sigma E_{\text{typical}}} = \frac{\epsilon_r \epsilon_0}{\sigma \mathcal{T}} \, , \tag{2.34}$$

where $\mathcal{T}$ is a *characteristic* time of the phenomenon observed. For sea water,

$$\frac{\epsilon_r \epsilon_0 |\partial \boldsymbol{E}/\partial t|_{\text{typical}}}{\sigma E_{\text{typical}}} \sim \frac{1.42 \times 10^{-10}}{\mathcal{T}} \, . \tag{2.35}$$

These results apply to a stationary medium. For non-stationary media, displacement currents are also negligible, at least in many circumstances, such as in a highly conducting plasma. See Exercise 3 at the end of Sect. A.8.

2.5 The Convection Current Density

In the $\boldsymbol{J}$ of Eq. 2.4 we also disregarded the convection current density $\widetilde{Q}\boldsymbol{v}$, where $\widetilde{Q}$ is the electric space charge density and $\boldsymbol{v}$ the velocity of the medium. Is the convection current density also negligible?

Conductors that move in a magnetic field carry a space charge (Sect. 7.2)

$$\widetilde{Q} = -\epsilon_0 \boldsymbol{\nabla} \cdot (\boldsymbol{v} \times \boldsymbol{B}) \, , \tag{2.36}$$

under steady conditions in a uniform medium. If we are only interested in the orders of magnitude of typical values of the fields, then, according to Eq. A.83 of Appendix A, we may write that

$$|\widetilde{Q}|_{\text{typical}} \lesssim \epsilon_0 \frac{|\boldsymbol{v} \times \boldsymbol{B}|_{\text{typical}}}{\mathcal{L}} \leq \epsilon_0 \frac{(vB)_{\text{typical}}}{\mathcal{L}} \, , \tag{2.37}$$

where $\mathcal{L}$ is a *characteristic* length of the phenomenon studied. For example, in the case of the Earth, $\mathcal{L}$ might be the Earth's radius.

Now assume that

$$(vB)_{\text{typical}} \sim v_{\text{typical}} B_{\text{typical}} , \tag{2.38}$$

$$|\widetilde{Q}v|_{\text{typical}} \sim |\widetilde{Q}|_{\text{typical}} v_{\text{typical}} . \tag{2.39}$$

For a discussion of this kind of equation, see Sect. A.7 of Appendix A.

For the typical value of the convection current $\widetilde{Q}\boldsymbol{v}$, then

$$|\widetilde{Q}v|_{\text{typical}} \lesssim \epsilon_0 \frac{B_{\text{typical}} v_{\text{typical}}^2}{\mathcal{L}} . \tag{2.40}$$

On the other hand, for the *net* current density, resulting from the movement of all charges, we have, from Maxwell's equation 2.4 and Eq. A.84,

$$J_{\text{net typical}} = \frac{1}{\mu_0} |\boldsymbol{\nabla} \times \boldsymbol{B}|_{\text{typical}} \sim \frac{B_{\text{typical}}}{\mu_0 \mathcal{L}} . \tag{2.41}$$

We have a $\sim$ sign here, instead of a $\lesssim$, because $\boldsymbol{B}$ is a so-called "transversal" field, since its divergence is zero (Sect. A.6.3).

From the two above equations, the ratio of the magnitudes of the convection and net current densities is

$$\frac{|\widetilde{Q}v|_{\text{typical}}}{J_{\text{net typical}}} \lesssim \frac{\epsilon_0 B_{\text{typical}} v_{\text{typical}}^2 / \mathcal{L}}{B_{\text{typical}} / (\mu_0 \mathcal{L})} = \epsilon_0 \mu_0 v_{\text{typical}}^2 = \frac{v_{\text{typical}}^2}{c^2} . \tag{2.42}$$

Here, v_{typical} will be, at most, 2 kilometers/second (solar plasma). Then

$$\frac{|\widetilde{Q}v|_{\text{typical}}}{J_{\text{net typical}}} \lesssim \frac{4 \times 10^6}{9 \times 10^{16}} \sim 10^{-10} . \tag{2.43}$$

We were therefore justified in neglecting both displacement and convection currents in the $\boldsymbol{J}$ of Eq. 2.4.

Now consider a conducting disk rotating in a magnetic field. Assume a steady state in which there are no conduction currents in the disk. There is a convection current, however, since there are space charges (Eq. 2.36), or at least surface charges. Does this contradict Eq. 2.42? In fact it does not, because the typical values in Eq. 2.42 are typical, not for the rotating disk alone, but for the total field. There are other currents, outside the disk, that produce the ambient $\boldsymbol{B}$. It is those currents that make the overall $J_{\text{net typical}}$ much larger than $|\widetilde{Q}v|_{\text{typical}}$.

2.6 Summary

We restate Maxwell's equations as follows. These equations apply to homogeneous, isotropic, linear, and stationary (HILS) media.

At a given point and at a given time, the *divergence* of $\boldsymbol{E}$ is proportional to the electric charge density $\widetilde{Q}$:

$$\nabla \cdot \boldsymbol{E} = \frac{\widetilde{Q}}{\epsilon_0} . \qquad (2.1)$$

As a consequence, the *integral of* $\boldsymbol{E}$ over the closed surface $\mathcal{A}$ enclosing a volume v is given by

$$\int_{\mathcal{A}} \boldsymbol{E} \cdot d\boldsymbol{\mathcal{A}} = \frac{1}{\epsilon_0} \int_v \widetilde{Q} \, dv . \qquad (2.17)$$

These are two equivalent statements of *Gauss's law for* $\boldsymbol{E}$.

The divergence of $\boldsymbol{B}$ is equal to zero:

$$\nabla \cdot \boldsymbol{B} = 0 \qquad (2.2)$$

and, as a consequence, the surface integral of $\boldsymbol{B}$ over a closed surface is equal to zero:

$$\int_{\mathcal{A}} \boldsymbol{B} \cdot d\boldsymbol{\mathcal{A}} = 0 . \qquad (2.18)$$

These are two equivalent forms of *Gauss's law for* $\boldsymbol{B}$.

At a given point and at a given time, the *curl* of $\boldsymbol{E}$ is equal to *minus* the time derivative of $\boldsymbol{B}$:

$$\nabla \times \boldsymbol{E} = -\frac{\partial \boldsymbol{B}}{\partial t} . \qquad (2.3)$$

It follows that the *line integral* of $\boldsymbol{E}$ around a closed curve C is equal to *minus* the time derivative of the magnetic flux Φ linked by C:

$$\oint_C \boldsymbol{E} \cdot d\boldsymbol{l} = -\frac{d}{dt} \int_{\mathcal{A}} \boldsymbol{B} \cdot d\boldsymbol{\mathcal{A}} . \qquad (2.19)$$

These are two equivalent statements of the *Law of Faraday.*

The curl of $\boldsymbol{B}$ is equal to μ_0 times the current density:

$$\nabla \times \boldsymbol{B} = \mu_0 \boldsymbol{J} . \qquad (2.4)$$

Then the line integral of $\boldsymbol{B}$ over a closed curve C is equal to the current through an open surface $\mathcal{A}$ bounded by C:

$$\oint_C \boldsymbol{B} \cdot d\boldsymbol{l} = \mu_0 \int_{\mathcal{A}} \boldsymbol{J} \cdot d\boldsymbol{\mathcal{A}} . \qquad (2.20)$$

These are two equivalent statements of *Ampère's circuital law.*

In isotropic, linear, and stationary conductors, the electric current density $\boldsymbol{J}$ is proportional to $\boldsymbol{E}$:

$$\boldsymbol{J} = \sigma \boldsymbol{E} . \qquad (2.10)$$

This is *Ohm's law.* This relation is only part of a more general one that we shall use extensively and that applies to moving conductors. See Sect. 6.4.

3 Electric Fields

Contents

Our objective in Chapters 2 to 5 is to set the stage for our study of magneto-fluid-dynamic phenomena. Here we think about electric fields in simple cases: for the moment there are no magnetic fields, and no moving conductors. We present an Example and a Case Study: the first concerns a proton beam, and the second the electrostatic potential at the surface of the Sun.

In this chapter we study the electric fields that result from accumulations of electric charge. By hypothesis, the charges move slowly, and accelerate slowly. We assume that $v^2 \ll c^2$, where v is the speed of the charge and c is the speed of light. Then the fields of the electric charges are essentially the same as if they were stationary.

We disregard magnetic fields, for the moment.

3.1 Electric Fields and Forces

Charged particles, whether positive or negative, carry multiples of the *electron charge*

$$e = 1.602176462 \times 10^{-19} \text{ coulomb} . \tag{3.1}$$

Electric charges possess electric fields: at a distance r from an isolated point charge of Q coulombs, in a vacuum,

$$\boldsymbol{E} = \frac{Q}{4\pi\epsilon_0 r^2}\hat{\boldsymbol{r}} , \tag{3.2}$$

where the unit vector $\hat{\boldsymbol{r}}$ points away from the charge Q.

Inside a uniform dielectric ϵ_r, at points near Q and remote from the surface,

$$\boldsymbol{E} = \frac{Q}{4\pi\epsilon_0\epsilon_r r^2}\hat{\boldsymbol{r}} . \tag{3.3}$$

Electric fields can exist in a vacuum, or in any material medium, whether solid, liquid, or gaseous, and whether insulating or conducting.

If the short vector $\boldsymbol{dl}$ is parallel to $\boldsymbol{E}$, then joining $\boldsymbol{dl}$ vectors end-to-end defines an *electric field line.*

If Q lies in a region where there is an electric field $\boldsymbol{E}$ because of the presence of other electric charges, then the electric force exerted on Q by these other charges is

$$\boldsymbol{F} = Q\boldsymbol{E} . \tag{3.4}$$

The force points in the direction of $\boldsymbol{E}$ if Q is positive.

Combining Eqs. 3.2 and 3.4, the force exerted *by* a point charge Q_1 *on* a point charge Q_2 a distance r away is

$$\boldsymbol{F} = \frac{Q_1 Q_2}{4\pi\epsilon_0 r^2}\hat{\boldsymbol{r}} , \tag{3.5}$$

where the unit vector $\hat{\boldsymbol{r}}$ points away from Q_1. The force is repulsive if the charges are of the same sign, and attractive if they are of different signs. This is *Coulomb's law* for charges in a vacuum.

Electric forces are huge: the force between two charges of one coulomb, separated by a distance of one meter would be about one megaton! Paradoxically, it is because electric forces are so large that they are usually unimportant in everyday life: an electrically charged macroscopic body attracts charges of opposite sign and quickly becomes neutral, or nearly so.

You guessed it: gravitational forces are completely negligible compared to electric forces. For example, at the surface of the Sun, $g = 273$ meters/second^2. The electric and gravitational forces on an electron at the surface of the Sun would be of equal magnitudes for an electric field of only about 1.6 nanovolts/meter. See Sect. 3.9.

A body that carries a net electric charge density $\widetilde{Q}$ and that lies in an electric field $\boldsymbol{E}$ is subjected to a force of density

$$\widetilde{\boldsymbol{F}} = \widetilde{Q}\boldsymbol{E} . \tag{3.6}$$

Electric forces are usually negligible compared to magnetic forces (Sect. 4.8).

3.2 Electric Currents

In *good conductors*, each atom, more or less, has one weakly bound *conduction electron* that is free to move about in an electric field. The positive ions remain fixed in the lattice.

Good conductors are usually solid, homogeneous, and isotropic.

Set n_e equal to the number density of the conduction electrons. If the electron cloud drifts at a velocity $\boldsymbol{v}$, then the current density is

$$\boldsymbol{J} = -n_\mathrm{e} e \boldsymbol{v} \,, \tag{3.7}$$

where e is the magnitude of the electronic charge, as in Eq. 3.1, and is a positive number. The charge density $-n_\mathrm{e} e$ is negative, and $\boldsymbol{J}$ and $\boldsymbol{v}$ point in opposite directions.

Now set

$$\boldsymbol{v} = -\mathcal{M} \boldsymbol{E} \,, \tag{3.8}$$

where the *mobility*

$$\mathcal{M} = |\boldsymbol{v}| / |\boldsymbol{E}| \tag{3.9}$$

is a positive quantity. Then

$$\boldsymbol{J} = n_\mathrm{e} e \mathcal{M} \boldsymbol{E} = \sigma \boldsymbol{E} \,. \tag{3.10}$$

This is *Ohm's law*, and the *conductivity*

$$\sigma = n_\mathrm{e} e \mathcal{M} \,. \tag{3.11}$$

However, see Sect. 6.4.

Liquid conductors, or *electrolytes*, conduct electricity because positive ions drift in the direction of $\boldsymbol{E}$, and negative ions and electrons in the opposite direction. The liquid remains neutral.

Ionized gases, or *plasmas*, can contain free electrons, negative ions, and positive ions that drift similarly. The electrons carry most of the current because they are much lighter and faster than the ions, and because their mean-free-path between collisions is much longer. Plasmas are normally neutral: their net electric charge density is zero.

The conductivity of a plasma is *anisotropic* in the presence of a magnetic field, electrons drifting more rapidly in the direction parallel to the local magnetic field than in perpendicular directions. The perpendicular conductivity is a function of B^2 (Cambel, 1963). The conductivity is then a tensor. The same effect also occurs in solid conductors, but it is then usually negligible. The distinction between parallel and perpendicular conductivity is not useful for our purposes here because, as a rule, the vectors $\boldsymbol{J}$ and $\boldsymbol{B}$ will be neither parallel nor perpendicular.

3.2.1 Example: The Drift Speeds of Charge Carriers

A current of 1 ampere flows in a copper wire having a cross-section of 1 square millimeter. What is the speed of the charge carriers?

Copper has an atomic weight of 64. Thus the mass of one atom is 64 times the mass of a proton, or 1.1×10^{-25} kilogram. Copper has a mass density of 8.9×10^3 kilograms/meter^3, and thus 8.1×10^{28} atoms/meter^3. With one conduction electron per atom, their charge density is

$$\widetilde{Q}_{\rm ce} = -8.1 \times 10^{28} \times 1.6 \times 10^{-19} = -1.3 \times 10^{10} \text{ coulombs/meter}^3 \ ! \quad (3.12)$$

There are surface charges but, inside, the *net* charge density is normally zero. (If the conductor moves in a magnetic field then, as rule, the charge density inside is not zero. See Chapter 7.)

From Eq. 3.7, the drift speed of the conduction electrons is low. For the above case,

$$v_{\rm d} = \frac{1/(1 \times 10^{-6})}{8.1 \times 10^{28} \times 1.6 \times 10^{-19}} = 7.7 \times 10^{-5} \text{ meter/second} , \quad (3.13)$$

or about 0.1 millimeter/second, or about 6 centimeters/minute!

Does that make sense? After all, light bulbs turn on instantaneously! Upon turning on a light bulb, an electromagnetic *wave* travels along the wires at about the speed of light and applies a longitudinal electric field both to the wires and to the bulb. So the cloud of conduction electrons has hardly moved at all by the time the light bulb is on, even kilometers away. In the above case, the drift speed is lower than the speed of light by a factor of

$$3 \times 10^8/(7.7 \times 10^{-5}) \approx 4 \times 10^{12} \ !$$

3.3 The Conservation of Electric Charge

It is an experimental fact that electric charge is indestructible. Given a volume v bounded by a closed surface $\mathcal{A}$, the net efflux of charge through $\mathcal{A}$ is equal to the rate of depletion of the charge within v:

$$\int_{\mathcal{A}} \boldsymbol{J} \cdot \boldsymbol{d\mathcal{A}} = -\frac{d}{dt}\int_v \widetilde{Q}\, dv = -\int_v \frac{\partial \widetilde{Q}}{\partial t}\, dv \, . \quad (3.14)$$

As usual, the vector $\boldsymbol{d\mathcal{A}}$ is normal to $\mathcal{A}$ and points *out*ward.

Applying the divergence theorem (see the page facing the front cover),

$$\boldsymbol{\nabla} \cdot \boldsymbol{J} = -\partial \widetilde{Q}/\partial t \, . \quad (3.15)$$

This is the *law of conservation of charge* in integral and in differential form.

3.3.1 Example: The Relaxation Time of a Conductor

Usually, conductors carry surface charges, but any net charge injected inside, say with a high-energy electron beam, disappears rapidly. However, as we shall see in Chapter 7, conductors that move in magnetic fields do carry permanent space, or volume, charges as a rule.

Imagine a spherical surface S of radius a inside a block of conductor. The net charge density inside S is uniform. The free charge enclosed is Q and, at the surface, from Eqs. 3.3 and 3.10,

$$\boldsymbol{E} = \frac{Q}{4\pi\epsilon_r\epsilon_0 a^2}\hat{\boldsymbol{r}}\,, \quad \boldsymbol{J} = \sigma\boldsymbol{E} = \frac{\sigma Q}{4\pi\epsilon_r\epsilon_0 a^2}\hat{\boldsymbol{r}}\,. \tag{3.16}$$

Then, from Eq. 3.14,

$$\frac{dQ}{dt} = -4\pi a^2 J = -\sigma\frac{Q}{\epsilon_r\epsilon_0}\,, \tag{3.17}$$

$$Q = Q_0 \exp\left(-\frac{\sigma}{\epsilon_r\epsilon_0}t\right)\,. \tag{3.18}$$

The charge Q decreases by a factor of e in a *relaxation time*

$$\tau = \epsilon_r\epsilon_0/\sigma\,, \tag{3.19}$$

where we can set $\epsilon_r \approx 3$. This time is of the order of 10^{-19} second for a good conductor, *if* we assume that the conductivity σ is the same at all time scales. Real relaxation times are longer by a few orders of magnitude, but are still very short.

The relaxation times for electrostatic charges inside conductors is short, even for very poor conductors. For example, to achieve a τ of one second, the conductivity would have to be equal to $\epsilon_r\epsilon_0$, or about 10^{-11} siemens/meter, while the conductivity of copper at room temperature is 5.8×10^7 siemens/meter.

Here is a more general proof of Eq. 3.18: take the divergence of both sides of Eq. 3.10, assuming a homogeneous medium:

$$\boldsymbol{\nabla}\cdot\boldsymbol{J} = \sigma\boldsymbol{\nabla}\cdot\boldsymbol{E}\,. \tag{3.20}$$

Now substitute Eq. 3.15 on the left, and Eq. 2.1 on the right:

$$-\frac{\partial\widetilde{Q}}{\partial t} = \frac{\sigma}{\epsilon_r\epsilon_0}\widetilde{Q}\,. \tag{3.21}$$

Then

$$\widetilde{Q} = \widetilde{Q}_0 \exp\left(-\frac{\sigma}{\epsilon_r\epsilon_0}t\right)\,. \tag{3.22}$$

3.4 The Electric Potential V

It is convenient to describe electric fields in terms of the *electric potential* V. This *scalar* quantity is expressed in *volts*. Contrary to a vector, a scalar has only a magnitude. Both $\boldsymbol{E}$ and V are usually functions of the coordinates, which makes them *point functions*, as well as functions of the time.

Near an isolated point charge Q in a dielectric ϵ_r,

$$V = \frac{Q}{4\pi\epsilon_0\epsilon_r r} \,. \tag{3.23}$$

In general, at a point P, in a vacuum,

$$V = \frac{1}{4\pi\epsilon_0}\int_\infty \frac{\widetilde{Q}}{r}\,dv \,, \tag{3.24}$$

where dv is an element of volume at the point P' where the electric charge density is $\widetilde{Q}$, and where r is the distance between P and P'. This equation sets V equal to zero at points infinitely remote from any charge. We are concerned here with static electric fields, and $\widetilde{Q}$ is constant.

In *electrostatic* fields,

$$\boldsymbol{E} = -\boldsymbol{\nabla} V \,. \tag{3.25}$$

This applies only in the absence of time-dependent magnetic fields, when $\boldsymbol{\nabla} \times \boldsymbol{E} = \boldsymbol{0}$ in Eq. 2.3. See Eq. 5.5.

There is an important difference between $\boldsymbol{E}$ and V. At a given point, $\boldsymbol{E}$ has a specific value that is measurable, at least in principle, but not V: one can add to V, for example to the integral of Eq. 3.24, any quantity that is independent of the coordinates, say K, without affecting $-\boldsymbol{\nabla} V = \boldsymbol{E}$. So the value of V remains undefined until one has chosen, arbitrarily, a value for K.

3.5 The Equations of Poisson and of Laplace

Substituting Eq. 3.25 into Eq. 2.1,

$$\boldsymbol{\nabla}\cdot\boldsymbol{\nabla} V = \boldsymbol{\nabla}^2 V = -\frac{\widetilde{Q}}{\epsilon_0} = -\frac{\widetilde{Q}_f}{\epsilon_r\epsilon_0} \,. \tag{3.26}$$

This is *Poisson's equation* for V. In regions where $\widetilde{Q}_f = 0$, *Laplace's equation* applies and

$$\boldsymbol{\nabla}^2 V = 0 \,. \tag{3.27}$$

These two equations apply on three conditions: the electric field must be constant, the medium must be homogeneous, and it must also be stationary if there is a magnetic field.

3.6 Joule Losses

In the absence of an electric field, the cloud of conduction electrons in a good conductor is in thermal equilibrium with the crystal lattice. Upon application of an electric field, the conduction electrons drift in the direction of $-\boldsymbol{E}$ and gain kinetic energy, which they share with the lattice through collisions. The power thus dissipated in the conductor, per cubic meter, is

$$\widetilde{P} = \boldsymbol{E} \cdot \boldsymbol{J} = \sigma E^2 = J^2/\sigma \,. \tag{3.28}$$

The same losses occur in electrolytes and in plasmas.

When a current I flows through a resistance R, it dissipates a power

$$P = I^2 R \,. \tag{3.29}$$

3.7 Electric Energy

Electric fields exert forces on electric charges, as in Eq. 3.4. Although electric forces are usually negligible, it is occasionally necessary to take into account the *electric energy* stored in an electric field, which is

$$\mathcal{E}_{\rm e} = \int_v \frac{\epsilon_{\rm r}\epsilon_0 E^2}{2}\, dv \,. \tag{3.30}$$

The electric energy density is $(\epsilon_{\rm r}\epsilon_0 E^2/2)$ joules/meter^3.

3.8 Example: A Proton Beam I

As a preparation for our investigation of the propagation of a broad proton beam in the solar atmosphere in Chapters 15 and 16, we discuss here the elementary case of propagation in a vacuum.

In Fig. 3.1, $V = 10^6$ volts, the beam current is 1 milliampere, the beam diameter 1 millimeter, and T has an inside diameter of 10 centimeters. A proton has a mass of $1.67262158 \times 10^{-27}$ kilogram.

a) CALCULATE THE PROTON SPEED.

Each proton has a kinetic energy of 10^6 electronvolts. If V is the accelerating voltage, 10^6 volts,

$$\tfrac{1}{2}mv^2 = eV \,, \tag{3.31}$$

$$v = \left(\frac{2eV}{m}\right)^{1/2} = 1.384 \times 10^7 \text{ meters/second} \,. \tag{3.32}$$

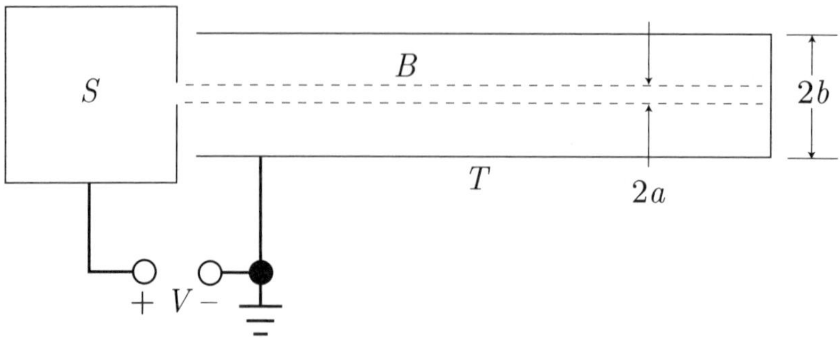

Fig. 3.1. Schematic diagram of a proton accelerator. A gas discharge within the source S ionizes hydrogen gas to produce protons. Some of the protons emerge through a hole and are focused into a beam B inside a conducting tube T. The source is at the potential V and the target, at the right-hand end, is grounded

Recall that the speed of light is approximately 3×10^8 meters/second, from Eq. 2.11. So $v^2 \ll c^2$ and the protons are non-relativistic.

b) Calculate the current density J, the charge density $\widetilde{Q}$, and the charge per meter λ.

Assume that the current density is uniform over the cross-section of the beam and call the beam radius a. Then

$$J = I/(\pi a^2) = 1.273 \times 10^3 \text{ amperes/meter}^2 , \tag{3.33}$$

$$\widetilde{Q} = J/v = 9.198 \times 10^{-5} \text{ coulomb/meter}^3 , \tag{3.34}$$

$$\lambda = I/v = 7.225 \times 10^{-11} \text{ coulomb/meter} . \tag{3.35}$$

c) Calculate the electric field strength outside the beam.

We use Eq. 2.17. Call the inside diameter of the copper tube b. Consider an imaginary cylinder of radius ρ, with $a \leq \rho \leq b$, that is coaxial with the beam and that has a length l. The electric charge inside the cylinder is λl. By symmetry, the electric field is radial. The beam is in a vacuum and $\epsilon_r = 1$. Then, from Eq. 2.17,

$$2\pi\rho l \epsilon_0 E_{\text{outside}} = \lambda l , \tag{3.36}$$

$$E_{\text{outside}} = \frac{\lambda}{2\pi\epsilon_0\rho} = \frac{1.299}{\rho} \text{ volts/meter} . \tag{3.37}$$

See Fig. 3.2.

d) Calculate the electric field strength inside the beam, assuming a uniform current density.

We again use Eq. 2.17:

$$E_{\text{inside}} = \frac{\pi\rho^2\widetilde{Q}}{2\pi\epsilon_0\rho} = 5.194 \times 10^6 \rho \text{ volts/meter} . \tag{3.38}$$

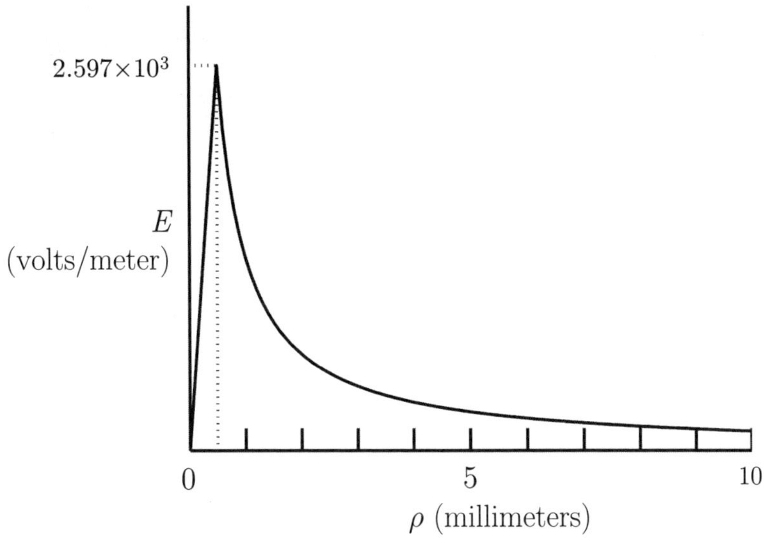

Fig. 3.2. The electric field strength E inside and outside the proton beam of Fig. 3.1

At the surface of the beam, $E_{\text{inside}} = E_{\text{outside}} = 2.593 \times 10^3$ volts/meter.

e) Calculate the electric potential outside the beam.
From Eq. 3.25,

$$\frac{\partial V_{\text{outside}}}{\partial \rho} = -E_{\text{outside}} = -\frac{\lambda}{2\pi\epsilon_0 \rho} , \tag{3.39}$$

$$V_{\text{outside}} = -\frac{\lambda}{2\pi\epsilon_0} \ln \rho + K , \tag{3.40}$$

where K is a constant of integration. Since, by hypothesis, $V_{\text{outside}} = 0$ at $\rho = b = 5$ centimeters,

$$V_{\text{outside}} = \frac{\lambda}{2\pi\epsilon_0} \ln \frac{b}{\rho} = 1.299 \ln \frac{0.05}{\rho} \text{ volts} . \tag{3.41}$$

At the surface of the beam, $\rho = 5 \times 10^{-4}$ meter and

$$V_{\text{outside}, \rho=a} = 5.982 \text{ volts} . \tag{3.42}$$

f) Calculate the electric potential inside the beam.
Inside,

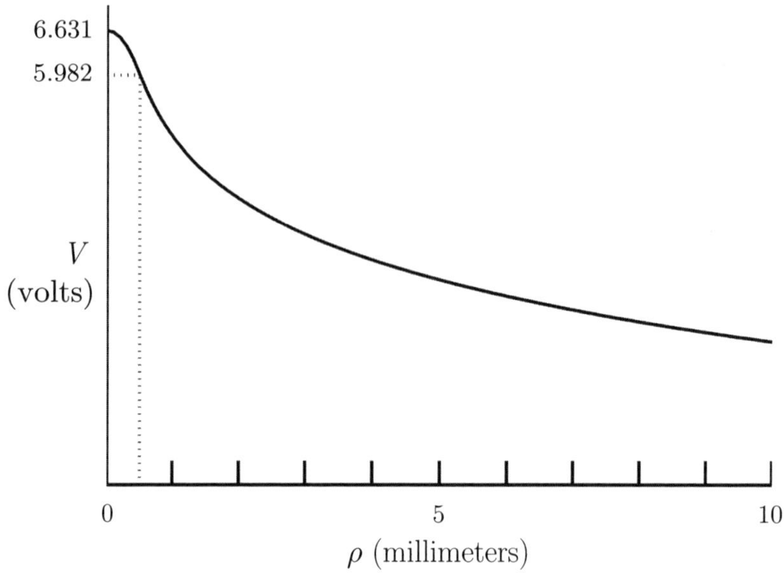

Fig. 3.3. The electric potential V inside and outside the proton beam of Fig. 3.1

$$\frac{\partial V_{\text{inside}}}{\partial \rho} = -E_{\text{inside}} = -\frac{\rho \widetilde{Q}}{2\epsilon_0} , \tag{3.43}$$

$$V_{\text{inside}} = -\frac{\widetilde{Q}\rho^2}{4\epsilon_0} + L , \tag{3.44}$$

where L is another constant of integration. Since, at the surface of the beam, $V_{\text{outside}} = V_{\text{inside}}$,

$$L = 5.982 + \frac{\widetilde{Q}a^2}{4\epsilon_0} , \quad V_{\text{inside}} = 5.982 + 2.597 \times 10^6 (a^2 - \rho^2) \text{ volts} , \tag{3.45}$$

$$V_{\rho=a} = 5.982 \text{ volts} , \quad V_{\rho=0} = 6.631 \text{ volts} . \tag{3.46}$$

See Fig. 3.3.

g) Check Laplace's equation 3.27 outside the beam, where $\widetilde{Q} = 0$.

We use cylindrical coordinates, with $\partial/\partial\phi = 0$, $\partial/\partial z = 0$. Then

$$\nabla^2 V_{\text{outside}} = \frac{1}{\rho}\frac{\partial}{\partial \rho}\left(\rho \frac{\partial V_{\text{outside}}}{\partial \rho}\right) \tag{3.47}$$

$$= \frac{1}{\rho}\frac{\partial}{\partial \rho}(-\rho E_{\text{outside}}) \tag{3.48}$$

$$= \frac{1}{\rho}\frac{\partial}{\partial \rho}\left[\rho\left(-\frac{\lambda}{2\pi\epsilon_0 \rho}\right)\right] \equiv 0 . \tag{3.49}$$

h) Check Poisson's equation inside the beam.
Similarly,

$$\nabla^2 V_{\text{inside}} = \frac{1}{\rho}\frac{\partial}{\partial\rho}\left(\rho\frac{\partial V_{\text{inside}}}{\partial\rho}\right) = \frac{1}{\rho}\frac{\partial}{\partial\rho}(-\rho E_{\text{inside}}) \tag{3.50}$$

$$= \frac{1}{\rho}\frac{\partial}{\partial\rho}\left[\rho\left(-\frac{\rho\widetilde{Q}}{2\epsilon_0}\right)\right] = -\frac{\widetilde{Q}}{\epsilon_0}\,. \tag{3.51}$$

i) Show that, in accordance with Eq. 2.3, the curl of $\boldsymbol{E}$ equals zero both outside and inside the beam.

Since $\boldsymbol{E}$ has only a radial component that is independent of both ϕ and z, $\nabla \times \boldsymbol{E} \equiv \mathbf{0}$.

j) Calculate the electric energy per meter outside the beam.
The electric energy density is $\epsilon_0 E^2/2$, from Sect. 3.7. Then

$$\widetilde{\mathcal{E}}_{\text{outside}} = \int_a^b \frac{\epsilon_0 E^2_{\text{outside}}}{2} 2\pi\rho\, d\rho = 2.162 \times 10^{-10} \text{ joule/meter}\,. \tag{3.52}$$

k) Calculate the electric energy per meter inside the beam.
Similarly,

$$\widetilde{\mathcal{E}}_{\text{inside}} = \int_0^a \frac{\epsilon_0 E^2_{\text{inside}}}{2} 2\pi\rho\, d\rho = 1.173 \times 10^{-11} \text{ joule/meter}\,. \tag{3.53}$$

l) Calculate the radial electrostatic force of repulsion on a proton at the periphery of the beam.

From Eq. 3.38,

$$F_{\text{el},\,\rho=a} = eE_{\text{in},\,\rho=a} = e\frac{\rho I}{2\pi\epsilon_0 a^2 v} \tag{3.54}$$

$$= 4.161 \times 10^{-16} \text{ newton}\,. \tag{3.55}$$

So what is the acceleration of a proton that is subjected to this force?

$$\text{Acceleration} = \frac{4.161 \times 10^{-16}}{1.673 \times 10^{-27}} \approx 2.5 \times 10^{11} \text{ meters/second}^2 \text{ !}$$

As we shall see in Sect. 4.11, there is also a magnetic pinching force that more or less opposes the electric force.

3.9 Case Study: V at the Surface of the Sun

The surface of the Sun that one sees in white light is known as the *photosphere*. Its temperature is about 6400 kelvins, and the pressure about 1.2×10^4

pascals, or about one eighth of the atmospheric pressure at the surface of the Earth. All the elements are present, but hydrogen (90%) and helium (9%) predominate. Only a small fraction of the hydrogen, about 0.04%, is ionized, and the number densities of free electrons and protons are about equal.

What is the electrostatic potential V at the surface of the Sun? The potential is not zero because electrons tend to escape more easily than the much heavier, and slower, ions. Must the net current be zero? Yes because, otherwise, the electrostatic charge on the Sun would drift... and the Sun has been around for quite a while! What is your guess? Is V positive or negative?

It is this evaporation that causes the *solar wind.* A similar phenomenon occurs on stars and galaxies.

This evaporation does not seem to have any appreciable astrophysical significance. Even giant galaxies have center-to-surface potential differences that are only of the order of one kilovolt, like the Sun.

The fraction F of the particles that possess enough energy to escape is given by

$$F = \exp\left(-\frac{mv^2/2}{kT}\right) , \tag{3.56}$$

where v is the *escape velocity*, k is the Boltzmann constant, and T the surface temperature. This is the *Maxwell velocity distribution function.*

a) First, what is the escape kinetic energy for an *uncharged* particle of mass m at the surface of a star of mass M and radius R?

The energy required to pull m away from the radius R to infinity, against the gravitational attraction, is GMm/R. So the escape *kinetic energy,*

$$\frac{1}{2}mv^2 = G\frac{Mm}{R} , \tag{3.57}$$

is proportional to the particle mass m, but the *escape velocity* v is independent of m.

If electrons and protons were uncharged, then the fractions F for electrons and for protons would be

$$F_{\text{e}} = \exp\left(-\frac{GMm_{\text{e}}}{RkT}\right) , \quad F_{\text{p}} = \exp\left(-\frac{GMm_{\text{p}}}{RkT}\right) , \tag{3.58}$$

and then we would have that

$$\frac{F_{\text{e}}}{F_{\text{p}}} = \exp\left[\frac{GM}{RkT}(m_{\text{p}} - m_{\text{e}})\right] \gg 1 . \tag{3.59}$$

This last exponent is about 4×10^3, and the ratio on the left is an even vastly larger number: electrons would escape much more easily than protons.

b) Now what is the escape kinetic energy for a *charged* particle?

The additional energy that is required to pull a particle of charge q out to infinity is $-qV$, where V is the electric potential at the radius R, or *minus* the electric potential energy.[1] The extra energy required, $-qV$, is positive if q and V have opposite signs, or if the electric force is attractive.

The escape kinetic energy for a particle of mass m and charge q is thus

$$\frac{1}{2}mv^2 = G\frac{Mm}{R} - qV \ . \tag{3.60}$$

c) So what is the equilibrium electric potential V at the surface of the star? Assume free electrons and protons at the same temperature, and zero net current.

The number densities of protons and electrons remain constant in a static situation. So $F_e = F_p$, and the escape *kinetic energies* for protons and electrons are equal. As usual, set $e = +1.6 \times 10^{-19}$ coulomb. That is the magnitude of the proton and electron charges. Then

$$G\frac{Mm_p}{R} - eV = G\frac{Mm_e}{R} + eV \ , \tag{3.61}$$

$$G\frac{M(m_p - m_e)}{R} = 2eV \ , \tag{3.62}$$

$$G\frac{Mm_p}{R} \approx 2eV \ , \quad V \approx G\frac{Mm_p}{2eR} \ , \tag{3.63}$$

and the electrostatic potential V at the surface of the star is *positive*: electrons escape somewhat more easily than protons.

d) For the case of the Sun, see the page facing the back cover for the values of the physical constants. In that case,

$$V \approx \frac{(6.7 \times 10^{-11})(2 \times 10^{30})(1.7 \times 10^{-27})}{2 \times (1.6 \times 10^{-19})(7 \times 10^{8})} \approx +1000 \text{ volts} \ . \tag{3.64}$$

Since $V = Q/(4\pi\epsilon_0 R)$,

$$Q \approx 4 \times 3.14 \times (8.85 \times 10^{-12})(7 \times 10^{8}) \times 1000 \approx +80 \text{ coulombs} \ . \tag{3.65}$$

There are slightly more free protons than free electrons on the Sun. The excess mass of protons is

$$80 \times \frac{1.7 \times 10^{-27}}{1.6 \times 10^{-19}} \approx 1 \times 10^{-6} \text{ kilogram} \ ,$$

or about 1 milligram... while the Sun has a mass of 2×10^{30} kilograms!

Since the escape *kinetic energies* are equal, the escape *velocities* are different, the escaping protons are much slower than the escaping electrons, and there is a net positive space charge density in the atmosphere of the Sun.

[1] The electric potential energy qV is equal to zero when q is at an infinite distance from the star, where V is zero, and its magnitude is proportional to $1/r$. It is positive if q and V have the same sign, and negative otherwise. In both cases, the electric force tries to *de*crease the potential energy. If you sketch a graph of the electric potential energy qV as a function of r, you will check that the electric force is repulsive if q and V have the same sign, and attractive otherwise.

3.10 Summary

Electric charge is conserved: any charge that appears within a given volume v bounded by a closed surface $\mathcal{A}$ results from a net electric current crossing the surface $\mathcal{A}$:

$$\int_{\mathcal{A}} \boldsymbol{J} \cdot \boldsymbol{d\mathcal{A}} = -\int_v \frac{\partial \widetilde{Q}}{\partial t} dv\,. \qquad (3.14)$$

At a given point, Eq. 3.15 applies:

$$\boldsymbol{\nabla} \cdot \boldsymbol{J} = -\partial \widetilde{Q} / \partial t\,. \qquad (3.15)$$

Normally, any net electric charge resides at the surface of a conducting body: any net charge deposited inside escapes to the surface within an exceedingly short time. We show in Chapter 7 that, on the contrary, conductors that move in magnetic fields usually carry a *permanent* net space charge inside.

In the absence of a time-dependent magnetic field,

$$\boldsymbol{E} = -\boldsymbol{\nabla} V\,. \qquad (3.25)$$

In time-independent fields and in homogeneous media,

$$\boldsymbol{\nabla} \cdot \boldsymbol{\nabla} V \equiv \boldsymbol{\nabla}^2 V = -\frac{\widetilde{Q}_{\mathrm{f}}}{\epsilon_{\mathrm{r}} \epsilon_0}\,. \qquad (3.26)$$

A medium of conductivity σ that carries a current of density $\boldsymbol{J}$ dissipates energy through the *Joule effect* and the power density is

$$\widetilde{P} = \boldsymbol{E} \cdot \boldsymbol{J} = \sigma E^2 = J^2/\sigma\,. \qquad (3.28)$$

When a current I flows through a resistance R,

$$P = I^2 R\,. \qquad (3.29)$$

The energy density in an electric field is $\epsilon_{\mathrm{r}} \epsilon_0 E^2/2$, as in Eq. 3.30.

4 Constant Magnetic Fields

Contents

After Chapters 2 and 3, this is our third step towards MFD. In this chapter we need not worry about electric fields; that was the object of the preceding chapter. We now have constant magnetic fields, generated by constant currents. Our application to solenoids will serve later in relation with solar magnetic elements and sunspots. We then have a glimpse at the Earth's magnetic field, and we return to the proton beam.

IN DISCUSSING MAGNETIC FIELDS IT IS ESSENTIAL TO REFER CONSTANTLY TO THE ASSOCIATED CURRENTS.

Although this point was repeatedly stressed by Alfvén (1975, 1981), his warnings have been left unheeded. It is of course pointless to discuss the shapes of imaginary magnetic field lines, unless the required electric current distribution can be justified.

Also, since we are concerned here largely with magnetic fields generated by convecting conducting fluids, it is equally important to ask what type of self-excited dynamo can generate the required electric current distribution. See Chapters 10 and following.

As stated in Sect. 2.2, we reserve the term *constant* for quantities that do not vary with time, and the term *uniform* for quantities that have the same value at every point in a given region. Many authors unfortunately use the term *constant* for both properties.

This chapter concerns time-*independent* magnetic fields; the next chapter explores time-dependent magnetic fields. As always, we disregard magnetic media.

4.1 Magnetic Field Lines

Define $\boldsymbol{dl}$ as a short vector parallel to $\boldsymbol{B}$. Then, joining $\boldsymbol{dl}$ vectors end-to-end defines a *magnetic field line.*[1] We return to magnetic field lines in Chapter 6.

Magnetic field lines serve to visualize the configuration of a magnetic field. They also serve to either explain or predict, *qualitatively*, magnetic forces, as we shall see in Sect. 4.8.2. It has long been realized that, otherwise, magnetic field lines have no physical significance (Slepian, 1951). They are of course invisible because they are not material objects. They are an invaluable "thinking crutch", but no more. Unfortunately, many authors treat them, implicitly, as material objects.

4.2 The Magnetic Flux Density $\boldsymbol{B}$ at a Point

At a point P in space,

$$\boldsymbol{B} = \frac{\mu_0}{4\pi} \int_v \frac{\boldsymbol{J} \times \hat{\boldsymbol{r}}}{r^2} \, dv \,, \tag{4.1}$$

where $\boldsymbol{J}$ is the electric current density at the point P', dv is the element of volume at P', $\hat{\boldsymbol{r}}$ is the unit vector that points *from* P' *to* P, and r is the distance between P' and P. The unit of B is the *tesla* (1 tesla $= 10^4$ gausses).

Equation 4.1 is known as the *law of Bio–Savart.* It shows that a magnetic field can exist only in the presence of electric currents.

[1] The ancient term *magnetic line of force* should have been abandoned long ago. It is copied from "electric line of force", which is appropriate because an electric force $Q\boldsymbol{E}$ points in the direction of the vector $\boldsymbol{E}$, but a magnetic force $Q\boldsymbol{v} \times \boldsymbol{B}$ is orthogonal to the vector $\boldsymbol{B}$, as we shall see in Sect. 4.8.

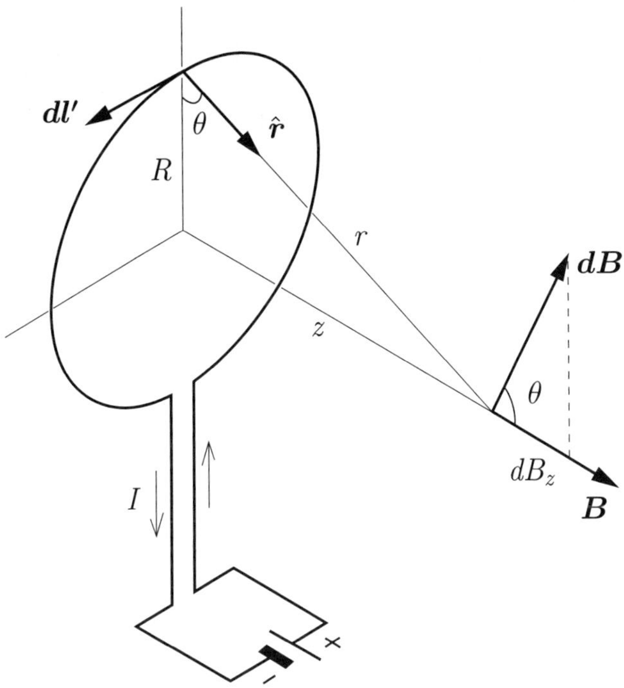

Fig. 4.1. Coil of wire of radius R carrying a current I. The element of length $\boldsymbol{dl}'$ of the wire contributes a field $\boldsymbol{dB}$ on the axis of symmetry, whose component along the z-axis is dB_z. The net magnetic field $\boldsymbol{B}$ of the coil lies on the axis of symmetry

4.2.1 Example: Calculating the Field of a Coil

Figure 4.1 shows a coil of wire of radius R that carries a current I. Clearly, on the axis, the vector $\boldsymbol{B}$ lies along the axis. Let us calculate the value of $\boldsymbol{B}$ at a distance z. Calculating $\boldsymbol{B}$ at a point off the axis is much more difficult. From Eq. 4.1, at a point $(0, 0, z)$ on the axis,

$$dB_z = \frac{\mu_0}{4\pi} \frac{I\, dl'}{r^2} \cos\theta \tag{4.2}$$

and

$$B = \frac{\mu_0}{4\pi} \frac{2\pi R I}{r^2} \cos\theta = \frac{\mu_0 I R^2}{2(R^2 + z^2)^{3/2}} \ . \tag{4.3}$$

The field at the center of the coil, at $z = 0$, is $\mu_0 I/(2R)$. Along the axis, B falls off as $1/z^3$ for $z^2 \gg R^2$.

4.3 Magnetic Flux Φ

By definition, the magnetic flux Φ that links the closed curve C is

$$\Phi = \int_{\mathcal{A}} \boldsymbol{B} \cdot \boldsymbol{d\mathcal{A}} , \tag{4.4}$$

where $\mathcal{A}$ is any open surface bounded by C.

Magnetic flux Φ is expressed in *webers*, and

$$1 \text{ tesla} = 1 \text{ weber/meter}^2 . \tag{4.5}$$

4.4 Ampère's Circuital Law

The integral form of Ampère's circuital law, Eq. 2.20, serves to calculate $\boldsymbol{B}$ for simple geometries:

$$\oint_C \boldsymbol{B} \cdot \boldsymbol{dl} = \mu_0 I_C , \tag{4.6}$$

where I_C is the conduction current linked by the closed curve C. This is reminiscent of Eq. 2.17, which often serves to calculate $\boldsymbol{E}$.

Recall, from Sects. 2.4 and 2.5 that the displacement current and the convection current are both negligible in the context of this book.

The differential form of Ampère's circuital law is Eq. 2.4:

$$\boldsymbol{\nabla} \times \boldsymbol{B} = \mu_0 \boldsymbol{J} . \tag{4.7}$$

4.4.1 Example: The Thin Solenoid

Figure 4.2 shows a cross-section through a thin solenoid. There are $\widetilde{N}$ turns per meter and the current is I. Assume that the solenoid is close-wound so that, in effect, one has a current sheet carrying $\widetilde{N}I$ amperes per meter of length. Assume also that the solenoid is long, and disregard end-effects. Then, inside, $\boldsymbol{B}$ is parallel to the axis. The return magnetic flux outside spreads over a large area where $B \approx 0$. The direction of $\boldsymbol{B}$ inside, and the direction of the current in the winding, satisfy the right-hand-screw rule.

To calculate the axial B inside the solenoid, consider the imaginary path C of height l of the figure. Since $B \approx 0$ outside, and since $\boldsymbol{B}$ is vertical inside,

$$Bl = \mu_0(\widetilde{N}Il) , \quad \text{and} \quad B = \mu_0 \widetilde{N} I . \tag{4.8}$$

The radial position of that portion of the path that lies inside being arbitrary, the B inside a long solenoid is independent of the radial position ρ. Then the enclosed magnetic flux in a thin, long solenoid of radius a is

$$\Phi = \mu_0 \widetilde{N} I \pi a^2 . \tag{4.9}$$

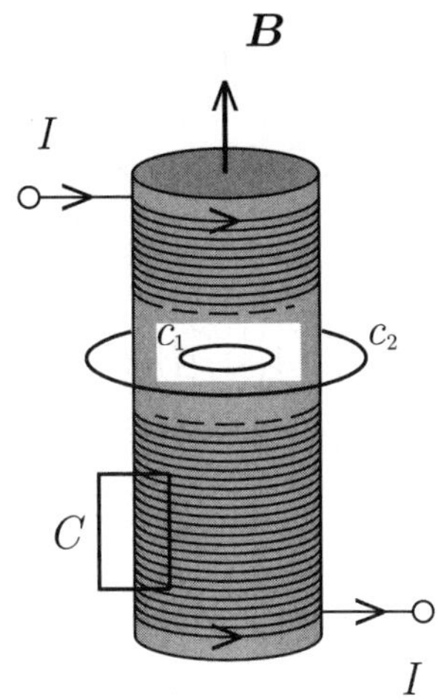

Fig. 4.2. Thin solenoid. The path C has a height l. We refer to paths c_1 and c_2 in Sect. 4.5.2

See Fig. 4.3. Note how quickly the field lines diverge on emerging from the solenoid. On the axis, at a distance of only one diameter from the end, the spacing has increased by a factor of 3.3. So the magnitude of the field decreases rapidly away from its source. This is a general rule that also applies to electric fields. We shall return to this fact when we shall discuss solar coronal loops in Chapters 15 and 16. This rapid decrease applies only to *static* fields: if the current fluctuates rapidly, then the field decreases only as $1/r$. So the field of a loop radio antenna can propagate over large distances.

4.4.2 Example: The Thick Solenoid

A thick solenoid has inner and outer radii a and b. It has $\widetilde{N}$ turns per meter. At the radius ρ such that $a \le \rho \le b$, it is only the current outside the radius ρ that contributes to B and

$$B = \mu_0 \widetilde{N} \frac{b - \rho}{b - a} I \ . \tag{4.10}$$

More generally, at the radius ρ,

$$B = \int_\rho^b \mu_0 \widetilde{N} I \frac{d\rho}{b - a} = \int_\rho^b \mu_0 J \, d\rho \ , \tag{4.11}$$

where J is the current density in the winding, in amperes/meter^2. The current density $\boldsymbol{J}$ is azimuthal. We shall often have occasion to use this last integral.

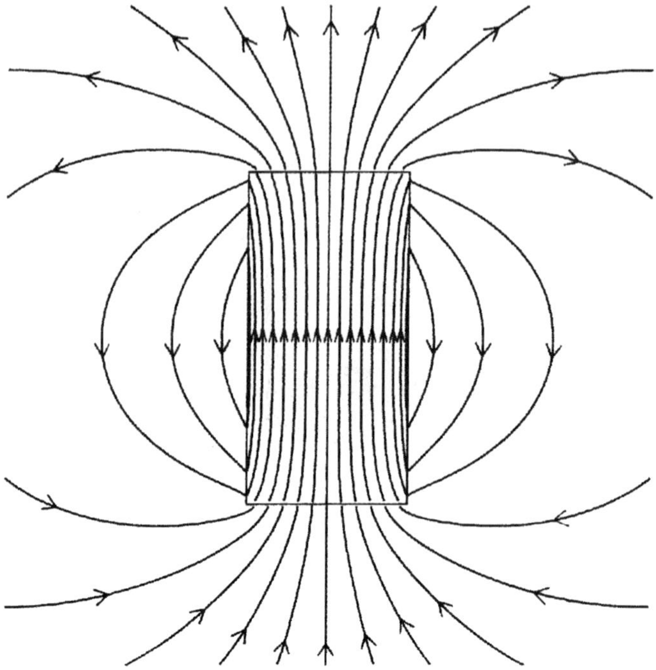

Fig. 4.3. Magnetic field lines for a thin solenoid

4.5 The Vector Potential

Recall that we stated in Sect. 3.4 that it is useful to define an electric potential V for electric fields, with $\boldsymbol{E} = -\boldsymbol{\nabla} V$.

Similarly, it is useful to describe magnetic fields in terms of the *vector potential* $\boldsymbol{A}$, defined by

$$\boldsymbol{B} = \boldsymbol{\nabla} \times \boldsymbol{A} \,. \tag{4.12}$$

This relation is allowed because $\boldsymbol{\nabla} \cdot \boldsymbol{B} = 0$, and the divergence of a curl is identically equal to zero.

Recall also that we found in Sect. 3.4 that, whereas $\boldsymbol{E}$ is measurable, any uniform quantity can augment V without affecting $\boldsymbol{E}$. We have a similar situation here in that $\boldsymbol{B}$ is measurable, while adding to $\boldsymbol{A}$ any vector whose curl is zero leaves $\boldsymbol{B}$ unaffected.

If C is a closed curve that bounds an arbitrary open surface $\mathcal{A}$, then, from Stokes's theorem (see the page facing the front cover), the magnetic flux linking C is

$$\Phi = \int_{\mathcal{A}} \boldsymbol{B} \cdot \boldsymbol{d\mathcal{A}} = \int_{\mathcal{A}} (\boldsymbol{\nabla} \times \boldsymbol{A}) \cdot \boldsymbol{d\mathcal{A}} = \oint_C \boldsymbol{A} \cdot \boldsymbol{dl} \,. \tag{4.13}$$

Given a current distribution in a volume v, then, at a point P either inside or outside v,

$$\boldsymbol{A} = \frac{\mu_0}{4\pi} \int_v \frac{\boldsymbol{J}}{r}\, dv \,, \tag{4.14}$$

where $\boldsymbol{J}$ and dv are at P', and r is the distance between P and P'. Recall, from the Introduction to Chapter 2 that we disregard magnetic materials.

4.5.1 Example: The Vector Potential near a Circuit

Note the $1/r$ factor in the above integral: it shows that, at a point close to a circuit, $\boldsymbol{A}$ is parallel to the local $\boldsymbol{J}$ and points in the same direction.

The time derivative $\partial \boldsymbol{A}/\partial t$ is important in MFD, as we shall see in the next chapter.

4.5.2 Example: The Thin Solenoid

Refer to Fig. 4.2. We found that, inside the solenoid, the magnetic field is uniform and $B = \mu_0 \widetilde{N} I$, where $\widetilde{N}$ is the number of turns per meter and I the current. The circular paths c_1 and c_2 are normal to the axis of symmetry and centered on the axis.

The vector $\boldsymbol{A}$ is tangential to c_1 and points in the direction of the current. From Eq. 4.13,

$$2\pi\rho A_{\text{inside}} = \pi\rho^2 \mu_0 \widetilde{N} I \,, \quad A_{\text{inside}} = \mu_0 \widetilde{N} I \frac{\rho}{2} \,. \tag{4 15}$$

Outside the solenoid, $\boldsymbol{A}$ is tangential to c_2 and also points in the direction of the current. Then

$$2\pi\rho A_{\text{outside}} = \pi a^2 \mu_0 \widetilde{N} I \,, \quad A_{\text{outside}} = \mu_0 \widetilde{N} I \frac{a^2}{2\rho} \,. \tag{4 16}$$

Observe that, outside a long solenoid,

$$\boldsymbol{A} \neq \boldsymbol{0} \,, \quad \text{but} \quad \boldsymbol{B} = \boldsymbol{\nabla} \times \boldsymbol{A} = \boldsymbol{0} \,. \tag{4.17}$$

Figure 4.3 shows magnetic field lines for a thin solenoid.

4.6 Magnetic Multipoles

Imagine a finite volume v that carries an arbitrary current distribution. The origin of coordinates O is inside v, and the point P is outside, at a distance $\boldsymbol{r}$ from O. Since the current distribution is not specified, $\boldsymbol{B}$ is an unknown function of $\boldsymbol{r}$, but a general discussion is nonetheless illuminating.

In the case of the Earth, v is the liquid part of the core, where dynamos, self-excited or not (Chapters 10 to 13), generate electric currents, and the point P is at the surface of the Earth.

It is possible to express $\boldsymbol{B}$ at the surface of the Earth as the sum of the $\boldsymbol{B}$'s of simple current distributions called *magnetic multipoles* (Durand, 1968; Owen, 1963; van Bladel, 1985). This procedure is analogous to Fourier analysis, where we consider a complex waveform of frequency f to be the sum of an infinite series of sine and cosine waves of frequencies f, $2f$, $3f$, etc.

This procedure is also similar to the calculation of the electric field $\boldsymbol{E}$ of a complex charge distribution by adding the fields of *electric multipoles*: a point charge, an electric dipole, an electric quadrupole, an electric octupole, etc. The calculation of $\boldsymbol{B}$ is however much more complex than the calculation of $\boldsymbol{E}$, and it is carefully avoided by most authors. The only satisfactory discussion seems to be that of Durand (1968, pp. 255 ff.).

The simplest source of magnetic field (assuming the non-existence of magnetic monopoles) is the current loop, or *magnetic dipole*. From Sect. 4.2.1, on the axis of symmetry, the magnetic field is axial and

$$B = \frac{\mu_0 I a^2}{2(a^2 + r^2)^{3/2}} \propto \frac{1}{r^3} \quad \text{if} \quad r^2 \gg a^2 \,. \tag{4.18}$$

Off the axis, the expression for $\boldsymbol{B}$ is more complex but, far away from the source, it is proportional to $1/r^3$.

The next simplest source is the *magnetic quadrupole*, which is a pair of dipoles, some distance apart, of opposite polarities, and not necessarily coaxial. Its $\boldsymbol{B}$ varies as $1/r^4$, at some distance from the source. One could go on indefinitely in this way, with pairs of quadrupoles, pairs of octupoles, etc. The field is inversely proportional to a power of r that increases from 3, to 4, to 5, etc., as the source becomes more and more complex.

Outside an arbitrary current distribution, $\boldsymbol{B}$ can thus be expressed as an infinite series of terms corresponding to the fields of a dipole, of a quadrupole, of an octupole, etc., proportional to $1/r^3$, $1/r^4$, $1/r^5$, etc. Close to the current distribution, the field can be quite complex because many terms in the series can be important. But, as we go further and further away, r becomes larger and larger and the higher-order terms become smaller and smaller. A long distance away, only the lowest order term, usually the dipole term, remains.

One might say that remoteness from the currents "filters out" complicated magnetic fields, just as remoteness from electric charges "filters out" complicated electric fields.

4.7 Case Study: The Earth's Magnetic Field

Figure 4.4 shows a cross-section of the Earth. Because of its high temperature, the iron core is non-magnetic: the Curie temperature, above which iron is non-

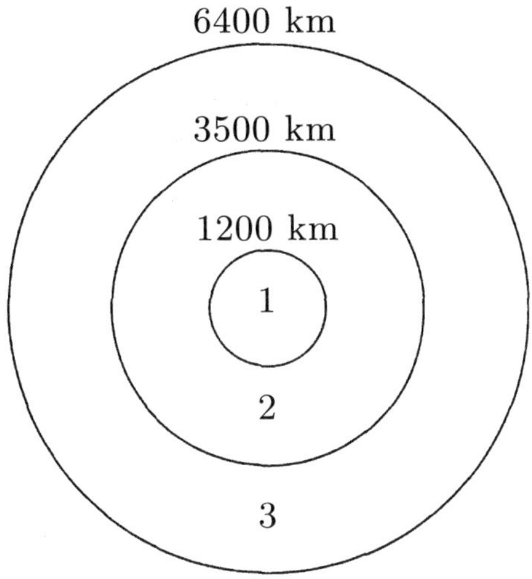

Fig. 4.4. Cross-section through the Earth, drawn to scale. 1. Solid inner core, mostly iron, radius $\approx$ 1200 kilometers, temperature $\approx$ 5000 kelvins, pressure at the center $\approx 3.6 \times 10^5$ atmospheres. 2. Liquid outer core, also mostly iron, outer radius $\approx$ 3500 kilometers, temperature $\approx$ 4000 kelvins. 3. Mantle. The crust has a thickness of only about 30 kilometers

magnetic, is about 800 kelvins. We return to the Earth's magnetic field in Chapter 9.

The heat generation in the core, in the mantle, and in the crust is attributed mostly to the disintegration of K^{40}, which has a half-life of 1.25×10^9 years. The resulting outward heat flux at the surface is roughly 5×10^{-2} watt/meter^2 (Stacey, 1992). This is negligible when compared to solar radiation, which amounts to about 1.3 kilowatts/meter^2 above the atmosphere.

The core temperature is about two-thirds of the temperature at the surface of the Sun, which is 6400 kelvins. So the core is nearly as bright as the Sun!

The electrical conductivity of the core's shell is estimated to be about 3×10^5 siemens/meter. That of the mantle is believed to vary widely from about 10^{-2} siemens/meter near the surface to about 10^3 near the outer core (Stacey, 1992). For comparison, the electrical conductivity of room-temperature iron is 1×10^7 siemens/meter, while that of room-temperature copper is 5.8×10^7 siemens/meter.

The liquid outer shell of the core acts as a dynamo: assuming a seed magnetic field, the convection of the conducting fluid in that field generates currents, and thus a further magnetic field that, eventually, becomes more or less constant (see Chapter 10).

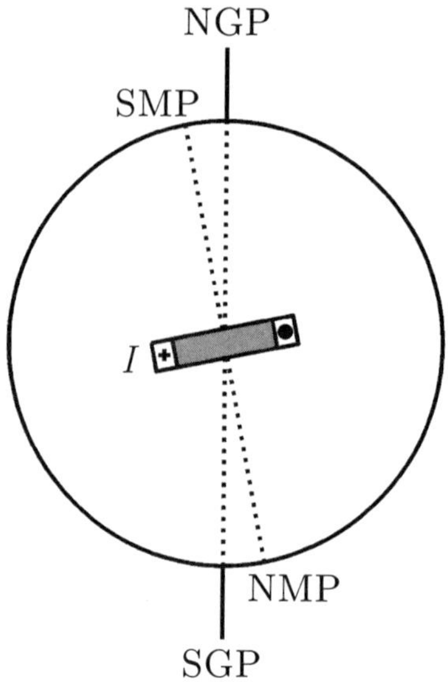

Fig. 4.5. Cross-section of the Earth and imaginary coil that could generate its dipolar magnetic field. We calculate that the current required in the coil would be about 8×10^{11} amperes, or 800 giga-amperes. The geographic poles (GP) and the magnetic poles (MP) do not coincide

Since the temperature and the chemical composition of the liquid core are not uniform, it convects under the combined action of gravitational, centrifugal, and Coriolis forces.

Imagine that the Earth's dipole field originates in a coil as in Fig. 4.5. That figure is not drawn to scale. Say the coil lies half-way down in the liquid part of the core; its radius a is about 2300 kilometers. What is the current in the coil? Now, at either one of the Earth's magnetic poles, or at one Earth radius z from the coil, $B \approx 10^{-4}$ tesla, or one gauss. From Sect. 4.2.1,

$$I = \frac{2(a^2 + z^2)^{3/2} B}{\mu_0 a^2} . \tag{4.19}$$

With

$$a = 2.3 \times 10^5 \text{ meters} \quad \text{and} \quad z = 6.4 \times 10^6 \text{ meters} , \tag{4.20}$$

$$I \approx 8 \times 10^{11} \text{ amperes} . \tag{4.21}$$

The power source is probably just the cooling of the core.

The complexity of the Earth's magnetic field at the surface (Glatzmaier and Roberts, 1995; Ladbury, 1996) results from the fact that the flow in the liquid part of the core is hugely complex, possibly as much so as the flow of air in the atmosphere, but on a vastly different time scale. There is probably a swarm of self-excited dynamos, all more or less coupled together. Moreover, the polarity of the dipolar component at the surface has changed many times over the Earth's history.

Disregarding the much weaker currents that flow in the mantle and in the ionosphere, the magnetic field that we observe at the surface of the Earth is that of the core currents, about 3 megameters away. The surface field has a large dipole component, but it also has quadrupole, octupole, etc. components.

The axis of symmetry of the dipole component forms an angle of a dozen degrees with the axis of rotation, and the axes do not lie in the same plane. The **North magnetic** pole is close to the **South geographic** pole, and inversely, as in Fig. 4.5. So $\boldsymbol{B}$ at the North magnetic pole points from the Earth outward. At the Equator, the dipole component of the magnetic field is horizontal and points North. The angle between true geographic North and the position of a compass is called the *declination*. Roughly, at the Equator declination varies from 20 degrees West to 10 degrees East, and, at a latitude of 45 degrees North it varies between 20 degrees West and 20 degrees East (Jacobs, 1987).

In the northern hemisphere the $\boldsymbol{B}$ of the dipolar component is nearly vertical (about 75 degrees at a latitude of 45 degrees) and points down. So a compass needle needs a small weight at one end or the other to keep it horizontal. There are compasses that have a horizontal axis of rotation to measure the *inclination* of the field.

Both magnetic poles wander slowly with time, and about 100 polarity reversals have occurred over the past 80 million years (De Bremaeker, 1985). See Sect. 10.2.1.

Both the dipolar and the multipolar fields vary, with time. For example, the magnitude of the dipolar field decreases by about 0.05% per year, and the multipolar field moves west by about 0.2 degree of longitude per year, or by about one millimeter per second. The speed at which these changes occur is proof that the source of the magnetic field does not lie in the mantle, which is solid, and that at least part of the core is liquid.

The field in the core is vastly more complex than the field at the surface because the multipolar components attenuate rapidly with distance. Also, the mantle filters out rapid changes. Indeed, if it were possible to survey the field at the surface of the core, it would be impossible to discern a dipole component. There exists a multitude of self-excited dynamos in the core.

The ionosphere, that surrounds the Earth at a height of about 60 kilometers and over, carries electric currents that affect somewhat the magnetic field measured at the surface. Above about 60 kilometers, there are enough free electrons to render the gas conducting. The nature of the solar radiation that ionizes the gas varies with altitude because the atmosphere filters out both long (infrared) and short (ultraviolet) wavelengths. Also, the chemical composition of the atmosphere varies with altitude and, at great heights, hydrogen predominates. Because of lunar tides, and because of the intermittent solar heating, the ionosphere moves up and down in the Earth's magnetic field, which causes electric currents to flow.

Electrons and protons oscillate North and South along lines of $\boldsymbol{B}$ with a period of about one half second. Protons drift westward, while electrons drift eastward, which gives a net westward ring current and a magnetic field that opposes the Earth's dipolar component. These oscillating charges form the *van Allen belt.*

In principle, it is possible to calculate the magnetic field at the core by interpolating the field at the surface, but that is not advisable because the higher-order multipole components, which must be multiplied by large factors, are poorly known.

For an extensive discussion of geomagnetism, see Jacobs (1987).

We return to the Earth's magnetic field in Chapter 9.

4.8 The Magnetic Force

The *magnetic force* on a particle of charge Q that moves at a velocity $\boldsymbol{v}$ in a magnetic field $\boldsymbol{B}$ is

$$\boldsymbol{F} = Q(\boldsymbol{v} \times \boldsymbol{B}) \,. \tag{4.22}$$

The magnetic force is perpendicular to the velocity and thus deflects a particle without affecting its speed.

If there is also an electric field $\boldsymbol{E}$, then

$$\boldsymbol{F} = Q[\boldsymbol{E} + (\boldsymbol{v} \times \boldsymbol{B})] \,. \tag{4.23}$$

This equation follows from Eq. 4.22, but it is really a consequence of Special Relativity, as we shall see in Sect. 6.2.2.

The magnetic force on an electric *current* follows from Eq. 4.22. As we saw in Sect. 3.2, we may assume that, in plasmas and in good conductors, the charge carriers are conduction electrons. Say the charge density of the conduction electrons at a given point is $\widetilde{Q}_{ce}$, a negative quantity, and that their mean drift velocity is $\boldsymbol{v}$. Then $\boldsymbol{J} = \widetilde{Q}_{ce}\boldsymbol{v}$, and the magnetic force per unit volume exerted on the conductor is

$$\widetilde{\boldsymbol{F}} = \widetilde{Q}_{ce}(\boldsymbol{v} \times \boldsymbol{B}) = \boldsymbol{J} \times \boldsymbol{B} \,. \tag{4.24}$$

That is the magnetic force density exerted on the conduction electrons. What is the force density on the *conductor*? The conduction electrons are embedded in the conductor and continually collide with the atoms and molecules that stand in their way. So $\widetilde{\boldsymbol{F}}$ is also the force per unit volume on the conductor.

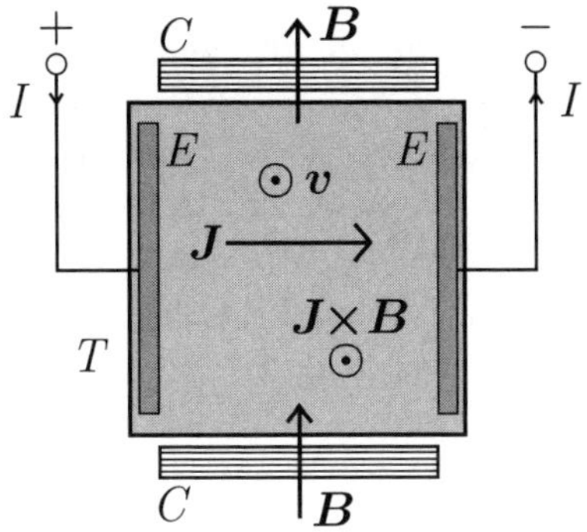

Fig. 4.6. Schematic diagram of an electromagnetic pump for a conducting fluid. The fluid flows out of the paper at the velocity $\boldsymbol{v}$ through the insulating tube T. Coils C apply a vertical magnetic field $\boldsymbol{B}$, and a current I is made to flow between a pair of electrodes E, on the sides of the tube. The force density $\boldsymbol{J} \times \boldsymbol{B}$ on the fluid points out of the paper

We saw above that the magnetic force of Eq. 4.22 deflects a moving charged particle without changing its speed, and hence without changing its kinetic energy. The magnetic force can nonetheless produce useful work by acting on current-carrying conductors, as in electric motors.

Whenever a current flows in a magnetic field, the charge carriers drift sideways. This is the *Hall effect*. We return to the Hall effect in Sect. 6.4.

4.8.1 Example: The Electromagnetic Pump

Figure 4.6 shows the principle of operation of an electromagnetic pump (Shercliff, 1965) With $\boldsymbol{B}$ pointing up and I flowing to the right, the fluid is made to flow out of the paper.

Figure 4.7 shows a rough approximation for the magnetic field of the pump of Fig. 4.6. A proper calculation taking into account the non-uniformity of the magnetic field of coils C and edge effects around the electrodes, would be excessively complex. The conducting fluid flows to the right. Note how the magnetic field lines shift *upstream*. The field lines are stationary. We return to these two aspects of the field in Chapter 8. The device is particularly suited for handling corrosive, conducting fluids such as the liquid sodium that serves as coolant in certain nuclear reactors.

4.8.2 Magnetic Field Lines Again

Figure 4.8 shows a cross-section of two parallel current-carrying wires, together with their magnetic field lines. In part (a) of the figure, the currents

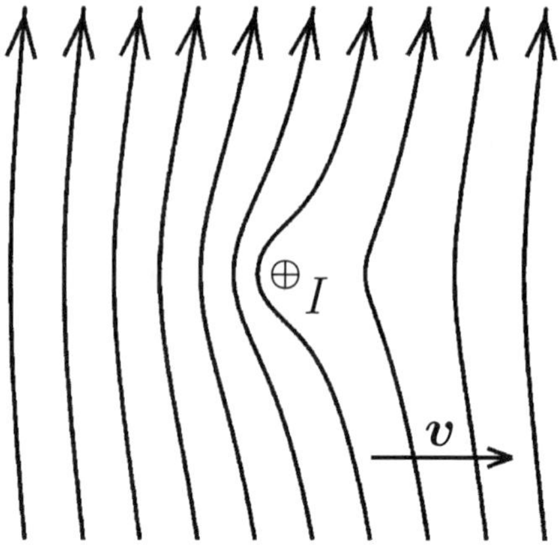

Fig. 4.7. Magnetic field lines for a tube carrying a current I, perpendicular to the paper and situated in a uniform vertical magnetic field. This is a rough approximation for the magnetic field of the electromagnetic pump of Fig. 4.6

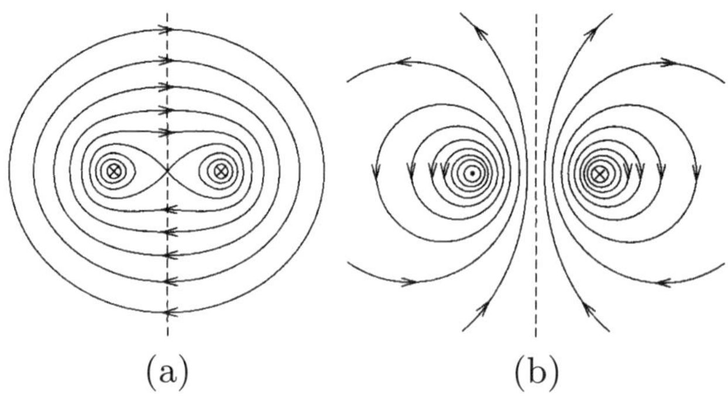

Fig. 4.8. Magnetic field lines for two parallel currents. Parallel currents attract, and antiparallel currents repel

flow in the same direction, and the force is attractive. Intuitively, the field lines "are under tension". In part (b), the currents flow in opposite directions and the force is repulsive. Again intuitively, the field lines "repel laterally". So we have the following rule:

Magnetic field lines "are under tension" and "repel laterally".

A similar rule applies to electric field lines.

4.9 Magnetic Pressure

A thin long solenoid has $\widetilde{N}$ turns per meter and carries a current I. The wire has a square cross-section of side a. Then $J = I/a^2$ and $\widetilde{N} = 1/a$. Inside the solenoid, from Sect. 4.4.1, $B = \mu_0 \widetilde{N} I$ and, outside, $B = 0$. Inside the winding,

$$B_{\text{av}} = \frac{\mu_0 \widetilde{N} I}{2}, \quad I = \frac{2B_{\text{av}}}{\mu_0 \widetilde{N}} . \tag{4.25}$$

The outward magnetic force on the wire, per cubic meter, is

$$\widetilde{F} = JB_{\text{av}} = \frac{I}{a^2} B_{\text{av}} = \frac{2B_{\text{av}}}{\mu_0 \widetilde{N} a^2} B_{\text{av}} \tag{4.26}$$

$$= \frac{2B_{\text{av}}^2}{\mu_0 \widetilde{N} a^2} = \frac{B^2}{2\mu_0 a} , \tag{4.27}$$

since $\widetilde{N} a = 1$.

What is the force per square meter on the winding, or the magnetic pressure? A short length l of the wire has a volume la^2 and has an area la. Then the force per unit area is

$$p_{\text{mag}} = \frac{\widetilde{F} l a^2}{la} = \frac{B^2}{2\mu_0} . \tag{4.28}$$

The outward magnetic force on the solenoid winding is the same as if there were a gas pressure of $B^2/(2\mu_0)$ pascals inside.

4.9.1 Example

Inside a certain solenoid, $B = 100$ teslas. Then the magnetic pressure is

$$\frac{100^2}{2 \times 4\pi \times 10^{-7}} \approx 4 \times 10^9 \text{ pascals} \approx 40\,000 \text{ atmospheres} .$$

Clearly, designing such a solenoid is a major engineering challenge.

Some solenoids are designed to operate only during a small fraction of a second, before exploding.

4.10 Magnetic Energy

The magnetic energy per unit volume is equal to the magnetic pressure $B^2/(2\mu_0)$. Thus the energy stored in a magnetic field occupying a volume v is

$$\mathcal{E}_{\text{mag}} = \int_v \frac{B^2}{2\mu_0} dv . \tag{4.29}$$

In the above case, the energy density is 4×10^9 joules/meter^3.

Note that μ_0 appears here in the *denominator* while, in Sect. 3.7, ϵ_0 appears in the *numerator* of the integrand for the *electric* energy.

4.11 Example: A Proton Beam II

We return to the 1-milliampere 1-million electronvolt proton beam of Sect. 3.8. Recall that this Example is in preparation for our study of the channeling of spicules and coronal loops in the solar atmosphere (Chapters 14, 15, and 16).

From Sect. 4.5, the vector potential $\boldsymbol{A}$ is parallel to the beam, both inside and outside, and $\boldsymbol{B}$ is azimuthal.

a) CALCULATE B OUTSIDE THE BEAM.
At a radius ρ, from Ampère's circuital law, Eq. 2.20,

$$2\pi\rho B = \mu_0 I\ , \quad B = \frac{\mu_0 I}{2\pi\rho} = \frac{2 \times 10^{-10}}{\rho}\ \text{tesla}\ , \tag{4.30}$$

$$B_{\rho=a} = 4.000 \times 10^{-7}\ \text{tesla}\ . \tag{4.31}$$

b) CALCULATE B INSIDE THE BEAM.
At a radius ρ, again from Eq. 2.20,

$$2\pi\rho B = \mu_0 \pi \rho^2 J\ , \quad B = \mu_0 \rho J/2 = 8.000 \times 10^{-4} \rho\ \text{tesla}\ . \tag{4.32}$$

The magnetic field vanishes on the axis, and increases linearly with ρ. At the surface of the beam, $B = B_{\rho=a}$.

c) CALCULATE THE MAGNETIC FORCE EXERTED ON A PROTON AT THE PERIPHERY OF THE BEAM.
From Eq. 4.22, the force points *inward*:

$$F_{\text{mag}} = -ev\frac{\mu_0 I}{2\pi a} = -8.870 \times 10^{-19}\ \text{newton}\ . \tag{4.33}$$

d) COMPARE THIS PINCHING MAGNETIC FORCE WITH THE REPULSIVE FORCE OF EQ. 3.55.
Recall, from Eq. 3.38 that, at the periphery,

$$F_{\text{el}} = e\frac{I}{2\pi\epsilon_0 a v}\ . \tag{4.34}$$

Then

$$\frac{F_{\text{mag}}}{F_{\text{el}}} = -v^2\epsilon_0\mu_0 = -\frac{v^2}{c^2}\ , \tag{4.35}$$

from Eq. 2.11. Thus

$$F_{\text{net}} = F_{\text{el}} + F_{\text{mag}} = F_{\text{el}}\left(1 - \frac{v^2}{c^2}\right)\ . \tag{4.36}$$

In this case the magnetic pinching force is negligible compared to the repulsive electrostatic force, since we found in Sect. 3.8 that $v = 1.384 \times 10^7$

meters/second and $v^2/c^2 \approx 0.002$. If $v^2 \approx c^2$, there is zero net radial force on a peripheral proton.

We have been considering the propagation of a proton beam in a vacuum. The situation is different if the beam propagates in a low-pressure gas. Then the beam ionizes the gas, and ambient electrons fall into the beam, more or less neutralizing the space charge and the electric repulsive force.

e) CALCULATE THE MAGNETIC ENERGY PER UNIT LENGTH INSIDE THE BEAM.

From Eq. 4.29,

$$\widetilde{\mathcal{E}}_{\text{mag, in}} = \int_0^a \frac{B^2}{2\mu_0} 2\pi\rho\, d\rho = \int_0^a \frac{(\mu_0 \rho J/2)^2}{2\mu_0} 2\pi\rho\, d\rho \tag{4.37}$$

$$= \frac{\pi\mu_0 J^2 a^4}{16} = \frac{\mu_0}{16} I^2 \tag{4.38}$$

$$= 7.854 \times 10^{-14} \text{ joule/meter} . \tag{4.39}$$

f) CALCULATE THE MAGNETIC ENERGY PER UNIT LENGTH OUTSIDE THE BEAM, ASSUMING THAT THE BEAM CURRENT RETURNS THROUGH THE COPPER TUBE.

Then there is no magnetic field outside the tube and

$$\widetilde{\mathcal{E}}_{\text{mag, out}} = \int_a^b \frac{B^2}{2\mu_0} 2\pi\rho\, d\rho = \int_a^b \left(\frac{\mu_0 I}{2\pi\rho}\right)^2 \frac{2\pi\rho\, d\rho}{2\mu_0} \tag{4.40}$$

$$= \frac{\mu_0 I^2}{4\pi} \ln\frac{b}{a} = 4.605 \times 10^{-13} \text{ joule/meter} . \tag{4.41}$$

Most of the magnetic energy lies outside the beam.

4.12 Summary

A *magnetic field line* is a fictitious entity that is useful for visualizing magnetic fields. A magnetic field line is everywhere parallel to the local $\boldsymbol{B}$ vector. To visualize magnetic forces, it is useful to remember the rule: "magnetic field lines are under tension, and repel laterally."

The *magnetic flux* that links a closed curve C is the surface integral of $\boldsymbol{B}$ through any open surface $\mathcal{A}$ bounded by C:

$$\Phi = \int_{\mathcal{A}} \boldsymbol{B} \cdot \boldsymbol{d\mathcal{A}} . \tag{4.4}$$

Ampère's circuital law states that the integral of $\boldsymbol{B} \cdot \boldsymbol{dl}$ around a closed curve C is proportional to the net electric current linking C:

$$\oint_C \boldsymbol{B} \cdot \boldsymbol{dl} = \mu_0 I_C , \tag{4.6}$$

and

$$\boldsymbol{\nabla} \times \boldsymbol{B} = \mu_0 \boldsymbol{J} \,. \qquad (4.7)$$

The *vector potential* $\boldsymbol{A}$ is defined by

$$\boldsymbol{B} = \boldsymbol{\nabla} \times \boldsymbol{A} \,. \qquad (4.12)$$

It follows that the magnetic flux linking a closed curve C is given by the line integral of $\boldsymbol{A} \cdot \boldsymbol{dl}$:

$$\Phi = \int_{\mathcal{A}} \boldsymbol{B} \cdot \boldsymbol{d\mathcal{A}} = \int_{\mathcal{A}} (\boldsymbol{\nabla} \times \boldsymbol{A}) \cdot \boldsymbol{d\mathcal{A}} = \oint_C \boldsymbol{A} \cdot \boldsymbol{dl} \,. \qquad (4.13)$$

Outside an arbitrary current distribution, $\boldsymbol{B}$ can be written as a series of terms corresponding to that of a dipole, of a quadrupole, of an octupole, etc., varying with distance respectively as $1/r^3$, $1/r^4$, $1/r^5$, etc.

The *magnetic force* per unit volume at a given point in a conductor is $\boldsymbol{J} \times \boldsymbol{B}$.

If a current is distributed over a surface, as in a thin solenoid, with a magnetic field $\boldsymbol{B}$ on one side and no magnetic field on the other, the magnetic field exerts an outward *magnetic pressure* of $B^2/(2\mu_0)$.

The *magnetic energy density* in a magnetic field is equal to $B^2/(2\mu_0)$, like the magnetic pressure. Hence the magnetic energy is given by

$$\mathcal{E}_{\text{mag}} = \int_v \frac{B^2}{2\mu_0} \, dv \,. \qquad (4.29)$$

5 Time-dependent Magnetic Fields: The Law of Faraday

Contents

In this short chapter, the magnetic field is a function of the time, and this brings us back to electric fields because time-dependent magnetic fields are always accompanied by electric fields, as we saw in Chapter 2. We return to the solenoid and to the Earth's magnetic field.

5.1 The Electric Field Strength

If $\boldsymbol{B}$ is either constant or zero, then Eqs. 2.12 and 3.25 apply:

$$\boldsymbol{\nabla} \times \boldsymbol{E} = \boldsymbol{0}\,, \quad \boldsymbol{E} = -\boldsymbol{\nabla} V\,. \tag{5.1}$$

If $\boldsymbol{B}$ fluctuates, then it is Eq. 2.3 that applies. From Eq. 4.12,

$$\boldsymbol{\nabla} \times \boldsymbol{E} = -\frac{\partial \boldsymbol{B}}{\partial t} = -\frac{\partial}{\partial t}(\boldsymbol{\nabla} \times \boldsymbol{A}) = \boldsymbol{\nabla} \times \left(-\frac{\partial \boldsymbol{A}}{\partial t}\right)\,. \tag{5.2}$$

Then

$$\boldsymbol{\nabla} \times \left(\boldsymbol{E} + \frac{\partial \boldsymbol{A}}{\partial t}\right) = \boldsymbol{0} \tag{5.3}$$

and we can set the parenthesis equal to the gradient of a scalar function because the curl of a gradient is identically equal to zero. Now, if $\partial \boldsymbol{A}/\partial t = \boldsymbol{0}$, then $\boldsymbol{E} = -\boldsymbol{\nabla} V$, as in Eq. 3.25. It follows that

$$\boldsymbol{E} + \frac{\partial \boldsymbol{A}}{\partial t} = -\boldsymbol{\nabla} V \,, \tag{5.4}$$

or

$$\boldsymbol{E} = -\boldsymbol{\nabla} V - \frac{\partial \boldsymbol{A}}{\partial t} \,. \tag{5.5}$$

THIS FUNDAMENTAL EQUATION ALWAYS APPLIES. We shall use it repeatedly.

It means that an electric field can arise from two distinct phenomena: there can be charge accumulations, and then $\boldsymbol{\nabla} V \neq \boldsymbol{0}$, or the magnetic field can be time-dependent, and then $\partial \boldsymbol{A}/\partial t \neq \boldsymbol{0}$. In this chapter we are concerned with the latter phenomenon.

We delay the important case of conductors that move in magnetic fields to Chapters 6 and 8.

5.2 The Law of Faraday[1]

Disregard the $-\boldsymbol{\nabla} V$ term in Eq. 5.5. Then

$$\boldsymbol{E} = -\frac{\partial \boldsymbol{A}}{\partial t} \,. \tag{5.6}$$

If we take the curl of both sides, we find the differential form of the *law of Faraday*, Eq. 2.3. If we take the line integral of both sides over a closed curve C and use Eq. 4.13, we find the integral form of the same law, Eq. 2.19.

Note the difference between the two forms of the law of Faraday. Equation 5.6 is a *local* relation between the values of $\boldsymbol{E}$ and of the time derivative of $\boldsymbol{A}$ *at a given point* and at a given time. But the integral form, Eq. 2.19,

$$\oint_C \boldsymbol{E} \cdot \boldsymbol{dl} = -\frac{d}{dt} \int_{\mathcal{A}} \boldsymbol{B} \cdot \boldsymbol{d\mathcal{A}} \,, \tag{5.7}$$

is *non-local*: it applies over a finite area $\mathcal{A}$ bounded by the curve C.

5.2.1 Lenz's Law

Note the negative sign in Eq. 5.6. Imagine a loop of wire whose axis of symmetry is the z-axis. The loop carries a current I in the $+\hat{\boldsymbol{\phi}}$-direction, and $\boldsymbol{B}$ points in the $+z$-direction. If I *de*creases, $\boldsymbol{A}$ decreases proportionally, and $\partial \boldsymbol{A}/\partial t$ points in the $-\hat{\boldsymbol{\phi}}$-direction. Then the induced electric field $\boldsymbol{E}$ points in the $+\hat{\boldsymbol{\phi}}$-direction, like the current I. So the induced electric field $\boldsymbol{E}$ opposes a *change* in $\boldsymbol{A}$ and $\boldsymbol{B}$. *This is Lenz's law.*

[1] Remember, from Sect. 1.3, that the only education that Faraday had was in Sunday school!

5.2.2 Example: The Thin Solenoid

A single-layer solenoid of radius a and length l has $\widetilde{N}$ turns per meter. As we saw in Sect. 4.5.2, the vector potential is parallel to the current and in the same direction. Then, if the current increases, $\boldsymbol{A}$ increases and the electric field $-\partial\boldsymbol{A}/\partial t$ inside the wire *opposes* the increase of current.

The electromotance $\mathcal{V}_1$ induced by the increasing current *over one turn* of the solenoid follows from Eq. 5.6:

$$\mathcal{V}_1 = \oint_C \boldsymbol{E} \cdot \boldsymbol{dl} = -\frac{d}{dt}\oint_C \boldsymbol{A} \cdot \boldsymbol{dl} = -\frac{d\Phi}{dt} , \tag{5.8}$$

where Φ is the magnetic flux $\pi a^2 B$ linking the solenoid. Thus the opposing electromotance over one turn is

$$\mathcal{V}_1 = -\frac{d}{dt}\left(\pi a^2 \mu_0 \widetilde{N} I\right) = -\mu_0 \pi a^2 \widetilde{N}\frac{dI}{dt} \tag{5.9}$$

and the total electromotance induced in the winding is $\widetilde{N}l$ larger:

$$\mathcal{V} = -\mu_0 \pi a^2 \widetilde{N}^2 l\,\frac{dI}{dt} = -N\frac{d\Phi}{dt} = -L\frac{dI}{dt} , \tag{5.10}$$

where

$$L = \mu_0 \pi a^2 \widetilde{N}^2 l = \frac{\mu_0 \pi N^2 a^2}{l} , \tag{5.11}$$

and N is the total number of turns.

We return to the self-inductance L below.

5.2.3 Example: A Secondary on a Long Solenoid

Outside a long solenoid, $\boldsymbol{B} \approx \boldsymbol{0}$ because the return flux extends over a large volume. But, if you wind a secondary on the solenoid as in Fig. 2.2, and if you vary the current in the solenoid, you find that there is a voltage induced in the secondary! How can there be an induced voltage, if there is no magnetic field? Many an author has been baffled by this fact. French (1994) asked the question in the American Journal of Physics, and Walstad (1997) suggested far-fetched explanations, but the problem remains, in that journal.

The explanation is simple: true, outside a long solenoid, $\boldsymbol{B} \approx \boldsymbol{0}$, but there is an $\boldsymbol{A}$ field, and the voltage induced on one turn of the secondary is given by Eq. 5.8! Outside a long solenoid, $\boldsymbol{B} = \boldsymbol{\nabla} \times \boldsymbol{A} = \boldsymbol{0}$, but $\boldsymbol{A} \neq \boldsymbol{0}$.

5.3 Self-inductance

The quantity L in Eq. 5.11 is the *self-inductance* of the solenoid, expressed in *henrys*. Note that

$$L \equiv N\Phi/I \,, \tag{5.12}$$

where Φ is the magnetic flux that links the solenoid when it carries a current I. Equation 5.11 gives a value of L that is somewhat too large because we have neglected end-effects: the axial B is weaker than $\mu_0 \widetilde{N} I$ near the ends.

Upon application of a constant voltage V to the winding,

$$V - L\frac{dI}{dt} = RI \,, \quad \text{or} \quad L\frac{dI}{dt} + RI = V \,, \tag{5.13}$$

and

$$I = \frac{V}{R}\left[1 - \exp\left(-\frac{R}{L}t\right)\right] . \tag{5.14}$$

The factor $T = L/R$ is the *time constant* of the solenoid. The current reaches about two thirds of its final value V/R within one time constant.

5.3.1 Magnetic Energy in a Self-inductance

According to Eq. 5.13, the source that maintains the voltage V at the terminals of the self-inductance L of resistance R supplies a power

$$VI = I^2 R + LI\frac{dI}{dt} . \tag{5.15}$$

This is the sum of the power dissipated in R, plus the power fed into the magnetic field. Integrating over the time t,

$$\int_0^t VI\,dt = \int_0^t I^2 R\,dt + \int_0^I LI\,dI = \int_0^t I^2 R\,dt + \tfrac{1}{2}LI^2 \,, \tag{5.16}$$

where I is the current at time t. The first term on the right is the energy dissipated in the resistor up to the time t, and the second is the magnetic energy stored in the field at t.

The magnetic energy stored in the field of a circuit whose self-inductance is L and that carries a current I is therefore

$$\mathcal{E}_{\text{mag}} = \tfrac{1}{2}LI^2 . \tag{5.17}$$

Equation 5.17 serves as the definition of the *self-inductance of a volume distribution of current*:

$$L = \frac{2\mathcal{E}_{\text{mag}}}{I^2} . \tag{5.18}$$

5.4 Mutual Inductance

Imagine two close-by circuits, 1 and 2, that carry currents I_1 and I_2. Circuit 2 is linked by the magnetic flux $\Phi_{1,2}$ due to I_1, and circuit 1 by the magnetic flux $\Phi_{2,1}$ due to I_2. Then the *mutual inductance* between the two circuits is defined as

$$M = \frac{\Phi_{1,2}}{I_1} = \frac{\Phi_{2,1}}{I_2} . \tag{5.19}$$

The unit of mutual inductance is also the *henry*.

5.5 Electromagnetic Waves

Say we have a time-dependent magnetic field $\boldsymbol{B}(t)$ in a region R. What will be the magnetic field at points remote from R? From the Maxwell equation 2.3, a time-dependent $\boldsymbol{B}$ generates an $\boldsymbol{E}$. So the disturbance in R will propagate as an electromagnetic wave with both $\boldsymbol{B}$ and $\boldsymbol{E}$ components.

If the medium of propagation is air, or a vacuum, the wave propagates at the speed of light c, except close to the source. If the medium is a dielectric, then the wave speed is less than c.

If the medium is a conductor, the speed of propagation depends on both the conductivity and the frequency, and the amplitude of a plane wave decreases exponentially with distance. If a plane wave of frequency f and amplitude B travels over a distance z in a medium of conductivity σ (P. Lorrain et al., 1988, p. 537),

$$B \propto \exp\left[-(\pi\sigma\mu_0 f)^{1/2} z\right] . \tag{5.20}$$

If we have two magnetic fields of initially equal amplitudes but of frequencies f and $2f$, then the ratio of their amplitudes after travelling a distance z is

$$\frac{B_{2f}}{B_f} = \frac{\exp\left[-(2\pi\sigma\mu f)^{1/2} z\right]}{\exp\left[-(\pi\sigma\mu f)^{1/2} z\right]} \tag{5.21}$$

$$= \exp\left[-0.414(\pi\sigma\mu f)^{1/2} z\right] . \tag{5.22}$$

Setting the attenuation factor

$$\alpha_f = \exp\left[-(\pi\sigma\mu_0 f)^{1/2} z\right] , \tag{5.23}$$

$$\alpha_{2f} = \alpha_f^{\sqrt{2}} . \tag{5.24}$$

For example, if, under given conditions, a wave is attenuated by a factor of 2 at the frequency f, then, at a frequency $2f$, it is attenuated by a factor of $2^{\sqrt{2}} = 2.665$.

5.5.1 Case Study: The Earth's Magnetic Field

Refer to Sect. 4.7. The Earth's magnetic field fluctuates. Disregarding high-frequency fluctuations ascribable to electric currents in the ionosphere, the fluctuations result from unsteady currents in the Earth's core. The fluctuating magnetic field at the core propagates as an electromagnetic wave through the mantle, which is 2.8 megameters thick. The conductivity of the mantle varies by orders of magnitude with depth: it is much more conductive near the core, where it is very hot.

Let us do a *very* rough calculation. Assume that the disturbance propagates through the mantle as a plane wave, and disregard magnetic ores. Assume that the conductivity is uniform and equal to its value half way down, ≈ 10 siemens/meter. That is by far the worst approximation because, as we saw in Sect. 4.7, the conductivity of the mantle is believed to vary by 5 orders of magnitude! And the conductivity appears in the exponent of e in Eq. 5.20!

For fluctuations with a period of one year, $f \approx 3.2 \times 10^{-8}$ hertz and, with $z = 2.8 \times 10^6$ meters,

$$-(\pi \sigma \mu_0 f)^{1/2} z \approx -3.1 \,. \tag{5.25}$$

If ΔB is the amplitude of the fluctuation at the surface, and ΔB_0 the amplitude at the core's surface,

$$\frac{\Delta B}{\Delta B_0} = \exp(-3.1) \approx 0.045 \,. \tag{5.26}$$

The attenuation of six-month fluctuations is four times larger. These attenuations are certainly much too small, but it is not worthwhile to do a more refined calculation of this type because the wave is not plane. It is not spherical either, because of the complexity of the source, and because the medium of propagation is not uniform.

Spatial and temporal fluctuations in the Earth's magnetic field at the surface are therefore very much weaker than at the core.

5.6 Summary

An electric field $\boldsymbol{E}$ can result from the accumulation of electric charges, which gives a field $-\boldsymbol{\nabla} V$, and it can result from a time-dependent magnetic field, which gives a $-\partial \boldsymbol{A}/\partial t$ field:

$$\boldsymbol{E} = -\boldsymbol{\nabla} V - \frac{\partial \boldsymbol{A}}{\partial t} \,. \tag{5.5}$$

The *law of Faraday* concerns only the second source of $\boldsymbol{E}$:

$$\boldsymbol{E} = -\frac{\partial \boldsymbol{A}}{\partial t} \,. \tag{5.6}$$

The value of V is immaterial, here.

If a single-turn circuit carrying a current I is linked by its own magnetic flux Φ, then the ratio Φ/I is the *self-inductance* L of the circuit:

$$L \equiv \Phi/I \, . \qquad (5.12)$$

If the circuit has N turns, all linked by the same flux Φ, L is N times larger.

The *magnetic energy* stored in a self inductance is

$$\mathcal{E}_{\text{mag}} = \tfrac{1}{2}LI^2 \, . \qquad (5.17)$$

If the current is distributed over a volume, then

$$L = \frac{2\mathcal{E}_{\text{mag}}}{I^2} \, . \qquad (5.18)$$

The mutual inductance between two circuits, with circuit 1 linked by the magnetic flux $\Phi_{2,1}$ of a current I_2 flowing in circuit 2, is $\Phi_{2,1}/I_2$. The symmetrical expression is also equal to M:

$$M = \frac{\Phi_{1,2}}{I_1} = \frac{\Phi_{2,1}}{I_2} \, . \qquad (5.19)$$

An electromagnetic disturbance travels through a medium of conductivity σ as an attenuated wave and, for a plane wave of amplitude B,

$$B \propto \exp\left[-(\pi\sigma\mu f)^{1/2} z\right] \, . \qquad (5.20)$$

Part III

Moving Conductors

6 Ohm's Law for Moving Conductors

Contents

We are now ready to discuss conductors that move in magnetic fields. An important question comes to mind: do you look at the electric and magnetic fields sitting down with the conductor moving in front of you, or do you ride on the conductor with your measuring instruments? It is important to tell between the two points of view because the fields can be radically different. As a rule it is best to stand still, but not always. Chapters 6, 7, and 8 are fundamental for all the rest of the book.

We have now established the basic principles of electromagnetic fields, on the assumption that the media involved are stationary. We now enter the field of Magneto-Fluid-Dynamics by allowing conductors to move. This will lead us to the fundamental equation of MFD, which is a more general form of the Ohm's law of Eq. 2.10, and to modifications to the Maxwell equations of Chapter 2.

The next chapter provides two Examples related to moving conductors.

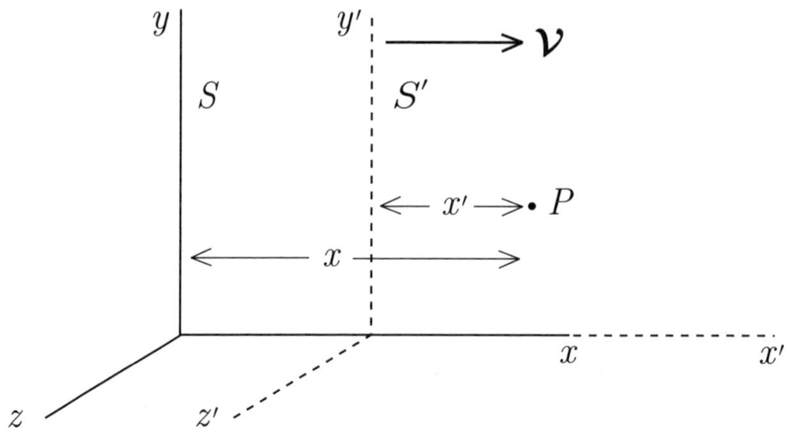

Fig. 6.1. Reference frames S and S', with S' moving to the right with respect to S at the constant velocity $\boldsymbol{\mathcal{V}} = \mathcal{V}\hat{\boldsymbol{x}}$. The origins coincide at the time $t = 0$. An object that has a velocity $\boldsymbol{v}$ with respect to frame S, has a velocity $\boldsymbol{v}'$ with respect to S'

6.1 Reference Frames

A *reference frame* is a coordinate system with respect to which, in a given situation, we refer all the variables: the coordinates x, y, z, the time t, the vectors $\boldsymbol{E}$, $\boldsymbol{B}$, $\boldsymbol{J}$, etc. If an object moves, then we imagine two reference frames, a fixed frame S, and a moving frame S' that follows the moving object. We also imagine two observers, O in S, and O' in S'. Figure 6.1 shows a simple case where the relative velocity $\boldsymbol{\mathcal{V}}$ is along the common x-axis.

Most of the time we shall use the fixed reference frame S but, occasionally, it will be helpful to think also of observations made in S'. If the moving object is a fluid, then S' applies to a given point that follows the fluid.

Now, strictly speaking, *nothing* is stationary. After all, the Earth is a space vehicle! None the less, the concept of a "fixed" reference frame is both useful and necessary.

6.2 Special Relativity

Special relativity concerns observations made by observers O and O', with S' moving at a *constant* velocity $\boldsymbol{\mathcal{V}}$ with respect to S. Both reference frames S and S' are *inertial*, or unaccelerated: they are either stationary or they move at a constant velocity, and they do not rotate. We deal briefly with rotating frames of reference below.

Strictly speaking again, the concept of unaccelerated frame is not realistic either because our space vehicle rotates, not only about its own axis, but also around the Sun... and the Sun is not stationary!

Because observers O and O' work in different reference frames, their measurements can yield different results. Say observer O on S notes that a light turned on at (x, y, z, t). Observer O' on S' disagrees: he maintains that the light turned on at (x', y', z', t'). Similarly, if O measures the vectors $\boldsymbol{E}$, $\boldsymbol{B}$, and $\boldsymbol{J}$, at P, then O', who measures the *same* quantities at P', could measure *different* vectors $\boldsymbol{E}'$, $\boldsymbol{B}'$, $\boldsymbol{J}'$.

6.2.1 The Kinematic Transformation Equations

A *transformation equation* expresses a variable in one reference frame in terms of variables in the other frame. So transformation equations come in pairs.

In *pre-relativistic mechanics*, at the time t, the origin of S' in Fig. 6.1 has reached $x = \mathcal{V}t$, and the transformation equations for the coordinates x, y, z, t and x', y', z', t' are trivial:

$$x = x' + \mathcal{V}t' , \quad y = y' , \quad z = z' , \quad t = t' , \tag{6.1}$$

$$x' = x - \mathcal{V}t , \quad y' = y , \quad z' = z , \quad t' = t . \tag{6.2}$$

Pre-relativistic *mechanics* applies whenever all speeds v satisfy the condition

$$v^2 \ll c^2 , \tag{6.3}$$

where c is the speed of light, about 3×10^8 meters/second. The condition applies to natural phenomena. For example, the orbital speed of the Earth around the Sun is about 3×10^4 meters/second, and $v^2/c^2 \approx 10^{-8}$. The tangential speed ωR at the Equator is much smaller, about 400 meters/second. Speeds in the solar atmosphere are also very much smaller that c.

From Eqs. 6.1 and 6.2, velocities transform simply:

$$\boldsymbol{v} = \boldsymbol{v}' + \boldsymbol{\mathcal{V}} . \tag{6.4}$$

Rotating reference frames are troublesome in Relativity because they are not Euclidean, and because clock time depends on the distance to the axis of rotation. Fortunately, Special Relativity continues to apply in the region close to a given point in a rotating frame (Møller, 1974). Chapter 9 concerns the rotating reference frame of the Earth.

6.2.2 The Transformation Equations of Electromagnetic Quantities

The transformation equations for electromagnetic quantities are not as simple as those for kinematic quantities, even if $\mathcal{V}^2 \ll c^2$.

Transformation of $\boldsymbol{E}$ and $\boldsymbol{B}$

The electric fields and magnetic flux densities observed in S' are not the same as those observed in S. If $\mathcal{V}^2 \ll c^2$, then we have the following *approximate* equations (P. Lorrain et al., 1988, p. 303):

$$\boldsymbol{E} = \boldsymbol{E}' - \boldsymbol{\mathcal{V}} \times \boldsymbol{B}' , \qquad \boldsymbol{E}' = \boldsymbol{E} + \boldsymbol{\mathcal{V}} \times \boldsymbol{B} , \tag{6.5}$$

$$\boldsymbol{B} = \boldsymbol{B}' + \frac{\boldsymbol{\mathcal{V}} \times \boldsymbol{E}'}{c^2} , \qquad \boldsymbol{B}' = \boldsymbol{B} - \frac{\boldsymbol{\mathcal{V}} \times \boldsymbol{E}}{c^2} . \tag{6.6}$$

Invariance of an Electric Charge Q

Electric charge is invariant. If a point-charge Q is observed in reference frame S, then, even for large $\mathcal{V}$'s, the same charge is observed in reference frame S':

$$Q' = Q . \tag{6.7}$$

Transformation of $\widetilde{Q}$ and $\boldsymbol{J}$

How do volume charge densities $\widetilde{Q}$ and conduction current densities $\boldsymbol{J}_{\mathrm{cond}}$ transform? Is $\widetilde{Q}' = \widetilde{Q}$ and $\boldsymbol{J}'_{\mathrm{cond}} = \boldsymbol{J}_{\mathrm{cond}}$?

Consider a small element of volume in a conducting medium. Let this element of volume be at rest in the fixed frame S. In S, the charge density of conduction electrons is $\widetilde{Q}_{\mathrm{ce}}$ and the other charges, at rest in the medium, have a net density $\widetilde{Q}_{\mathrm{other}}$. The net charge density in S is

$$\widetilde{Q}_{\mathrm{net}} = \widetilde{Q}_{\mathrm{ce}} + \widetilde{Q}_{\mathrm{other}} . \tag{6.8}$$

Call $\boldsymbol{v}_{\mathrm{d}}$ the drift velocity of the conduction electrons. The current densities in S are

$$\boldsymbol{J}_{\mathrm{ce}} = \widetilde{Q}_{\mathrm{ce}} \boldsymbol{v}_{\mathrm{d}} , \quad \boldsymbol{J}_{\mathrm{other}} = \boldsymbol{0} , \quad \boldsymbol{J}_{\mathrm{net}} = \widetilde{Q}_{\mathrm{ce}} \boldsymbol{v}_{\mathrm{d}} . \tag{6.9}$$

The moving reference frame S' has the velocity $\boldsymbol{\mathcal{V}} = \mathcal{V}\hat{\boldsymbol{x}}$ relative to S. Let $\boldsymbol{v}_{\mathrm{d}\|}$ be the component of $\boldsymbol{v}_{\mathrm{d}}$ parallel to $\boldsymbol{\mathcal{V}}$, and $\boldsymbol{v}_{\mathrm{d}\perp}$ the component of $\boldsymbol{v}_{\mathrm{d}}$ perpendicular to $\boldsymbol{\mathcal{V}}$. The relativistic equations for the current and charge densities, as measured in S', are the following, where γ equals $(1-\mathcal{V}^2/c^2)^{-1/2}$ and $\boldsymbol{v}_{\mathrm{d}\|} = v_{\mathrm{d}\|}\hat{\boldsymbol{x}}$ (P. Lorrain et al., 1988, p. 287):

$$\boldsymbol{J}'_{\mathrm{ce}} = \gamma(\widetilde{Q}_{\mathrm{ce}} \boldsymbol{v}_{\mathrm{d}\|} - \widetilde{Q}_{\mathrm{ce}} \boldsymbol{\mathcal{V}}) + \widetilde{Q}_{\mathrm{ce}} \boldsymbol{v}_{\mathrm{d}\perp} , \tag{6.10}$$

$$\boldsymbol{J}'_{\mathrm{other}} = -\gamma \widetilde{Q}_{\mathrm{other}} \boldsymbol{\mathcal{V}} , \tag{6.11}$$

$$\boldsymbol{J}'_{\mathrm{net}} = \widetilde{Q}_{\mathrm{ce}}(\gamma \boldsymbol{v}_{\mathrm{d}\|} + \boldsymbol{v}_{\mathrm{d}\perp}) - \gamma \widetilde{Q}_{\mathrm{net}} \boldsymbol{\mathcal{V}} , \tag{6.12}$$

$$\widetilde{Q}'_{\mathrm{ce}} = \gamma \left(\widetilde{Q}_{\mathrm{ce}} - \frac{\mathcal{V} \widetilde{Q}_{\mathrm{ce}} v_{\mathrm{d}\|}}{c^2} \right) , \tag{6.13}$$

$$\widetilde{Q}'_{\text{other}} = \gamma \widetilde{Q}_{\text{other}} , \tag{6.14}$$

$$\widetilde{Q}'_{\text{net}} = \gamma \left(\widetilde{Q}_{\text{net}} - \widetilde{Q}_{\text{ce}} \frac{\mathcal{V} v_{\text{d}\|}}{c^2} \right) . \tag{6.15}$$

Then, with $\mathcal{V}^2 \ll c^2$,

$$\boldsymbol{J}'_{\text{net}} = \boldsymbol{J}_{\text{net}} - \widetilde{Q}_{\text{net}} \boldsymbol{\mathcal{V}} , \tag{6.16}$$

$$\widetilde{Q}'_{\text{net}} = \widetilde{Q}_{\text{net}} - \widetilde{Q}_{\text{ce}} \frac{\mathcal{V} v_{\text{d}\|}}{c^2} . \tag{6.17}$$

If $\widetilde{Q}_{\text{net}} = 0$,

$$\boldsymbol{J}'_{\text{net}} = \boldsymbol{J}_{\text{net}} , \tag{6.18}$$

$$\widetilde{Q}'_{\text{net}} = -\widetilde{Q}_{\text{ce}} \frac{\mathcal{V} v_{\text{d}\|}}{c^2} . \tag{6.19}$$

In general, the last term of Eqs. 6.17 and 6.19 cannot be neglected, for, although here $|\mathcal{V} v_{\text{d}\|}/c^2| \ll 1$, the density $\widetilde{Q}_{\text{ce}}$ is *very* large, compared with $\widetilde{Q}_{\text{net}}$. This is an example of a relativistic effect measurable at so-called "non-relativistic" speeds. See the example and comments of Sect. 6.2.3.

Invariance of *Conduction* Currents

The conduction current $\boldsymbol{J}'_{\text{cond}}$ is equal to the charge density $\widetilde{Q}'_{\text{ce}}$ (measured in S') of the conduction electrons, times the velocity $\boldsymbol{v}'_{\text{d relative}}$ (measured in S') of the electrons *relative to the conductor*:

$$\boldsymbol{J}'_{\text{cond}} = \widetilde{Q}'_{\text{ce}} \boldsymbol{v}'_{\text{d relative}} . \tag{6.20}$$

We consider only non-relativistic speeds. So the above product will be hardly affected if, *in this product*, we set

$$\widetilde{Q}'_{\text{ce}} = \widetilde{Q}_{\text{ce}} \quad \text{and} \quad \boldsymbol{v}'_{\text{d relative}} = \boldsymbol{v}_{\text{d}} \tag{6.21}$$

(Eqs. 6.13 and 6.4). Finally,

$$\boldsymbol{J}'_{\text{cond}} = \widetilde{Q}'_{\text{ce}} \boldsymbol{v}'_{\text{d relative}} = \widetilde{Q}_{\text{ce}} \boldsymbol{v}_{\text{d}} = \boldsymbol{J}_{\text{cond}} . \tag{6.22}$$

Conduction currents are invariant.

Convection Currents

Relative to S, there is no convection current. However, relative to S', there is, from Eq. 6.17, a convection current density

$$\boldsymbol{J}'_{\text{conv}} = \widetilde{Q}'_{\text{net}}(-\boldsymbol{\mathcal{V}}) = -\widetilde{Q}_{\text{net}}\boldsymbol{\mathcal{V}} + \widetilde{Q}_{\text{ce}}\frac{\mathcal{V}v_{\text{d}\|}}{c^2}\boldsymbol{\mathcal{V}} \tag{6.23}$$

$$= -\widetilde{Q}_{\text{net}}\boldsymbol{\mathcal{V}} + \frac{\mathcal{V}^2}{c^2}\widetilde{Q}_{\text{ce}}\boldsymbol{v}_{\text{d}\|} \; . \tag{6.24}$$

Note that the last term above is negligible, compared to $\widetilde{Q}_{\text{ce}}\boldsymbol{v}_{\text{d}} = \boldsymbol{J}_{\text{net}} = \boldsymbol{J}_{\text{cond}}$. Therefore, for all practical purposes, from Eqs. 6.22–6.24,

$$\boldsymbol{J}'_{\text{net}} = \boldsymbol{J}'_{\text{cond}} + \boldsymbol{J}'_{\text{conv}} = \boldsymbol{J}_{\text{cond}} - \widetilde{Q}_{\text{net}}\boldsymbol{\mathcal{V}}, \tag{6.25}$$

as in Eq. 6.16.

Convection currents, however, are negligible in all our Case Studies (Sect. 2.5).

Invariance of the *B* of *Conduction* Currents

The equality of *conduction* currents in both frames (Eq. 6.22) implies that the magnetic flux densities due to conduction currents only will, at non-relativistic speeds, be the same in both frames:

$$\boldsymbol{B}'_{\text{cond}} = \boldsymbol{B}_{\text{cond}} \; . \tag{6.26}$$

We neglect retardation in the calculation of both fields, since our emphasis is on steady states.

In contrast, *convection* currents are different in the two frames. Therefore the magnetic flux densities due to convection currents are also different in the two frames.

Invariance of the Conductivity σ

Let us finally consider the conductivities σ and σ'. They concern only the conduction currents, not the convection current. Relative to S, the conduction current is

$$\boldsymbol{J}_{\text{cond}} = \boldsymbol{J}_{\text{net}} = \sigma\boldsymbol{E} \; . \tag{6.27}$$

Relative to S', the conducting medium has velocity $-\boldsymbol{\mathcal{V}}$ and the conducting current is

$$\boldsymbol{J}'_{\text{cond}} = \sigma'(\boldsymbol{E}' - \boldsymbol{\mathcal{V}} \times \boldsymbol{B}') = \sigma'\boldsymbol{E} \; , \tag{6.28}$$

from Eq. 6.5. However, from Eqs. 6.22 and 6.27,

$$\boldsymbol{J}'_{\text{cond}} = \boldsymbol{J}_{\text{cond}} = \sigma\boldsymbol{E} \; . \tag{6.29}$$

Finally, from Eqs. 6.28 and 6.29, $\sigma'\boldsymbol{E} = \sigma\boldsymbol{E}$ and, at non-relativistic speeds, conductivity is invariant:

$$\sigma' = \sigma \; . \tag{6.30}$$

6.2.3 Example: Straight Wire Carrying a Current

Imagine a straight wire, at rest in S and carrying a constant current density $\boldsymbol{J}_{\text{net}}$. Let $\widetilde{Q}_{\text{net}} = 0$.

A charge q travels at velocity $\boldsymbol{\mathcal{V}}$ relative to the wire (with $\mathcal{V}^2 \ll c^2$), at a point where $\boldsymbol{E} = \boldsymbol{0}$. An observer moving with q observes in the wire the $\boldsymbol{J}'_{\text{net}}$ and $\widetilde{Q}'_{\text{net}}$ of Eqs. 6.18 and 6.19.

For the observer in S, q is subjected to a magnetic force $q(\boldsymbol{\mathcal{V}} \times \boldsymbol{B})$. However, for the observer in S', in which q is at rest, the force felt by q is an *electric* force $q\boldsymbol{E}'$, which happens to be equal to $q\boldsymbol{\mathcal{V}} \times \boldsymbol{B}$ (Eq. 6.5). The *electrostatic* field $\boldsymbol{E}'$ observed in S' comes from the charge density $\widetilde{Q}'_{\text{net}} = -\widetilde{Q}_{\text{ce}}\mathcal{V}v_{\text{d}\|}/c^2$ observed in S' along the wire.

The charge density $\widetilde{Q}'_{\text{net}}$ is due to the fact that the conduction electrons have, relative to S', a slightly different velocity than the other charges in the wire, which entails a slightly different Lorentz contraction for both sets of charges. Admittedly, with $|\mathcal{V}v_{\text{d}\|}| \ll c^2$, this difference in Lorentz contractions is extremely small. However, in a conductor, the charge density $\widetilde{Q}_{\text{ce}}$ is so large, that the relativistic charge density $Q'_{\text{net}} = -\widetilde{Q}_{\text{ce}}\mathcal{V}v_{\text{d}\|}/c^2$ cannot be neglected, even at "non-relativistic" speeds (P. Lorrain et al., 1988, pp. 287–288).

6.3 The Lorentz Force

The electric force on a charge Q that lies in an electric field $\boldsymbol{E}$ is $Q\boldsymbol{E}$. If Q moves at a velocity $\boldsymbol{v}$ with respect to a fixed reference frame S in superposed electric and magnetic fields $\boldsymbol{E}$ and $\boldsymbol{B}$, the force on Q (Sect. 4.8) is

$$\boldsymbol{F} = Q[\boldsymbol{E} + (\boldsymbol{v} \times \boldsymbol{B})] \,. \tag{6.31}$$

This is the *Lorentz force*. In the reference frame S' moving with Q the charge is stationary and

$$\boldsymbol{F}' = Q'\boldsymbol{E}' = Q\boldsymbol{E}' \,, \tag{6.32}$$

the charge having the same value in both frames, from Eq. 6.7.

As a rule, charges move either in a solid, or in a liquid, or in a plasma. Since their mean-free-path between collisions with the ambient atoms and molecules is finite, *a force on the charge carriers is also a force on the ambient medium.*

6.4 Ohm's Law for Moving Conductors

Consider a conductor that moves in superposed $\boldsymbol{E}$ and $\boldsymbol{B}$ fields. At a given point P, the conductor has a velocity $\boldsymbol{v}$, the conductivity is σ, and the electric current density is $\boldsymbol{J}$.

The average velocity of the conduction electrons $\boldsymbol{v}_{ce}$ is essentially equal to the conductor velocity.

We measure all the variables with respect to the same, fixed, reference frame S. The charge carriers are subjected to the Lorentz force and

$$\boldsymbol{J} = \sigma[\boldsymbol{E} + (\boldsymbol{v} \times \boldsymbol{B})] \tag{6.33}$$

$$= \sigma\left[-\boldsymbol{\nabla} V - \frac{\partial \boldsymbol{A}}{\partial t} + (\boldsymbol{v} \times \boldsymbol{B})\right] . \tag{6.34}$$

THIS GENERALIZED OHM'S LAW IS THE FUNDAMENTAL EQUATION OF MAGNETO-FLUID-DYNAMICS. WE SHALL USE IT THROUGHOUT THE REST OF THIS BOOK.

This equation is not as general as one might wish because σ is anisotropic in the presence of a magnetic field, the reason being that conduction electrons flow more easily in directions parallel or anti-parallel to $\boldsymbol{B}$. Also, there are more complete forms of Ohm's law. See, for example, Miyamoto (1989). Equation 2.10 is the more usual form of Ohm's law, which applies to stationary conductors.

Equation 6.33 is not general, also in another way: we have neglected the Hall effect of Sect. 4.8. We have assumed that the drift velocity of the conduction electrons relative to the moving conductor is negligible, relative to the velocity $\boldsymbol{v}$ of the conductor.

If the conductor is stationary, however, $\boldsymbol{v} = \boldsymbol{0}$ and the drift velocity $\boldsymbol{v}_{ce}$ of the conduction electrons in the conductor is not negligible, relative to $\boldsymbol{v}$. Then

$$\boldsymbol{J} = \sigma[\boldsymbol{E} + (\boldsymbol{v}_{ce} \times \boldsymbol{B})] = \widetilde{Q}_{ce}\boldsymbol{v}_{ce} , \tag{6.35}$$

where $\widetilde{Q}_{ce}$ is the space charge density of the conduction electrons. The term between parentheses is orthogonal to $\boldsymbol{v}_{ce}$, and thus orthogonal to $\boldsymbol{J}$; the field $\boldsymbol{E}$ adjusts accordingly. See Fig. 6.2. That is the origin of the Hall effect of Sect. 4.8.

Consider once again a conductor moving at such a velocity $\boldsymbol{v}$ that the Hall effect is negligible. In the moving frame S' in which the conductor is at rest, because of Eq. 6.30,

$$\boldsymbol{J}' = \sigma'\boldsymbol{E}' = \sigma\boldsymbol{E}' = \sigma[\boldsymbol{E} + (\boldsymbol{v} \times \boldsymbol{B})] = \boldsymbol{J} . \tag{6.36}$$

Within our approximation $v^2 \ll c^2$, the conduction current densities in the two frames are the same, as in Eq. 6.22.

The above $\boldsymbol{B}$ is the *net* $\boldsymbol{B}$: it is the magnetic field originating in external sources, plus the magnetic field of the induced currents:

$$\boldsymbol{B}_{\text{net}} = \boldsymbol{B}_{\text{ext}} + \boldsymbol{B}_{\text{ind}} . \tag{6.37}$$

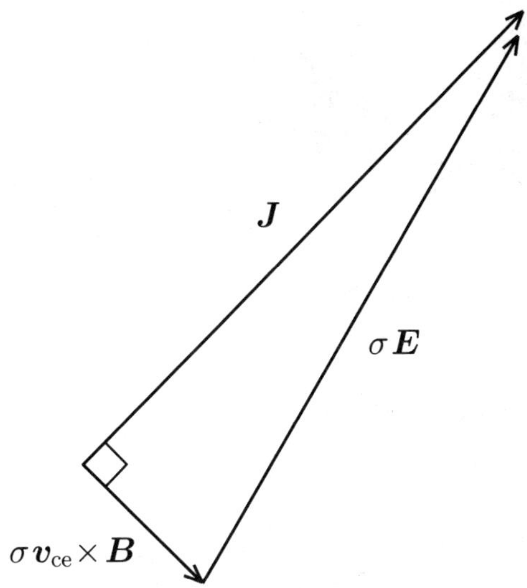

Fig. 6.2. The vectors $\boldsymbol{J}$, $\sigma\boldsymbol{v}_{ce}\times\boldsymbol{B}$, and $\sigma\boldsymbol{E}$

Equations 6.33 and 6.34 are far more complex than they seem because each one of the six variables depends on the five others. In other words, the six variables are all tightly knitted together. See Fig. 6.3.

So Ohm's law for moving conductors is hopelessly complex in the general case. Nonetheless, in specific cases, its solution is straightforward, as we shall see in the following chapters.

With an ideal non-conductor, $\boldsymbol{J}$ and σ would both be zero, and Ohm's law would be meaningless.

The $\boldsymbol{J}$ wheel of Fig. 6.3 illustrates how $\boldsymbol{J}$ is related to the other variables discussed here. Note the following points.

1. The conductivity σ is in reality a tensor whose value is a function of the magnitude of $\boldsymbol{B}$ and of the orientation of $\boldsymbol{J}$ with respect to $\boldsymbol{B}$: the conductivity parallel or antiparallel to $\boldsymbol{B}$ is larger than in orthogonal directions.

2. From Eq. 2.4, $\boldsymbol{J}$ is closely related to $\boldsymbol{B}$:

$$\nabla\times\boldsymbol{B}=\mu_0\boldsymbol{J}\,. \qquad (6.38)$$

3. The electric potential V is a function of the charge density $\widetilde{Q}$, and, from Eq. 7.10,

$$\widetilde{Q}_{\rm f}=-\epsilon_0\nabla\cdot(\boldsymbol{v}\times\boldsymbol{B})\,. \qquad (6.39)$$

4. Both $\boldsymbol{B}$ and $\boldsymbol{A}$ are *non-local* functions of $\boldsymbol{J}$. This means that the values of $\boldsymbol{B}$ and $\boldsymbol{A}$ at P depend on the value of $\boldsymbol{J}$, as in Eqs. 4.1 and 4.14.

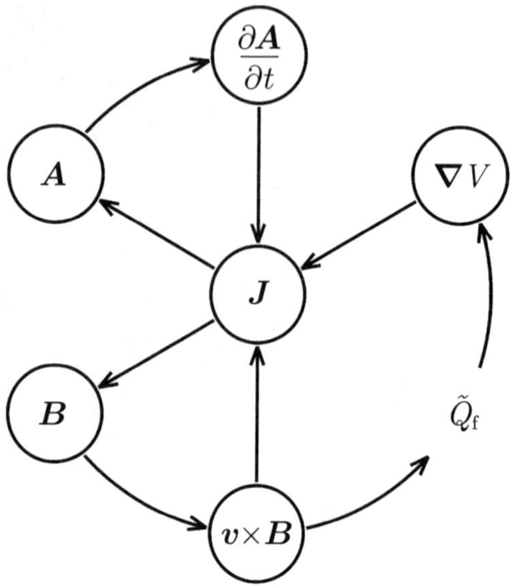

Fig. 6.3. The $\boldsymbol{J}$ wheel. The value of $\boldsymbol{J}$ yields both $\boldsymbol{A}$ and $\boldsymbol{B}$ at every point (Eqs. 4.1 and 4.14). If the field is time-dependent, then $-\sigma\partial\boldsymbol{A}/\partial t$ at P provides one component of $\boldsymbol{J}$ at P. If the local velocity of the medium is $\boldsymbol{v}$, then $\sigma\boldsymbol{v}\times\boldsymbol{B}$ gives us another component of $\boldsymbol{J}$ at P. Finally, the field $\boldsymbol{v}\times\boldsymbol{B}$ yields the value of $\widetilde{Q}_{\mathrm{f}}$ (Eq. 7.10), and then $-\sigma\boldsymbol{\nabla}V$ at P (Eq. 3.24), which is the remaining component of $\boldsymbol{J}$. Further major complication: the velocity $\boldsymbol{v}$ at P depends on the magnetic force, of density $\boldsymbol{J}\times\boldsymbol{B}$ (Sect. 4.8)

5. The velocity $\boldsymbol{v}$ of the conducting medium depends on the external forces that drive it, but also on the magnetic force, of density $\boldsymbol{J}\times\boldsymbol{B}$ as in Eq. 4.24. We return to the magnetic force in Sect. 6.7.

6.5 The Induction Equation

Assume a uniform σ and take the curl of Eq. 6.33:

$$\boldsymbol{\nabla}\times\boldsymbol{J}=\sigma[\boldsymbol{\nabla}\times\boldsymbol{E}+\boldsymbol{\nabla}\times(\boldsymbol{v}\times\boldsymbol{B})]\,. \tag{6.40}$$

Then, from Eqs. 2.4 and 2.3,

$$\frac{1}{\mu_0}\boldsymbol{\nabla}\times(\boldsymbol{\nabla}\times\boldsymbol{B})=\sigma\left[-\frac{\partial\boldsymbol{B}}{\partial t}+\boldsymbol{\nabla}\times(\boldsymbol{v}\times\boldsymbol{B})\right] \tag{6.41}$$

and

$$\frac{\partial \boldsymbol{B}}{\partial t} = \boldsymbol{\nabla} \times (\boldsymbol{v} \times \boldsymbol{B}) - \frac{1}{\mu_0 \sigma} \boldsymbol{\nabla} \times (\boldsymbol{\nabla} \times \boldsymbol{B}) \, . \tag{6.42}$$

This is the induction equation. Its only dependent variable is $\boldsymbol{B}$.

It is the custom to substitute the Laplacian of $\boldsymbol{B}$ for $-\boldsymbol{\nabla} \times (\boldsymbol{\nabla} \times \boldsymbol{B})$ (see the page facing the front cover), using Eq. 2.2. That is not advisable because that transformation applies only in Cartesian coordinates, which are of little use in MFD, because the most common geometries have cylindrical symmetry.

From Eq. 2.4,

$$\frac{\partial \boldsymbol{B}}{\partial t} = \boldsymbol{\nabla} \times (\boldsymbol{v} \times \boldsymbol{B}) - \frac{1}{\sigma} \boldsymbol{\nabla} \times \boldsymbol{J} \, . \tag{6.43}$$

It is also the custom to state that the last term of the induction equation is negligible when the conductivity σ is very large. That is incorrect because $\boldsymbol{J}$ is proportional to σ, which makes $\boldsymbol{\nabla} \times \boldsymbol{J}$ proportional to σ.

6.6 The Magnetic Reynolds Number

Let us estimate the relative orders of magnitude of the two terms on the right side of Eq. 6.42. More specifically, we wish to know under what conditions the second term is negligible.

Let $\mathcal{L}$ be a characteristic length of the phenomenon. Considering only typical values of the fields as in Sect. A.8 of Appendix A, and using the fact that $\boldsymbol{B}$ is transversal since its divergence is zero (Sect. A.6.3 and Eq. A.87), we have, for the second term,

$$\frac{|\boldsymbol{\nabla} \times (\boldsymbol{\nabla} \times \boldsymbol{B})|_{\text{typical}}}{\mu_0 \sigma} \sim \frac{B_{\text{typical}}}{\mathcal{L}^2 \mu_0 \sigma} \, . \tag{6.44}$$

The order of magnitude of the term $\boldsymbol{\nabla} \times (\boldsymbol{v} \times \boldsymbol{B})$ is more difficult to estimate. An inequality such as Eq. A.83 would be useless here. Let us first assume that the field $\boldsymbol{v} \times \boldsymbol{B}$ is not too longitudinal; this is just another way of saying that its curl is, overall, not close to zero (Sect. A.6.3). This implies that

$$|\boldsymbol{\nabla} \times (\boldsymbol{v} \times \boldsymbol{B})|_{\text{typical}} \sim \frac{|\boldsymbol{v} \times \boldsymbol{B}|_{\text{typical}}}{\mathcal{L}} \, . \tag{6.45}$$

Now suppose that $\boldsymbol{v}$ and $\boldsymbol{B}$ are, overall, not too parallel. Equation 6.45 can then be rewritten as follows:

$$|\boldsymbol{\nabla} \times (\boldsymbol{v} \times \boldsymbol{B})|_{\text{typical}} \sim \frac{(vB)_{\text{typical}}}{\mathcal{L}} \, . \tag{6.46}$$

Now assume that

$$(vB)_{\text{typical}} \sim v_{\text{typical}} B_{\text{typical}} \, . \tag{6.47}$$

For a discussion of this kind of equation, see Sect. A.7. From Eqs. 6.46 and 6.47,

$$|\boldsymbol{\nabla} \times (\boldsymbol{v} \times \boldsymbol{B})|_{\text{typical}} \sim \frac{v_{\text{typical}} B_{\text{typical}}}{\mathcal{L}} . \tag{6.48}$$

Finally, from Eqs. 6.44 and 6.48,

$$\frac{|\boldsymbol{\nabla} \times (\boldsymbol{v} \times \boldsymbol{B})|_{\text{typical}}}{|\boldsymbol{\nabla} \times (\boldsymbol{\nabla} \times \boldsymbol{B})|_{\text{typical}}/(\mu_0 \sigma)} \sim \mu_0 \sigma \mathcal{L} v_{\text{typical}} . \tag{6.49}$$

The quantity

$$R_{\text{m}} = \mu_0 \sigma \mathcal{L} v_{\text{typical}} \tag{6.50}$$

is the *magnetic Reynolds number.*

If

$$R_{\text{m}} \gg 1 , \quad \text{or if} \quad \mu_0 \sigma \mathcal{L} v_{\text{typical}} \gg 1 , \tag{6.51}$$

then

$$\frac{\partial \boldsymbol{B}}{\partial t} \approx \boldsymbol{\nabla} \times (\boldsymbol{v} \times \boldsymbol{B}) . \tag{6.52}$$

In this case

$$|\boldsymbol{\nabla} \times (\boldsymbol{v} \times \boldsymbol{B})| \gg \left| \frac{1}{\mu_0 \sigma} \boldsymbol{\nabla} \times (\boldsymbol{\nabla} \times \boldsymbol{B}) \right| , \tag{6.53}$$

$$|\boldsymbol{\nabla} \times (\boldsymbol{v} \times \boldsymbol{B})| \gg \left| \frac{1}{\sigma} \boldsymbol{\nabla} \times \boldsymbol{J} \right| . \tag{6.54}$$

If, instead,

$$R_{\text{m}} \ll 1 , \quad \text{or if} \quad \mu_0 \sigma \mathcal{L} v_{\text{typical}} \ll 1 , \tag{6.55}$$

then

$$\frac{\partial \boldsymbol{B}}{\partial t} \approx -\frac{1}{\sigma} \boldsymbol{\nabla} \times \boldsymbol{J} . \tag{6.56}$$

If $\boldsymbol{B}$ is time-*in*dependent, then

$$\boldsymbol{\nabla} \times (\boldsymbol{v} \times \boldsymbol{B}) = \boldsymbol{0} \tag{6.57}$$

for large Reynolds numbers.

6.6.1 Examples

For the Earth's core, we can set

$$\mu_0 \sim 10^{-6} \text{ henry/meter} , \quad \sigma \sim 3 \times 10^5 \text{ siemens/meter} ,$$
$$\mathcal{L} \sim 10^6 \text{ meters} , \quad v_{\text{typical}} \sim 10^2 \text{ meters/second} ,$$

and $R_{\text{m}} \sim 3 \times 10^7$.

In the case of a sunspot (Chapter 13),

$$\sigma \sim 0.1 \text{ siemens/meter} , \quad \mathcal{L} \sim 10^5 \text{ meters} , \quad v_{\text{typical}} \sim 300 \text{ meters/second} ,$$

and $R_{\text{m}} \sim 3$, so neither Eq. 6.51 nor Eq. 6.55 applies.

6.7 Magnetic Forces on Moving Conductors

Section 4.8.1 described how, in an electromagnetic pump, the $\boldsymbol{J} \times \boldsymbol{B}$ force propels a conducting fluid in a prescribed direction.

But what is the role of magnetic forces in a convecting conducting fluid, such as the liquid part of the Earth's core? As we shall see immediately, these forces seriously disturb the flow pattern.

From Eq. 4.24, the magnetic force density is

$$\widetilde{\boldsymbol{F}}_{\text{mag}} = \boldsymbol{J} \times \boldsymbol{B} = \sigma[\boldsymbol{E} + (\boldsymbol{v} \times \boldsymbol{B})] \times \boldsymbol{B} \tag{6.58}$$

$$= \sigma[\boldsymbol{E} \times \boldsymbol{B} + (\boldsymbol{v}_\perp \times \boldsymbol{B}) \times \boldsymbol{B}] \tag{6.59}$$

$$= \sigma \left(\boldsymbol{E} \times \boldsymbol{B} - \boldsymbol{v}_\perp B^2\right) , \tag{6.60}$$

where $\boldsymbol{v}_\perp$ is the velocity component of the moving medium that is orthogonal to the local $\boldsymbol{B}$.

The magnetic force density on a moving conductor therefore has two components. The first one, $\sigma \boldsymbol{E} \times \boldsymbol{B}$, is orthogonal to $\boldsymbol{B}$, and can point in any direction with respect to the local fluid velocity $\boldsymbol{v}$. The second component of the magnetic force density is proportional to $-\boldsymbol{v}_\perp$, and is thus somewhat equivalent to a viscous braking force: it tries to cancel the velocity component that is orthogonal to $\boldsymbol{B}$. That component is proportional to the local B, squared.

Magnetic forces on a convecting fluid affect not only the velocity $\boldsymbol{v}$; they also affect $\boldsymbol{v} \times \boldsymbol{B}$, and thus the electric charge density $\widetilde{Q}$ through Eq. 6.39, hence the electric potential V, and its gradient $\boldsymbol{\nabla} V$, which in turn alters the current density $\boldsymbol{J}$, and hence $\boldsymbol{B}$. Also, a change in $\boldsymbol{J}$ alters the stored magnetic energy and the Joule losses! See Fig. 6.3 for the $\boldsymbol{J}$ wheel.

In the absence of an electric field,

$$\widetilde{\boldsymbol{F}}_{\text{mag}} = \sigma(\boldsymbol{v} \times \boldsymbol{B}) \times \boldsymbol{B} = \sigma \left[\boldsymbol{B}(\boldsymbol{B} \cdot \boldsymbol{v}) - B^2 \boldsymbol{v}\right] \tag{6.61}$$

$$= \sigma \left(Bv \cos\theta\, \boldsymbol{B} - B^2 \boldsymbol{v}\right) , \tag{6.62}$$

where θ is the angle between $\boldsymbol{v}$ and $\boldsymbol{B}$.

If $\theta = 0$, the conductor moves parallel to the field. Then $\boldsymbol{v} \times \boldsymbol{B} = \boldsymbol{0}$, $\boldsymbol{J} = \boldsymbol{0}$, and there is no magnetic force. If $\theta = \pi/2$, the conductor moves in the direction perpendicular to the magnetic field. Then the magnetic force is proportional to B^2, and it brakes the motion, as above.

Many authors write the magnetic force density as

$$\widetilde{\boldsymbol{F}}_{\text{mag}} = \boldsymbol{J} \times \boldsymbol{B} = \frac{1}{\mu_0}(\boldsymbol{\nabla} \times \boldsymbol{B}) \times \boldsymbol{B} , \tag{6.63}$$

and then apply Identity 5 (see the page facing the front cover):

$$\widetilde{\boldsymbol{F}}_{\text{mag}} = \frac{1}{\mu_0} \left[(\boldsymbol{B} \cdot \boldsymbol{\nabla})\boldsymbol{B} - \frac{1}{2}\boldsymbol{\nabla}\left(B^2\right)\right] . \tag{6.64}$$

That is legitimate, but the relation is valid only in Cartesian coordinates, which are next to useless in MFD, as we saw in Sect. 6.5.

6.8 Summary

It is often useful to think in terms of two *reference frames*, a fixed reference frame S, and a second frame S' that moves at a velocity $\boldsymbol{\mathcal{V}} = \mathcal{V}\hat{\boldsymbol{x}}$ with respect to S. If $\mathcal{V}^2 \ll c^2$, then x, y, z, t transform as in pre-relativistic mechanics:

$$x = x' + \mathcal{V}t' \,, \quad y = y' \,, \quad z = z' \,, \quad t = t' \,, \qquad (6.1)$$

$$x' = x - \mathcal{V}t \,, \quad y' = y \,, \quad z' = z \,, \quad t' = t \,. \qquad (6.2)$$

A point-charge Q, a conduction current $\boldsymbol{J}_{\text{cond}}$, the magnetic flux density $\boldsymbol{B}_{\text{cond}}$ due to conduction currents, and the conductivity σ are invariant, as in pre-relativistic Physics:

$$Q' = Q \,, \qquad (6.7)$$

$$\boldsymbol{J}'_{\text{cond}} = \boldsymbol{J}_{\text{cond}} \,, \qquad (6.22)$$

$$\boldsymbol{B}'_{\text{cond}} = \boldsymbol{B}_{\text{cond}} \,, \qquad (6.26)$$

$$\sigma' = \sigma \,. \qquad (6.30)$$

The $\boldsymbol{B}$ due to convection currents is in general not invariant, because such currents are different, relative to different reference frames. However, convection currents are negligible here (Sect. 2.5).

For the electric field $\boldsymbol{E}$, *even at non-relativistic speeds,*

$$\boldsymbol{E}' = \boldsymbol{E} + (\boldsymbol{\mathcal{V}} \times \boldsymbol{B}) \,. \qquad (6.5)$$

Ohm's law for moving conductors states that the conduction current is

$$\boldsymbol{J} = \sigma[\boldsymbol{E} + (\boldsymbol{v} \times \boldsymbol{B})] \,. \qquad (6.33)$$

THIS IS *the* FUNDAMENTAL EQUATION OF MAGNETO-FLUID-DYNAMICS.

The *induction equation* can be written in two equivalent forms:

$$\frac{\partial \boldsymbol{B}}{\partial t} = \boldsymbol{\nabla} \times (\boldsymbol{v} \times \boldsymbol{B}) - \frac{1}{\mu_0 \sigma} \boldsymbol{\nabla} \times (\boldsymbol{\nabla} \times \boldsymbol{B}) \,, \qquad (6.42)$$

or

$$\frac{\partial \boldsymbol{B}}{\partial t} = \boldsymbol{\nabla} \times (\boldsymbol{v} \times \boldsymbol{B}) - \frac{1}{\sigma} \boldsymbol{\nabla} \times \boldsymbol{J} \,. \qquad (6.43)$$

The *magnetic force density* exerted on a conductor that moves in a magnetic field $\boldsymbol{B}$ is

$$\widetilde{\boldsymbol{F}}_{\text{mag}} = \sigma \left(\boldsymbol{E} \times \boldsymbol{B} - \boldsymbol{v}_{\perp} B^2 \right) \,, \qquad (6.60)$$

where $\boldsymbol{v}_{\perp}$ is the component of the conductor velocity that is perpendicular to the local $\boldsymbol{B}$.

7 Charges Inside Moving Conductors

Contents

This chapter concerns a little-known, but crucially important phenomenon: when a conductor moves in a magnetic field, it nearly always carries, INSIDE, a space charge. And the electric field of that space charge cancels part, or even all of, the $\boldsymbol{v} \times \boldsymbol{B}$ field.[1]

Our first Example will be the Faraday disk, which is the standard model for natural dynamos in convecting conducting fluids, that we shall discuss at length in Chapters 10 to 14. Our second Example will be the solid rotating conducting sphere, which has mystified many an author, and which concerns the magnetic field of the Earth (Chapter 9).

[1] See P. Lorrain, *Charges in moving conductors*, Eur. Jour. Physics **11**, 94–98 (1990); but the value of $\widetilde{Q}_f$ deduced there is incorrect. See Redžić (2001), P. Lorrain (2001), and below.

7.1 Introduction

Conductors usually carry *surface* electric charges, and any extra charge deposited inside moves out to the periphery almost instantaneously, as we saw in Sect. 3.3.1.

However, if the conductor moves in a magnetic field, then it most probably carries a net charge density both on the surface and INSIDE. Let $\boldsymbol{E}_{\text{sc}}$ be the electric field of these space and surface charges.

Now let $\boldsymbol{E}_{\text{ext}}$ be the electric field of all other charges. The net electric field is $\boldsymbol{E} = \boldsymbol{E}_{\text{sc}} + \boldsymbol{E}_{\text{ext}}$.

These space and surface charges in conductors play an essential role in MFD because their electric field $\boldsymbol{E}_{\text{sc}}$ opposes, more or less, the $\boldsymbol{E}_{\text{ext}} + (\boldsymbol{v} \times \boldsymbol{B})$ field in Eq. 7.1 below.

The phenomenon seems to be unknown throughout the Geophysics and Astrophysics communities. A review of the literature showed that, out of a total of 22 authors, 18 assume, either implicitly or explicitly, that the electric charge density is equal to zero. The other four deduce an erroneous value, and then disregard the corresponding electric field. See, for example, Moffatt (1978) and Parker (1979).

Note that $\boldsymbol{E}_{\text{sc}}$ is not, in general, equal to the total electric field $\boldsymbol{E}$. That will be clear below.

7.2 Electric Charges in Moving Conductors

We assume isotropic media.

Imagine a fixed reference frame S where the electric field strength is $\boldsymbol{E}$ and the magnetic flux density $\boldsymbol{B}$. A conductor, whether solid or not, and of arbitrary conductivity σ, moves in that field. At a given point, the conductor velocity is $\boldsymbol{v}$. Then, from Sect. 6.4, at that point, the conduction current density $\boldsymbol{J}$ is a function of both $\boldsymbol{E}$ and $\boldsymbol{v} \times \boldsymbol{B}$:

$$\boldsymbol{J} = \sigma[\boldsymbol{E} + (\boldsymbol{v} \times \boldsymbol{B})] = \sigma[\boldsymbol{E}_{\text{sc}} + \boldsymbol{E}_{\text{ext}} + (\boldsymbol{v} \times \boldsymbol{B})] . \tag{7.1}$$

All variables can be both space- and time-dependent.

The $\boldsymbol{v} \times \boldsymbol{B}$ field acts like the electric field of a distributed source (P. Lorrain et al., 1988, p. 494): at each point, a charge feels, not only the electric field

$$\boldsymbol{E} = -\boldsymbol{\nabla} V - \frac{\partial \boldsymbol{A}}{\partial t} , \tag{7.2}$$

but also the $\boldsymbol{v} \times \boldsymbol{B}$ field. In other words, the $\boldsymbol{v} \times \boldsymbol{B}$ field gives rise to an electric *space* charge and a corresponding electric field.

Take the divergence of both sides of Eq. 7.1, assuming a uniform conductivity σ:

$$\boldsymbol{\nabla} \cdot \boldsymbol{J} = \sigma[\boldsymbol{\nabla} \cdot \boldsymbol{E} + \boldsymbol{\nabla} \cdot (\boldsymbol{v} \times \boldsymbol{B})] . \tag{7.3}$$

Applying the law of conservation of charge of Sect. 3.3 on the left, that term becomes equal to $-\partial\widetilde{Q}_{\rm f}/\partial t$, where $\widetilde{Q}_{\rm f}$ is the free space charge density. If $\partial/\partial t = 0$, then $\boldsymbol{\nabla}\cdot\boldsymbol{J} = 0$ and

$$\boldsymbol{\nabla}\cdot\boldsymbol{E} + \boldsymbol{\nabla}\cdot(\boldsymbol{v}\times\boldsymbol{B}) = 0\,. \tag{7.4}$$

As we saw in Sect. 2.2, the space charge density that appears on the RHS of Eq. 2.1 comprises not only the *free* charge density $\widetilde{Q}_{\rm f}$, but also the *bound* charge density $\widetilde{Q}_{\rm b}$, which comes from the distortion of atoms and molecules when subjected to a non-uniform electric field. So

$$\boldsymbol{\nabla}\cdot\boldsymbol{E} = \frac{\widetilde{Q}_{\rm f}+\widetilde{Q}_{\rm b}}{\epsilon_0} = \frac{\widetilde{Q}_{\rm f}-\boldsymbol{\nabla}\cdot\boldsymbol{P}}{\epsilon_0}\,, \tag{7.5}$$

where $\boldsymbol{P}$ is the polarization of the medium (P. Lorrain et al., 1988, p. 173).

Now we cannot use Eq. 2.8 here because atoms and molecules cannot tell between $\boldsymbol{E}$ and $\boldsymbol{v}\times\boldsymbol{B}$ fields. So, instead of the usual

$$\boldsymbol{P} = \epsilon_0(\epsilon_{\rm r}-1)\boldsymbol{E}\,, \tag{7.6}$$

we must set

$$\boldsymbol{P} = \epsilon_0(\epsilon_{\rm r}-1)[\boldsymbol{E}+(\boldsymbol{v}\times\boldsymbol{B})]\,. \tag{7.7}$$

Then, if the medium is homogeneous,

$$\boldsymbol{\nabla}\cdot\boldsymbol{E} = \frac{\widetilde{Q}_{\rm f}}{\epsilon_0} - (\epsilon_{\rm r}-1)\boldsymbol{\nabla}\cdot[\boldsymbol{E}+(\boldsymbol{v}\times\boldsymbol{B})]\,. \tag{7.8}$$

Under steady conditions, $\boldsymbol{\nabla}\cdot\boldsymbol{J} = -\partial\widetilde{Q}_{\rm f}/\partial t = 0$ and $\boldsymbol{\nabla}\cdot(\boldsymbol{E}+\boldsymbol{v}\times\boldsymbol{B}) = 0$ (Eq. 7.4). Then Eq. 7.8 reduces to

$$\boldsymbol{\nabla}\cdot\boldsymbol{E} = \frac{\widetilde{Q}_{\rm f}}{\epsilon_0} \tag{7.9}$$

Also, $\boldsymbol{\nabla}\cdot\boldsymbol{E} = -\boldsymbol{\nabla}\cdot(\boldsymbol{v}\times\boldsymbol{B})$ and the free charge density is given by

$$\widetilde{Q}_{\rm f} = -\epsilon_0\boldsymbol{\nabla}\cdot(\boldsymbol{v}\times\boldsymbol{B})\,. \tag{7.10}$$

Note that, if $\partial/\partial t = 0$, then $\widetilde{Q}_{\rm f}$ is independent of the conductivity σ, at least if σ is uniform as above. Of course the $\boldsymbol{B}$ and $\boldsymbol{J}$ of Eq. 7.1 are always related through Eq. 2.4:

$$\boldsymbol{\nabla}\times\boldsymbol{B} = \mu_0\boldsymbol{J}\,, \tag{7.11}$$

and $\boldsymbol{J}$ depends on σ.

As we shall see in the cases of the Faraday disk and of the rotating sphere below, if the conductor is isolated, then surface charges compensate for the volume charges.

This electric space charge makes sense because the $\boldsymbol{v} \times \boldsymbol{B}$ field sweeps charges around inside the conductor. For example, in a region where $\boldsymbol{\nabla} \cdot (\boldsymbol{v} \times \boldsymbol{B})$ is positive, electrons swarm in, making the region negative. This explains the negative sign in Eq. 7.10.

If $\boldsymbol{J} = \boldsymbol{0}$, this means that charges have distributed themselves so that

$$\boldsymbol{E} = -(\boldsymbol{v} \times \boldsymbol{B}) . \tag{7.12}$$

What if $\boldsymbol{J} \neq \boldsymbol{0}$? In a steady-state situation, as here, $\boldsymbol{\nabla} \times \boldsymbol{E} = \boldsymbol{0}$. Then $\boldsymbol{E}$ is longitudinal (Sect. A.6.3) and, from Eq. A.82,

$$|\boldsymbol{\nabla} \cdot \boldsymbol{E}|_{\text{typical}} \sim \frac{E_{\text{typical}}}{\mathcal{L}} , \tag{7.13}$$

where $\mathcal{L}$ is a characteristic length. On the other hand, from Eq. A.81,

$$|\boldsymbol{\nabla} \cdot (\boldsymbol{v} \times \boldsymbol{B})|_{\text{typical}} \lesssim \frac{|\boldsymbol{v} \times \boldsymbol{B}|_{\text{typical}}}{\mathcal{L}} . \tag{7.14}$$

Now $|\boldsymbol{\nabla} \cdot \boldsymbol{E}|_{\text{typical}} = |\boldsymbol{\nabla} \cdot (\boldsymbol{v} \times \boldsymbol{B})|_{\text{typical}}$ (Eq. 7.4). Then

$$E_{\text{typical}} \lesssim |\boldsymbol{v} \times \boldsymbol{B}|_{\text{typical}} . \tag{7.15}$$

This is all we can say in general, if $\boldsymbol{J} \neq \boldsymbol{0}$.

Whether $\boldsymbol{J} = \boldsymbol{0}$ or not, E_{typical} is, at most, of the same order of magnitude as $|\boldsymbol{v} \times \boldsymbol{B}|_{\text{typical}}$.

Parker (1979) deduces Eq. 7.10, but then he argues that the electric charge density in conducting media, such as plasmas in Space, is in fact zero because of the abundance of charge carriers! He concludes that, under steady conditions, $\boldsymbol{E} = \boldsymbol{0}$ inside conductors. That is of course a misconception: under steady conditions and with a uniform σ, the above equations for $\widetilde{Q}_{\mathrm{f}}$ always apply, and $\widetilde{Q}_{\mathrm{f}} \neq 0$ if $\boldsymbol{\nabla} \cdot (\boldsymbol{v} \times \boldsymbol{B}) \neq 0$.

7.2.1 Example: Rotating Solid Conductor

An axisymmetric conducting rigid body rotates around its axis of symmetry at a constant angular velocity ω, in a uniform axial field $\boldsymbol{B} = B\hat{\boldsymbol{z}}$. Then

$$\boldsymbol{v} \times \boldsymbol{B} = \omega\rho\hat{\boldsymbol{\phi}} \times (B\hat{\boldsymbol{z}}) = \omega\rho B\hat{\boldsymbol{\rho}} , \tag{7.16}$$

$$\boldsymbol{\nabla} \cdot (\boldsymbol{v} \times \boldsymbol{B}) = \boldsymbol{\nabla} \cdot (\omega\rho B\hat{\boldsymbol{\rho}}) = \omega B \boldsymbol{\nabla} \cdot (\rho\hat{\boldsymbol{\rho}}) = 2\omega B , \tag{7.17}$$

and

$$\widetilde{Q}_{\mathrm{f}} = -2\epsilon_0\omega B . \tag{7.18}$$

Note that, in this case, $\widetilde{Q}_{\mathrm{f}}$ is uniform and independent of the conductivity σ.

For example, if $\omega = 100$ radians/second and $B = 1$ tesla, then $\widetilde{Q}_{\mathrm{f}} \approx -2 \times 10^{-9}$ coulomb/meter^3.

How does this compare with the charge density of conduction electrons in copper? In copper, there are 8.1×10^{28} conduction electrons per cubic meter, and $\widetilde{Q}_{\mathrm{ce}} \approx -10^{10}$ coulombs/meter^3, or *nineteen* orders of magnitude larger than $\widetilde{Q}_{\mathrm{f}}$.

Therefore $|\widetilde{Q}_{\mathrm{f}}| \ll |\widetilde{Q}_{\mathrm{ce}}|$. Despite this, in normal matter, positive and negative charges are so finely balanced that $\widetilde{Q}_{\mathrm{f}}$ gives an all-important electric field. For more on rotating conductors, see Sect. 8.5.

7.2.2 Time-dependent Situations

If the phenomenon is time-dependent, or if the medium is not homogeneous, then there still exist space charges.

With a uniform conductivity and $\boldsymbol{P} \neq \mathbf{0}$, we obtain, from the equation for the conservation of charge 3.15 and Eq. 7.8,

$$-\frac{\partial \widetilde{Q}_{\mathrm{f}}}{\partial t} = \boldsymbol{\nabla} \cdot \boldsymbol{J}_{\mathrm{f}} = \sigma \left[\boldsymbol{\nabla} \cdot \boldsymbol{E} + \boldsymbol{\nabla} \cdot (\boldsymbol{v} \times \boldsymbol{B}) \right] \tag{7.19}$$

$$= \frac{\sigma \widetilde{Q}_{\mathrm{f}}}{\epsilon_{\mathrm{r}} \epsilon_0} + \frac{\sigma}{\epsilon_{\mathrm{r}}} \boldsymbol{\nabla} \cdot (\boldsymbol{v} \times \boldsymbol{B}) \,. \tag{7.20}$$

So, rearranging,

$$\frac{\partial \widetilde{Q}_{\mathrm{f}}}{\partial t} + \frac{\sigma}{\epsilon_{\mathrm{r}} \epsilon_0} \widetilde{Q}_{\mathrm{f}} = -\frac{\sigma}{\epsilon_{\mathrm{r}}} \boldsymbol{\nabla} \cdot (\boldsymbol{v} \times \boldsymbol{B}) \,. \tag{7.21}$$

Setting the right-hand side equal to zero gives the equation for the transient effect:

$$\frac{\partial \widetilde{Q}_{\mathrm{f}}}{\partial t} + \frac{\sigma}{\epsilon_{\mathrm{r}} \epsilon_0} \widetilde{Q}_{\mathrm{f}} = 0 \tag{7.22}$$

and

$$\widetilde{Q}_{\mathrm{f,\,transient}} = C \exp\left(-\frac{\sigma}{\epsilon_{\mathrm{r}} \epsilon_0} t \right) , \tag{7.23}$$

where C is a constant of integration.

The coefficient of t in the exponent is a *very* large number: for copper,

$$\frac{\sigma}{\epsilon_{\mathrm{r}} \epsilon_0} = \frac{5.8 \times 10^7}{8.85 \times 10^{-12}} \approx 7 \times 10^{18} \,. \tag{7.24}$$

So, in copper, the transient effect for $\widetilde{Q}_{\mathrm{f}}$ is totally negligible, except for time intervals of the order of 10^{-18} second or less.

Disregarding the time derivative in Eq. 7.21,

$$\widetilde{Q}_{\mathrm{f}} = -\epsilon_0 \boldsymbol{\nabla} \cdot (\boldsymbol{v} \times \boldsymbol{B}) \,, \tag{7.25}$$

which is just Eq. 7.10. However, we now know that, in this equation, both $\boldsymbol{v}$ and $\boldsymbol{B}$ can be time-dependent.

7.2.3 The Magnetic Field of the Convection Current

Since a moving conductor usually carries a space charge, there is usually a convection current. We saw in Sect. 2.5 that convection currents are negligible, compared to conduction currents, but is the *magnetic field* $\boldsymbol{B}_{\text{conv}}$ of the convection current negligible, compared to the net magnetic field $\boldsymbol{B}_{\text{net}}$, due to all currents? Set $v^2/c^2 \ll 1$. From Eq. 2.4,

$$\boldsymbol{\nabla} \times \boldsymbol{B}_{\text{conv}} = \mu_0 \widetilde{Q}_{\text{f}} \boldsymbol{v} , \quad |\boldsymbol{\nabla} \times \boldsymbol{B}_{\text{conv}}| = \mu_0 |\widetilde{Q}_{\text{f}} v| . \tag{7.26}$$

Call $\mathcal{L}$ a characteristic length of the system. Then, for typical values of the fields, and from Eq. A.84,

$$|\boldsymbol{\nabla} \times \boldsymbol{B}_{\text{conv}}|_{\text{typical}} \sim \frac{B_{\text{conv typical}}}{\mathcal{L}} . \tag{7.27}$$

We have a $\sim$ sign here instead of $\lesssim$ because $\boldsymbol{B}_{\text{conv}}$ is transversal, since its divergence is zero (Sect. A.6.3). The above two equations imply that

$$B_{\text{conv typical}} \sim \mu_0 |\widetilde{Q}_{\text{f}} v|_{\text{typical}} \mathcal{L} . \tag{7.28}$$

There is a similar equation for $\boldsymbol{B}_{\text{net}}$ and $\boldsymbol{J}_{\text{net}}$:

$$B_{\text{net typical}} \sim \mu_0 J_{\text{net typical}} \mathcal{L} . \tag{7.29}$$

From Eq. 2.42, then,

$$\frac{B_{\text{conv typical}}}{B_{\text{net typical}}} \lesssim \frac{v^2}{c^2} \ll 1 . \tag{7.30}$$

So, as long as $v^2 \ll c^2$, the magnetic field of the convection current associated with $\widetilde{Q}_{\text{f}}$ is negligible, at least if the conditions for Eq. 7.10 are satisfied (steady state and uniform σ).

7.3 Example: The Faraday Disk

Figure 7.1 shows a *Faraday disk* used as a generator. This device is the classical model for natural dynamos (Chapters 10 to 14). It is simple-minded if one thinks, for example, of the vastly complex flows in the liquid part of the Earth's core that generate its magnetic field. Nonetheless, the disk dynamo is a useful model if analyzed correctly. The discussions of the disk dynamo in the literature are erroneous, particularly those by Bullard (1955, 1978). We return to the Faraday disk in Chapters 8 and 10. See also P. Lorrain et al. (1988).

The conducting disk rotates in air at the angular velocity ω in the constant and uniform axial magnetic field $\boldsymbol{B}$. We assume that the condition $v^2 \ll c^2$

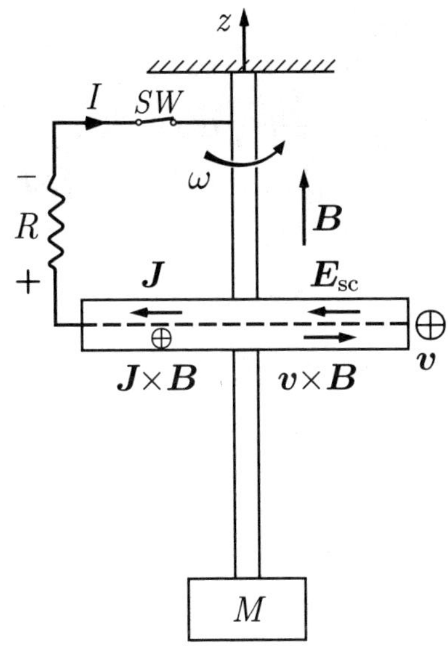

Fig. 7.1. Faraday disk connected as a generator. A motor M rotates the disk in the axial magnetic field $\boldsymbol{B}$

applies everywhere on the disk. Sliding contacts on the periphery of the disk and on the axle connect to an external circuit that draws a current I.

Industrial versions of Faraday disks are called *homopolar motors*, or *homopolar generators*. They have continuous brushes all around the disk, and all around the axle. The magnetic field is often supplied by superconducting coils.

7.3.1 The Faraday Disk with the Switch *SW* Open

Let us first open switch SW of Fig. 7.1. An electric charge in the disk feels the two fields $\boldsymbol{v} \times \boldsymbol{B}$ and $\boldsymbol{E}_{sc}$. With ω and $\boldsymbol{B}$ as in the figure, $\boldsymbol{v} \times \boldsymbol{B}$ points outward, in the direction of the rim. The space charge density in the disk being negative, as in Eq. 7.18, $\boldsymbol{E}_{sc}$ points inward, toward the axle. Then, since $\boldsymbol{J} = \boldsymbol{0}$ with the switch SW open, the two fields cancel. Let us check this. So as to avoid end-effects, we replace the disk by a long cylinder.

At a distance ρ from the center of the disk,

$$\boldsymbol{v} \times \boldsymbol{B} = \omega \rho B \hat{\boldsymbol{\rho}} , \tag{7.31}$$

while the field of the space charge, of uniform density $-2\epsilon_0 \omega B$ (Eq. 7.18), is the charge within a cylinder of radius ρ, divided by $2\pi\epsilon_0\rho$:

$$\boldsymbol{E}_{sc} = -\frac{\pi \rho^2 \widetilde{Q}_f}{2\pi\epsilon_0 \rho} \hat{\boldsymbol{\rho}} = -\omega \rho B \hat{\boldsymbol{\rho}} . \tag{7.32}$$

So the field of the space charge cancels $\boldsymbol{v} \times \boldsymbol{B}$ everywhere within the disk, when the switch SW is open.

At the circumference, of radius b,

$$|\boldsymbol{E}_{\rm sc}| = |\boldsymbol{v} \times \boldsymbol{B}| = \omega b B \,, \tag{7.33}$$

and the surface charge density is $\epsilon_0 \omega b B$. If the length of the cylinder, or the thickness of the disk, is s, then the charge at the periphery is

$$Q_{\rm sc\,periphery} = 2\pi b s(\epsilon_0 \omega b B) = 2\pi \epsilon_0 \omega b^2 s B \,. \tag{7.34}$$

The total charge associated with $\widetilde{Q}_{\rm f}$ is

$$Q_{\rm f} = \pi b^2 s \widetilde{Q}_{\rm f} = \pi b^2 s(-2\epsilon_0 \omega B) \tag{7.35}$$

$$= -2\pi \epsilon_0 \omega b^2 s B \,. \tag{7.36}$$

The surface charge at the periphery compensates for the space charge over the volume of the disk.

7.3.2 The Faraday Disk as a Generator

Now close SW in Fig. 7.1. The current in the generator depends on its characteristics, on the applied torque, and on the load resistance R. The radial current density in the disk is $\boldsymbol{J}$. Assume again that $\boldsymbol{B}$ is uniform over the surface of the disk. This assumption is not realistic, but it is appropriate here. Disregard end effects at the faces of the disk. This is equivalent to considering the disk to be a cylinder, as above. For the moment, disregard the magnetic fields of the currents in the axle and in the disk. We return to them in a moment.

With the polarities shown in Fig. 7.1, the magnetic force density $\boldsymbol{J} \times \boldsymbol{B}$ is azimuthal, points in the direction shown, and brakes the motion. The vector $\boldsymbol{v} \times \boldsymbol{B}$ is radial and points *out*ward. Inverting the direction of rotation inverts the polarity of the output. There is also a magnetic force due to $\boldsymbol{J}$ and to the azimuthal magnetic fields of the currents in the axle and in the disk. We return to those currents below in Sect. 7.3.4.

The rotation of the space charge, of density $\widetilde{Q}_{\rm f}$, gives an azimuthal convection current of density $\widetilde{Q}_{\rm f}\, \boldsymbol{v}$ whose magnetic field $\boldsymbol{B}_{\rm conv}$ opposes the externally applied magnetic field $\boldsymbol{B}$, but it is negligible, as we saw in Sect. 7.2.3. With the above space charge density, at a distance of 1 meter from the axis of rotation, and with $v = 10$ meters/second, the convection current density at that radius is only about 2×10^{-9} ampere/meter^2: although the speed of a point on the disk is very much larger than the drift speed of the conduction electrons, the net space charge density is so small that the convection current is negligible.

With the switch SW in Fig. 7.1 closed, conduction electrons escape at the axle into the external circuit, where $\boldsymbol{v} \times \boldsymbol{B} = \boldsymbol{0}$, and return at the rim.

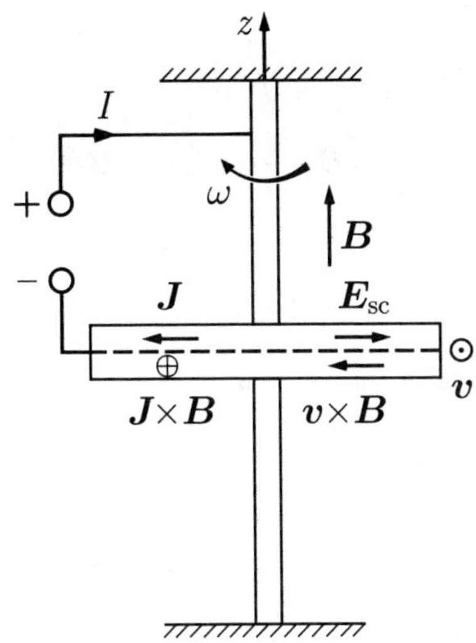

Fig. 7.2. Faraday disk connected as a motor

The electric field $\boldsymbol{J}/\sigma$ in the disk points in the direction of the rim, in the direction of $\boldsymbol{J}$. Then

$$\boldsymbol{E} = \boldsymbol{E}_{\text{sc}} + \frac{\boldsymbol{J}}{\sigma} \, . \tag{7.37}$$

The electric field has two components: $\boldsymbol{E}_{\text{sc}}$, which points in the direction of the axle, and $\boldsymbol{J}/\sigma$, which points in the direction of the rim.

As we shall see in Sect. 9.4.3, the Faraday generator also works without an external source of magnetic field, but with a disk magnetized in the direction parallel to the axle.

7.3.3 The Faraday Disk as a Motor

The current in a motor depends not only on the characteristics of the motor, but also on the source of current that feeds it and on its mechanical load.

To use the Faraday disk as a motor, connect it as in Fig. 7.2. Note that the direction of rotation is the opposite of that in Fig. 7.1, but that the direction of the current is the same. Inverting the polarity of the input inverses the direction of rotation.

Equation 7.37 again applies. The magnetic force $\boldsymbol{J} \times \boldsymbol{B}$ exerts a braking torque in the direction shown.

Let us calculate the potentials V and $\boldsymbol{A}$ for the motor. Assume that the magnetic field $\boldsymbol{B}$ is that of a long solenoid. For a long solenoid, from Sect. 4.5.2,

$$\boldsymbol{A} = \frac{B\rho}{2}\hat{\boldsymbol{\phi}} \quad \text{and} \quad \frac{\partial \boldsymbol{A}}{\partial t} = \boldsymbol{0} \, , \tag{7.38}$$

then

$$\boldsymbol{E} = -\boldsymbol{\nabla} V = -\frac{\partial V}{\partial \rho}\hat{\boldsymbol{\rho}} \tag{7.39}$$

$$= -\boldsymbol{v} \times \boldsymbol{B} + \frac{\boldsymbol{J}}{\sigma} = \left(-\omega\rho B + \frac{J}{\sigma}\right)\hat{\boldsymbol{\rho}} \tag{7.40}$$

and

$$\frac{\partial V}{\partial \rho} = \left(\omega\rho B - \frac{J}{\sigma}\right)\hat{\boldsymbol{\rho}}\,. \tag{7.41}$$

Since $J = I/(2\pi\rho s)$, where the current I is independent of the radius,

$$V = \frac{\omega\rho^2 B}{2} - \frac{I}{2\pi s\sigma}\ln\frac{\rho}{\rho_0}\,. \tag{7.42}$$

The quantity ρ_0 is a constant of integration; we can set the second term equal to zero at the radius $\rho = \rho_0$.

We now turn to the magnetic fields of the currents in the axle and in the disk.

7.3.4 The Currents in the Axle and in the Disk

Assume that the magnetic field of the current in the *axle* is the same as if the axle were infinitely long. Set the origin at the center of the disk, and assume that the current starts at $z = -\infty$ and ends at $z = 0$. At the point $\rho,\ z$, from Eq. 4.1,

$$\boldsymbol{B}_{\text{axle}} = \frac{\mu_0 I}{4\pi\rho}\left[1 + \frac{z}{(\rho^2 + z^2)^{1/2}}\right]\hat{\boldsymbol{\phi}}\,. \tag{7.43}$$

At $z = 0$ the field is half as large as if the current extended from $z = -\infty$ to $z = +\infty$. Since $\boldsymbol{v}$ and $\boldsymbol{B}_{\text{axle}}$ are both azimuthal,

$$\boldsymbol{v} \times \boldsymbol{B}_{\text{axle}} = \boldsymbol{0}\,. \tag{7.44}$$

The magnetic field of the radial current through the *disk* is also azimuthal. Then $B_{\text{disk},\,\rho} = B_{\text{disk},\,z} = 0$. We can show this by using Eq. 2.4 for the curl of $\boldsymbol{B}$, in cylindrical coordinates. Since the disk current density has only a radial component, and for a constant angular velocity $\partial/\partial\phi = 0$,

$$\frac{\partial B_{\text{disk},\,\phi}}{\partial z} = \mu_0 J_{\text{disk}}\,. \tag{7.45}$$

Calling the disk radius b, and the disk thickness s as above,

$$J_{\text{disk}} = \frac{I}{2\pi\rho s} \tag{7.46}$$

and

$$\frac{\partial B_{\text{disk},\,\phi}}{\partial z} = -\mu_0 \frac{I}{2\pi\rho s} \, . \tag{7.47}$$

So

$$B_{\text{disk},\,\phi} = -\frac{\mu_0 I z}{2\pi\rho s} + K \, , \tag{7.48}$$

where the constant of integration K is a function of ρ, but not of z.

Also, since the ϕ and z-components of the disk current density are zero, again from Eq. 2.4 in cylindrical coordinates,

$$\frac{\partial B_{\text{disk},\,\rho}}{\partial z} = \frac{\partial B_{\text{disk},\,z}}{\partial \rho} \, . \tag{7.49}$$

This is in agreement with our assumption of zero ρ and z-components of $\boldsymbol{B}$.

Setting now $K = 0$, which means that $B_{\text{disk},\,\phi}$ changes sign at $z = 0$,

$$B_{\text{disk},\,\phi} = -\frac{\mu_0 I}{2\pi\rho s} z \, . \tag{7.50}$$

The magnetic field of the current inside the disk is azimuthal, proportional to z, and inversely proportional to ρ. So, inside the disk,

$$\boldsymbol{v} \times \boldsymbol{B}_{\text{disk}} = \boldsymbol{0} \, . \tag{7.51}$$

Above and below the disk, $J = 0$ and $\boldsymbol{B}_{\text{disk}}$ is azimuthal and independent of z. This field is continuous at the top and bottom surfaces of the disk. We have disregarded edge effects at the rim of the disk.

7.4 Example: The Rotating Sphere

A solid sphere of radius R and uniform conductivity σ rotates at the constant angular velocity ω in a constant magnetic field $\boldsymbol{B}$. The magnetic field may have ρ, ϕ, and z-components, but it is axisymmetric. Figure 7.3 illustrates a situation in which this field is axial: $\boldsymbol{B} = B\hat{\boldsymbol{z}}$.

Take the curl of Eq. 7.1. As we shall see below, there is no induced current, and thus no induced magnetic field. For a uniform σ,

$$\boldsymbol{\nabla} \times \boldsymbol{J} = \sigma[\boldsymbol{\nabla} \times \boldsymbol{E} + \boldsymbol{\nabla} \times (\boldsymbol{v} \times \boldsymbol{B})] \, . \tag{7.52}$$

Since $\boldsymbol{B}$ is constant, $\boldsymbol{\nabla} \times \boldsymbol{E} = \boldsymbol{0}$, from Eq. 2.3, and

$$\boldsymbol{\nabla} \times \boldsymbol{J} = \sigma \boldsymbol{\nabla} \times (\boldsymbol{v} \times \boldsymbol{B}) = \sigma \boldsymbol{\nabla} \times (\omega\rho\hat{\boldsymbol{\phi}} \times \boldsymbol{B}) \tag{7.53}$$

$$= \sigma \boldsymbol{\nabla} \times [\omega\rho\hat{\boldsymbol{\phi}} \times (B_\rho\hat{\boldsymbol{\rho}} + B_z\hat{\boldsymbol{z}})] \tag{7.54}$$

$$= \sigma\omega\rho \left(\frac{B_\rho}{\rho} + \frac{\partial B_\rho}{\partial \rho} + \frac{\partial B_z}{\partial z} \right) \hat{\boldsymbol{\phi}} \tag{7.55}$$

$$= \sigma\omega\rho(\boldsymbol{\nabla} \cdot \boldsymbol{B})\hat{\boldsymbol{\phi}} = \boldsymbol{0} \, , \tag{7.56}$$

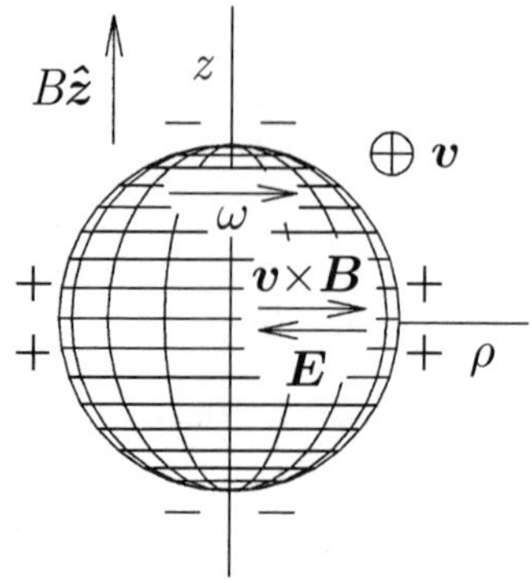

Fig. 7.3. Solid conducting sphere rotating in an axisymmetric axial magnetic field. There is a uniform space charge density inside, and the surface charges up as shown

from Eq. 2.2.

The fact that the line integral of $\boldsymbol{J} \cdot \boldsymbol{dl}$ around any arbitrary closed curve within the sphere is equal to zero, implies, in a steady-state situation, that $\boldsymbol{J} = \boldsymbol{0}$ everywhere inside the sphere.[2] Thus, in the sphere,

$$\boldsymbol{J} = \boldsymbol{0}\,, \quad \boldsymbol{E} = -(\boldsymbol{v} \times \boldsymbol{B}) \; : \tag{7.57}$$

the field of the space charge cancels the $\boldsymbol{v} \times \boldsymbol{B}$ field exactly at every point.

There is no induced current and the rotating sphere therefore has no effect on the magnetic field. This was noted by J.J. Thomson (1893) over one century ago.

Most authors are baffled by this classic result: how can there be no induced current when there is a $\boldsymbol{v} \times \boldsymbol{B}$ field? See, for example, Moffatt (1978), who says that "magnetic field lines disconnect and reconnect at the surface"! The reason is simple: the field $\boldsymbol{E}$ of the space and surface charges cancels the $\boldsymbol{v} \times \boldsymbol{B}$ field exactly at every point!

[2] Let the line integral of $\boldsymbol{J}$ around any arbitrary closed curve within a given *finite* conductor *of any shape* be equal to zero. There then exists a "potential" function ψ such that $\boldsymbol{J} = \boldsymbol{\nabla}\psi$. Suppose that $\boldsymbol{J} \neq \boldsymbol{0}$, that is, that ψ is not uniform in the conductor. Since the conductor is finite, ψ has both a minimum value $\psi_{\min}$ and a maximum value $\psi_{\max}$ in the conductor, with $\psi_{\min} < \psi_{\max}$. Current flows from regions with low values of ψ to regions with high values of ψ. Negative charges will accumulate at points where $\psi = \psi_{\min}$ and positive charges will accumulate at points where $\psi = \psi_{\max}$. This contradicts our usual assumption of a steady state, in which volume and surface charge densities are constant in time. It follows that ψ is uniform in the conductor and that $\boldsymbol{J} = \boldsymbol{\nabla}\psi \equiv \boldsymbol{0}$.

This problem was solved correctly by Mason and Weaver in their classic book *The Electromagnetic Field* that was republished by Dover in 1929 (Mason and Weaver, 1929).

The electric charge density *inside* the conducting sphere is

$$\widetilde{Q}_{\rm f} = -\epsilon_0 \boldsymbol{\nabla} \cdot (\boldsymbol{v} \times \boldsymbol{B}) \tag{7.58}$$

$$= -\epsilon_0 \boldsymbol{\nabla} \cdot [\omega\rho\hat{\boldsymbol{\phi}} \times (B_\rho\hat{\boldsymbol{\rho}} + B_\phi\hat{\boldsymbol{\phi}} + B_z\hat{\boldsymbol{z}})] \tag{7.59}$$

$$= \epsilon_0 \boldsymbol{\nabla} \cdot (-\omega\rho B_\rho\hat{\boldsymbol{z}} + \omega\rho B_z\hat{\boldsymbol{\rho}}) \tag{7.60}$$

$$= -\epsilon_0\omega\left(-\rho\frac{\partial B_\rho}{\partial z} + 2B_z + \rho\frac{\partial B_z}{\partial \rho}\right) \neq 0\,. \tag{7.61}$$

7.4.1 The Surface Charge

There is also a surface charge. The surface of the sphere is *not* an equipotential, and the lines of $\boldsymbol{E}$ outside are thus *not* normal to the surface (van Bladel, 1984; P. Lorrain, 1990).

If the applied axisymmetric magnetic field is a uniform field $B\hat{\boldsymbol{z}}$, then the space charge density is given by

$$\widetilde{Q}_{\rm f} = -2\epsilon_0\omega B\,, \tag{7.62}$$

and there results a radial electric field

$$\boldsymbol{E}_{\rm sc} = -\frac{(4/3)\pi r^3(2\epsilon_0\omega B)}{4\pi\epsilon_0 r^2} = -\frac{2}{3}\omega B r\hat{\boldsymbol{r}}\,. \tag{7.63}$$

In this case, the electric field of the space charge does *not* cancel the $\boldsymbol{v}\times\boldsymbol{B}$ field. Indeed, that electric field is along $-\hat{\boldsymbol{r}}$, while the $\boldsymbol{v}\times\boldsymbol{B}$ field is along $+\hat{\boldsymbol{\rho}}$, and the magnitudes of the two fields are not the same. The explanation is that there also exists a surface charge, and it is the combination of the electric fields of both the volume and surface charges that cancels the $\boldsymbol{v}\times\boldsymbol{B}$ field.

Let us calculate the surface charge density $\widetilde{S}_{\rm f}$. From Gauss's law,

$$\widetilde{S}_{\rm f} = \epsilon_0 E_{r,\,\rm outside} - \epsilon_0 E_{r,\,\rm inside}\,. \tag{7.64}$$

Inside the sphere, in cylindrical coordinates,

$$\boldsymbol{E} = -(\boldsymbol{v}\times\boldsymbol{B}) = -\omega\rho B\hat{\boldsymbol{\rho}} \tag{7.65}$$

or, in spherical coordinates,

$$\boldsymbol{E} = -\omega r(\sin\theta)B(\hat{\boldsymbol{r}}\sin\theta + \hat{\boldsymbol{\theta}}\cos\theta)\,, \tag{7.66}$$

$$E_{r,\,\rm inside} = -\omega r B\sin^2\theta\,. \tag{7.67}$$

Also, inside the sphere, $E_z = 0$ and the potential along the z-axis is uniform. Call that potential $V_{\rm A}$. Then, on the surface, from Eq. 7.65,

$$V_{\text{surface}} = V_{\text{A}} + \tfrac{1}{2}\omega B\rho^2 = V_{\text{A}} + \tfrac{1}{2}\omega BR^2\sin^2\theta\,. \qquad (7.68)$$

Outside the sphere, the space charge density is zero, Laplace's equation 3.27 applies, and we can expand V as a series of Legendre polynomials (P. Lorrain et al., 1988, p. 225):

$$V_{\text{outside}} = \frac{B_0}{r} + \frac{B_1}{r^2}\cos\theta + \frac{B_2}{r^3}\frac{3\cos^2\theta - 1}{2} + \cdots \qquad (7.69)$$

We neglect higher order terms. The first term, which is the monopole term, implies a net charge on the sphere. That term would apply only if the sphere carried a net electric charge. The second, dipole, term implies charge asymmetry between the upper and lower halves of the sphere. That term is also zero. We therefore set the third, or quadrupole term equal to the V of Eq. 7.68 at $r = R$:

$$V_{\text{A}} + \frac{1}{2}\omega BR^2\sin^2\theta = \frac{B_2}{R^3}\frac{3\cos^2\theta - 1}{2} = \frac{B_2}{R^3}\left(1 - \frac{3}{2}\sin^2\theta\right)\,. \qquad (7.70)$$

Upon matching corresponding terms, we find that

$$B_2 = -\omega\frac{BR^5}{3} \quad \text{and} \quad V_{\text{A}} = -\omega\frac{BR^2}{3}\,. \qquad (7.71)$$

Thus

$$V_{\text{outside}} = -\frac{\omega BR^5}{3r^3}\left(1 - \frac{3}{2}\sin^2\theta\right)\,, \qquad (7.72)$$

$$E_{r,\,\text{outside}} = -\frac{\partial V}{\partial r} = -\frac{\omega BR^5}{r^4}\left(1 - \frac{3}{2}\sin^2\theta\right)\,. \qquad (7.73)$$

Finally, from Eqs. 7.64, 7.67, and 7.73, the surface charge density is

$$\widetilde{S}_{\text{f}} = \epsilon_0\omega BR\left(\frac{5}{2}\sin^2\theta - 1\right) \text{ coulombs/meter}^2\,. \qquad (7.74)$$

The surface charge density is positive in the region of the equator, and negative near the poles, as in Fig. 7.3. The net surface charge is equal to minus the space charge.

The value of the potential inside the sphere follows from Eqs. 7.65 and 7.71:

$$V_{\text{inside}} = \omega B\left(\frac{\rho^2}{2} - \frac{R^2}{3}\right)\,. \qquad (7.75)$$

The potential is zero on a cylindrical surface of radius $(2/3)^{1/2}R$.

Both the space and surface charge densities are independent of the conductivity σ of the sphere.

We return to the rotating conducting sphere in Sect. 8.7.5.

7.5 Summary

Under steady conditions, a homogeneous conductor that moves at a velocity $\boldsymbol{v}$ in a magnetic field $\boldsymbol{B}$ carries a net electric space charge whose density is

$$\widetilde{Q}_{\mathrm{f}} = -\epsilon_0 \boldsymbol{\nabla} \cdot (\boldsymbol{v} \times \boldsymbol{B}) \; . \qquad (7.10)$$

The electric field of these charges plays a fundamental role in MFD because it cancels, more or less, the $\boldsymbol{v} \times \boldsymbol{B}$ *field.*

However, the magnetic field produced by the convection of these charges is negligible, compared to the total magnetic field.

The Faraday disk of Sect. 7.3 is the classical model for natural dynamos. It carries a uniform space charge density

$$\widetilde{Q}_{\mathrm{f}} = -2\epsilon_0 \omega B \; . \qquad (7.18)$$

With the switch *SW* of Fig. 7.1 open, $\boldsymbol{J} = \boldsymbol{0}$ and the electric field of $\widetilde{Q}_{\mathrm{f}}$ cancels the $\boldsymbol{v} \times \boldsymbol{B}$ field exactly at every point:

$$E_{\mathrm{sc}} \hat{\boldsymbol{\rho}} = -\omega \rho B \hat{\boldsymbol{\rho}} = -(\boldsymbol{v} \times \boldsymbol{B}) \; . \qquad (7.32)$$

If a current flows, then

$$\boldsymbol{E} = \boldsymbol{E}_{\mathrm{sc}} + \boldsymbol{J}/\sigma \; . \qquad (7.37)$$

If a solid conducting sphere rotates in an axisymmetric magnetic field, the field of the electric volume and surface charges cancels the $\boldsymbol{v} \times \boldsymbol{B}$ field at every point, there is no induced current, and the rotating sphere has no effect on the magnetic field.

8 Nine Examples: Magnetic Fields in Moving Conductors

Nine Examples:
Magnetic Fields in Moving Conductors

Contents

What happens if you move a conductor in a given magnetic field? The answer is simple, in principle: the conductor usually carries induced currents that generate an induced magnetic field, which adds to the original field: the net field is the sum of the two. To calculate the net field, one must therefore calculate the induced currents from Ohm's law for moving conductors (Sect. 6.4). One must of course take into account the electric field of the space charges associated with the $\boldsymbol{v} \times \boldsymbol{B}$ field.[1]

[1] This chapter extends the discussion in P. Lorrain, J. McTavish, F. Lorrain, *Magnetic fields in moving conductors: four simple examples*, Eur. J. Phys. **19**, 451–457 (1998).

8.1 Introduction

Conductors that move in magnetic fields are everywhere. For example, the liquid shell of the Earth's core, which is composed largely of liquid iron, and which is therefore conducting, convects in its own magnetic field (Chapter 9). The solar plasma is another example of a conductor that moves in a magnetic field: certain regions act as self-excited dynamos, for example sunspots (Chapter 13), while in other regions the plasma convects in magnetic fields generated elsewhere.

A convecting fluid can act as a self-excited dynamo: given a seed magnetic field, the motion of the fluid generates, under appropriate conditions, an *aiding* magnetic field, providing positive feedback, and the total field increases until it reaches an equilibrium value that is a function of the power that drives the conductor. See Chapters 10 to 14. Certain self-excited dynamos in convecting plasmas can also generate charged-particle beams, as in Chapter 14. Other examples of conductors that move in magnetic fields are of course electric motors and generators.

8.2 The Induced Currents

As a rule, a conductor that moves in a magnetic field carries an induced current. That is not always the case, as we saw when discussing the rotating conducting sphere in Sect. 7.4, and as we shall see in some of the Examples discussed below. If there is an induced current, then its magnetic field adds to any externally applied field. Indeed, modifying a magnetic field inside a moving conductor requires the addition of a further magnetic field, and this requires an induced current. If there is no induced current, then the moving conductor has no effect on the magnetic field.

The magnetic force density on the moving conductor is given by $\boldsymbol{J} \times \boldsymbol{B}_{\text{net}}$, where $\boldsymbol{B}_{\text{net}}$ is the sum of the externally applied magnetic field $\boldsymbol{B}_{\text{ext}}$ and of the magnetic field of the induced currents, $\boldsymbol{B}_{\text{ind}}$. If there is no induced current, then $\boldsymbol{J} = \boldsymbol{0}$ and there is no magnetic force.

We are concerned here with the net *magnetic field* $\boldsymbol{B}_{\text{net}}$.

We therefore calculate the *currents*, if any, induced in the moving conductors. Once the induced currents are known, we can calculate the induced magnetic field from either Eq. 2.4 or Eq. 2.20, and the net field.

There exists an abundant literature on the interaction between magnetic fields and moving conductors, particularly in the context of Magneto-Fluid-Dynamics. For a general discussion of electromagnetic fields in moving media, see Penfield and Haus (1967), even though those authors do not discuss the subject of this chapter.

Geophysicists and Astrophysicists have been particularly prolific in this field. See, for example, Parker (1979) and Russell et al. (1990). These discussions concern the shape of imaginary magnetic field lines, invariably disregarding the electric currents. It is of course pointless to investigate imaginary field lines without first making sure that the required electric currents make sense.

Also, discussions on magnetic fields in moving conductors never refer to the electrostatic space and surface charges that play an essential role in MFD, as we saw in the preceding chapter. Alfvén's repeated warnings against Pseudo-MHD, quoted in the Preface and in Sect. 8.3 below, have gone unheeded.

Moreover, most authors do not seem to realize the importance of distinguishing clearly between the two reference frames involved, one that is fixed and another one that follows the moving conductor.

8.3 Magnetic Field Lines Once More

The concept of magnetic field line of Sect. 4.1 is a useful "thinking crutch", but no more. See Slepian (1951) and McCune and Sears (1959). It is legitimate to plot magnetic field lines as we do, but it is of course impossible to identify a particular magnetic field line, and it does not make sense to discuss its motion.

According to the *frozen-flux principle* (Alfvén, 1943), magnetic field lines follow a perfect conductor as if they were frozen-in. If σ is not infinite, then it is said that the lines are not quite frozen because of "slippage".

After Alfvén proposed the principle, Lundquist (1952) attempted to illustrate the principle by calculating magnetic field lines for a moving conducting plate. This is a simple version of our Example 5 below (Sect. 8.7.4). He found that the field lines are STATIONARY, for any value of σ. But then he stated without proof that the frozen-flux principle applies nonetheless, for infinite conductivity!

Unfortunately, Alfvén, Lundquist, and their readers did not realize that the principle is a misconception. Otherwise, much time and effort, and paper and ink would have been saved over the past half-century!

The "principle" is commonly invoked even now, despite Alfvén's repeated disavowals (Alfvén, 1943, 1975, 1981; Alfvén and Fälthammar, 1963). It is worthwhile quoting Alfvén, again, as in the Preface:

> *It is meaningless to speak about a translational movement of magnetic field lines [...] An impression is developed that you understand a situation even if in reality you have misunderstood it [...] There is no need for 'frozen-in' field lines moving with the plasma, still less for 'field line reconnection' or 'magnetic merging' [...] Forgetting the 'frozen-in' picture [...] We have no use for [...] magnetic field line reconnection [...] this monstrous concept is a product of the frozen-in picture* in absurdum *[..., etc.]*

Curiously enough, the frozen-field principle is commonly applied to poor conductors. For example, some geophysicists apply it to the liquid part of the Earth's core, whose σ is thought to be smaller than that of room-temperature copper by three orders of magnitude.

If the frozen-field principle were true, then the electric field $\boldsymbol{E}'$ in the moving reference frame would be equal to zero, and the magnetic flux Φ' linking a closed curve C' would be constant. Both of these consequences prove to be wrong, as we shall see in the following Examples.

Now the frozen-field principle has been "proved" many times. So where is the pitfall? All the "proofs" rest, basically, on the following reasoning. According to Ohm's law for moving conductors (Sect. 6.4),

$$\boldsymbol{J} = \sigma[\boldsymbol{E} + (\boldsymbol{v} \times \boldsymbol{B})] . \tag{8.1}$$

If σ tends to infinity, then the bracket must tend to zero because $\boldsymbol{J}$ cannot be infinite. That is the pitfall. In all the examples investigated here, *both* $\boldsymbol{E}$ *and* $\boldsymbol{v} \times \boldsymbol{B}$ *are independent of* σ*, and* $\boldsymbol{J}$ *is proportional to* σ. If the bracket tended to zero when σ tends to infinity, then both $\boldsymbol{E}$ and $\boldsymbol{v} \times \boldsymbol{B}$ would be functions of σ, which proves to be wrong.

It is not difficult to show that $\boldsymbol{E}$ is independent of σ under steady-state conditions. The electric charges are then the only source of $\boldsymbol{E}$, and the electric space charge density is independent of σ, as in Eq. 7.10: with $\widetilde{Q}_{\mathrm{f}}$ everywhere constant, the net charge per unit time flowing out of any given volume is zero, and the value of σ is immaterial. The *surface* charge density is also independent of σ for the same reason: since it is constant, by hypothesis, the net charge per unit time flowing out of any given part of the surface is zero, and again the value of σ bears no importance.

Is $\boldsymbol{v} \times \boldsymbol{B}$ independent of σ under steady-state conditions? With a given $\boldsymbol{v}$ and a given externally applied $\boldsymbol{B}_{\mathrm{ext}}$, $\boldsymbol{v} \times \boldsymbol{B}_{\mathrm{ext}}$ is independent of σ. But is $\boldsymbol{v} \times \boldsymbol{B}_{\mathrm{ind}}$ also independent of σ, with $\boldsymbol{B}_{\mathrm{ind}}$ a function of $\boldsymbol{J}$, and thus of σ? The answer is yes, according to the Examples below.

Before the paper by P. Lorrain et al. (1998), only Lundquist (1952) had discussed the distortion of magnetic field lines by moving conductors in clear-cut cases. As stated above, Lundquist found, despite his assertion to the contrary, that the frozen-field principle does not apply.

The near-total absence of real tests of the behavior of magnetic field lines in moving conductors is both striking and regrettable. Typically, Newcomb (1958) wrote a 39-page theoretical paper without testing his statements.

8.4 The Net Magnetic Field

The net magnetic field depends solely on the electric currents, whether externally applied or induced. For example, the net magnetic field is unaffected if one replaces the moving conductors by their induced currents. The case of the Faraday disk of Sect. 8.7.8 below is striking: the magnetic field is unaffected if the disk stops rotating, and the current is restored with an external source.

The magnetic field of the induced currents, $\boldsymbol{B}_{\text{ind}}$, presents a problem. According to Ohm's law for moving conductors (Sect. 6.4),

$$\boldsymbol{J} = \sigma[\boldsymbol{E} + (\boldsymbol{v} \times \boldsymbol{B})] \, , \tag{8.2}$$

where $\boldsymbol{B}$ is the *net* field

$$\boldsymbol{B} \equiv \boldsymbol{B}_{\text{net}} = \boldsymbol{B}_{\text{ext}} + \boldsymbol{B}_{\text{ind}} \, , \tag{8.3}$$

$\boldsymbol{B}_{\text{ext}}$ being the externally applied field and $\boldsymbol{B}_{\text{ind}}$ the magnetic field of the induced currents. So

$$\boldsymbol{J} = \sigma\{\boldsymbol{E} + [\boldsymbol{v} \times (\boldsymbol{B}_{\text{ext}} + \boldsymbol{B}_{\text{ind}})]\} \, . \tag{8.4}$$

It would be appropriate here to add a subscript "ind" on $\boldsymbol{J}$, since it is an induced current density.

Let there be no externally applied current. Now $\boldsymbol{B}_{\text{ind}}$ depends on $\boldsymbol{J}$: from Eq. 2.4,

$$\boldsymbol{\nabla} \times \boldsymbol{B}_{\text{ind}} = \mu_0 \boldsymbol{J} \, . \tag{8.5}$$

Similarly, the space charge density of Eq. 7.10 is a function of the *net* magnetic field. Now the net charge density $\widetilde{Q}$ is the sum of the bound charge density $\widetilde{Q}_{\text{b}}$ resulting from the distortion of dielectric atoms or molecules, and the free charge density $\widetilde{Q}_{\text{f}}$ that results from the addition or subtraction of free electric charges as in Sect. 7.2. We are solely concerned here with conductors, and $\widetilde{Q} = \widetilde{Q}_{\text{f}}$. We use $\widetilde{Q}_{\text{f}}$, as in Chapter 7. Then

$$\widetilde{Q}_{\text{f}} = -\epsilon_0 \boldsymbol{\nabla} \cdot [\boldsymbol{v} \times (\boldsymbol{B}_{\text{ext}} + \boldsymbol{B}_{\text{ind}})] \, . \tag{8.6}$$

Also,

$$\widetilde{Q}_{\text{f}} = -\epsilon_0[\boldsymbol{B}_{\text{net}} \cdot (\boldsymbol{\nabla} \times \boldsymbol{v}) - \boldsymbol{v} \cdot (\boldsymbol{\nabla} \times \boldsymbol{B}_{\text{net}})] \tag{8.7}$$

$$= -\epsilon_0[\boldsymbol{B}_{\text{net}} \cdot (\boldsymbol{\nabla} \times \boldsymbol{v}) - \mu_0 \boldsymbol{v} \cdot \boldsymbol{J}] \tag{8.8}$$

$$= -\epsilon_0 \boldsymbol{B}_{\text{net}} \cdot (\boldsymbol{\nabla} \times \boldsymbol{v}) + \frac{\boldsymbol{v} \cdot \boldsymbol{J}}{c^2} \, . \tag{8.9}$$

So $\boldsymbol{E}$ in Eq. 8.2 depends on $\widetilde{Q}_{\text{f}}$, which depends on $\boldsymbol{B}_{\text{ind}}$. We avoid this difficulty here by investigating only cases where $\boldsymbol{B}_{\text{ind}} = \boldsymbol{0}$, or where $\boldsymbol{v} \times \boldsymbol{B}_{\text{ind}} = \boldsymbol{0}$. Then we can disregard the $\boldsymbol{B}_{\text{ind}}$ terms in Eqs. 8.4 and 8.6. Some readers might devise more general examples.

8.5 The Case of Rotating Conductors

If the velocity $\boldsymbol{v}$ of a point in the conductor is not constant, then its frame S' is not inertial. For instance, in Examples 2, 3, 5, and 6 below, the vector $\boldsymbol{v}$

rotates with the conductor. Then the above equations, as well as Maxwell's equations, apply only in the neighborhood of a given point P', as we saw in Sect. 6.2.1. The moving reference frame S' at P' then applies only at a given instant and is an inertial frame whose velocity is the instantaneous velocity of P' (Møller, 1974).

Consider a *finite* axisymmetric passive conductor that rotates around its axis at a constant angular velocity ω. The conductor is not connected to a stationary circuit through sliding contacts.

Let the externally applied magnetic field $\boldsymbol{B}_{\text{ext}}$ be symmetric around the same axis. We assume a steady state. Then the net magnetic field $\boldsymbol{B} = \boldsymbol{B}_{\text{net}}$ (Eq. 8.3) will also be axisymmetric.

The curl of $\boldsymbol{J}$ is zero for the following reason. From Sect. 6.4,

$$\boldsymbol{J} = \sigma[\boldsymbol{E} + (\boldsymbol{v} \times \boldsymbol{B})] \tag{8.10}$$

and

$$\boldsymbol{\nabla} \times \boldsymbol{J} = \sigma[\boldsymbol{\nabla} \times \boldsymbol{E} + \boldsymbol{\nabla} \times (\boldsymbol{v} \times \boldsymbol{B})] \tag{8.11}$$

$$= \sigma\left[-\frac{\partial \boldsymbol{B}}{\partial t} + \boldsymbol{\nabla} \times (\boldsymbol{v} \times \boldsymbol{B})\right] . \tag{8.12}$$

We have used Eq. 2.3. For a steady state, the time derivative vanishes and

$$\boldsymbol{\nabla} \times \boldsymbol{J} = \sigma\boldsymbol{\nabla} \times [\omega\rho\hat{\boldsymbol{\phi}} \times (B_\rho\hat{\boldsymbol{\rho}} + B_\phi\hat{\boldsymbol{\phi}} + B_z\hat{\boldsymbol{z}})] \tag{8.13}$$

$$= \sigma\omega\boldsymbol{\nabla} \times (-\rho B_\rho\hat{\boldsymbol{z}} + \rho B_z\hat{\boldsymbol{\rho}}) \tag{8.14}$$

$$= \sigma\omega\left[\frac{\partial}{\partial\rho}(\rho B_\rho) + \frac{\partial}{\partial z}(\rho B_z)\right]\hat{\boldsymbol{\phi}} \tag{8.15}$$

$$= \sigma\omega\rho\left[\frac{1}{\rho}\frac{\partial}{\partial\rho}(\rho B_\rho) + \frac{\partial B_z}{\partial z}\right]\hat{\boldsymbol{\phi}} \tag{8.16}$$

$$= \sigma\omega\rho(\boldsymbol{\nabla}\cdot\boldsymbol{B})\hat{\boldsymbol{\phi}} \equiv \boldsymbol{0} . \tag{8.17}$$

So the curl of $\boldsymbol{J}$ is zero inside our conductor.

Then, from Stokes's theorem, if C is any closed curve lying entirely inside the conductor, and if $\mathcal{A}$ is any surface that is inside the conductor and is bounded by C,

$$\oint_C \boldsymbol{J}\cdot\boldsymbol{dl} = \int_{\mathcal{A}} (\boldsymbol{\nabla}\times\boldsymbol{J})\cdot\boldsymbol{d\mathcal{A}} = 0 . \tag{8.18}$$

Since the shape of C is arbitrary, we conclude from p. 98, note 2, that

$$\boldsymbol{J} \equiv \boldsymbol{0} . \tag{8.19}$$

There are no induced currents and hence no induced magnetic field.

So, if an axisymmetric conductor rotates around the axis of an axisymmetric net magnetic field (Eq. 8.3), then $\boldsymbol{J} = \boldsymbol{0}$ and, from Eq. 8.10,

$$\boldsymbol{E} = -(\boldsymbol{v} \times \boldsymbol{B}) \,. \tag{8.20}$$

The field of the induced electric charges cancels the $\boldsymbol{v} \times \boldsymbol{B}$ field exactly everywhere. There are no induced currents and the magnetic field is therefore unaffected by the rotating conductor.

8.6 The Newcomb Criterion

Newcomb (1958) stated that, if a closed curve C moves and distorts with a conducting medium, then the magnetic flux Φ linking C is constant in time if, and only if,

$$\boldsymbol{\nabla} \times [\boldsymbol{E} + (\boldsymbol{v} \times \boldsymbol{B})] = \boldsymbol{0} \,. \tag{8.21}$$

This statement is known as the *Newcomb criterion.* As we shall see, this is, in fact, a sufficient, but not a necessary condition for the constancy of Φ. Also, Newcomb sets $\boldsymbol{v} \cdot \boldsymbol{B} = 0$, but that condition is unnecessary.

In a fixed reference frame S,

$$\boldsymbol{\nabla} \times \boldsymbol{E} = -\frac{\partial \boldsymbol{B}}{\partial t} \,, \tag{8.22}$$

as in Eq. 2.4. Similarly, in the moving frame S',

$$\boldsymbol{\nabla}' \times \boldsymbol{E}' = -\frac{\partial \boldsymbol{B}'}{\partial t'} \,, \tag{8.23}$$

where $\boldsymbol{B} = \boldsymbol{B}'$ and $t = t'$, as in Eq. 6.26 and Sect. 6.2.1. Then, from Stokes's theorem,

$$\oint_{C'} \boldsymbol{E}' \cdot d\boldsymbol{l}' = \frac{d}{dt'} \int_{\mathcal{A}'} \boldsymbol{B}' \cdot d\boldsymbol{\mathcal{A}}' = \frac{d\Phi'}{dt'} \,, \tag{8.24}$$

where $\mathcal{A}'$ is any open surface bounded by the closed curve C'.

Now we saw in Sect. 8.3 that $\boldsymbol{E}$ is in general not zero, even when σ tends to infinity. The same applies to $\boldsymbol{E}'$. It follows that the magnetic flux Φ' linked by the closed curve C' is *not* constant, in general. This will be confirmed by the Examples that follow.

Let us look at this result more closely.

If σ is uniform, the Newcomb criterion is equivalent to stating that Φ is constant if, and only if,

$$\boldsymbol{\nabla} \times \boldsymbol{J} = \boldsymbol{0} \,, \quad \text{or} \quad \boldsymbol{\nabla} \times (\boldsymbol{\nabla} \times \boldsymbol{B}) = \boldsymbol{0} \,. \tag{8.25}$$

We have used Eq. 2.4 and Identity 1 on the page facing the front cover. Here $\boldsymbol{B}$ is the induced magnetic field, assuming that $\boldsymbol{\nabla} \times \boldsymbol{B}_{\text{ext}}$ is zero.

Newcomb's argument, slightly restated as we do below, rests on a mathematical principle, Eq. 8.26 below, and on one of Maxwell's equations. It is therefore reliable.

Call $\mathcal{A}$ the area of any surface bounded by the closed curve C, $\hat{\boldsymbol{n}}$ the unit vector normal to $\mathcal{A}$, and $\boldsymbol{dl}$ an element of length along C. Then, for a stationary observer, with $\boldsymbol{v}$ the velocity of a point on the curve C, in reference frame S,

$$\frac{d\Phi}{dt} = \int_{\mathcal{A}} \frac{\partial \boldsymbol{B}}{\partial t} \cdot \hat{\boldsymbol{n}}\, d\mathcal{A} + \oint_C \boldsymbol{B} \cdot (\boldsymbol{v} \times \boldsymbol{dl}) \,. \tag{8.26}$$

The first term on the right is the time derivative of the magnetic flux through the surface $\mathcal{A}$, if the curve C is independent of the time. The second term is the flux swept by the element $\boldsymbol{dl}$ in one second. Then

$$\frac{d\Phi}{dt} = -\int_{\mathcal{A}} (\boldsymbol{\nabla} \times \boldsymbol{E}) \cdot \hat{\boldsymbol{n}}\, d\mathcal{A} - \oint_C (\boldsymbol{v} \times \boldsymbol{B}) \cdot \boldsymbol{dl} \,. \tag{8.27}$$

We have used Eq. 2.3 and Identity 16 (page facing the front cover). Applying now Stokes's theorem to the second term on the right,

$$\frac{d\Phi}{dt} = -\int_{\mathcal{A}} \{\boldsymbol{\nabla} \times [\boldsymbol{E} + (\boldsymbol{v} \times \boldsymbol{B})]\} \cdot \hat{\boldsymbol{n}}\, d\mathcal{A} \tag{8.28}$$

$$= -\int_{\mathcal{A}} \left(\boldsymbol{\nabla} \times \frac{\boldsymbol{J}}{\sigma}\right) \cdot \hat{\boldsymbol{n}}\, d\mathcal{A} \tag{8.29}$$

$$= -\int_C \frac{\boldsymbol{J}}{\sigma} \cdot \boldsymbol{dl} \,. \tag{8.30}$$

The time derivative of the magnetic flux Φ linked by C is zero if either one of these last two *integrals* is zero. That is a necessary condition. Equation 8.21 is a sufficient condition for this to be true.

In Examples 1 to 6 below, $\boldsymbol{E} + (\boldsymbol{v} \times \boldsymbol{B}) = \boldsymbol{0}$ and the magnetic flux Φ linked by a curve C that follows the moving conductor moving at a constant velocity, is constant. This is in agreement with the fact that the magnetic field is stationary in the reference frame S' of the moving conductor.

Example 8 concerns a sphere of fluid in which the angular velocity is a function of the z-coordinate as in Eq. 8.46. Equations 8.51–8.53 show that $\boldsymbol{\nabla} \times \boldsymbol{J}$ is in general not equal to zero. The magnetic flux linked by a closed curve C that moves and distorts with the fluid is not constant, in agreement with the Newcomb criterion.

8.7 Nine Examples

We calculate below the currents induced in moving conductors of uniform σ, and the net magnetic field. We find that, in all these Examples, the magnetic field lines are either *unaffected* by the moving conductor, or distorted *downstream*, or distorted *upstream*. In all cases the magnetic field lines are *stationary*.

Many readers will think of other examples, either just as simple, or more general.

8.7.1 Assumptions

We assume steady-state conditions: with respect to at least one inertial reference frame S, the field variables $\boldsymbol{E}$, $\boldsymbol{B}_{\text{ext}}$, $\boldsymbol{A}$, V, $\boldsymbol{J}$, as well as the electrostatic space and surface charge densities, are all independent of the time, so we set $\partial/\partial t = 0$ in that frame. The conductor velocities are given. All transient phenomena have become negligible, charges have moved to their equilibrium positions, and the conductors have reached their asymptotic velocities. We disregard dielectrics, and thus polarization charges and currents.

If the conductor moves in a straight line, then the velocity $\boldsymbol{v}$ of a given point inside is constant. If the conductor rotates, then $\boldsymbol{v}$ is a function of the time, but the angular velocity ω is constant.

The conductors move in a vacuum and there are no externally applied electric charges, or electric fields, or electric currents. This does not apply to the electromagnetic pump referred to below.

Under those conditions, $\boldsymbol{B}_{\text{ind}}$ and $\boldsymbol{B}_{\text{net}}$ are constant, and the configuration of the magnetic field lines is independent of the time in reference frame S.

We also assume non-relativistic speeds and non-magnetic materials:

$$v^2 \ll c^2 , \quad \mu_r = 1 , \tag{8.31}$$

where v is the speed of a point in the conductor with respect to reference frame S, and c is the speed of light.

Recall that $\boldsymbol{B}' = \boldsymbol{B}$, from Eq. 6.26, whatever the value of the conductivity σ. This relation is important here. It means that observers in S and S' measure *identical* magnetic fields. It also means that magnetic field lines in S' are *identical* to magnetic field lines in S, whatever the conductivity σ of the moving medium.

In the following Examples all the conductors are solid, save for Example 8, but our general conclusions apply to conducting fluids such as the liquid part of the Earth's core, and to plasmas such as those of the Sun, of the stars, and of Space (Peratt, 1991): the basic Physics is the same.

The induction equation of Sect. 6.5 applies: both sides are equal to zero.

8.7.2 Examples 1, 2, 3: The Three Shercliff Cases

Shercliff (1965) gives one example of a magnetic field that is *not* affected by a moving conductor, and two examples where the moving conductor shifts magnetic field lines *upstream*. In the first two cases, the magnetic flux Φ linked by a curve C that follows the moving conductor at a constant velocity, is constant. This is in agreement with the fact that the magnetic field is stationary in the reference frame S' of the moving conductor. Unfortunately, his discussion is brief and qualitative.

In his first example, a conducting fluid flows inside a non-conducting pipe, in the field of a magnet. The fluid velocity and the magnetic field are orthogonal. There is a Hall effect (Sect. 4.8):

> "Charges appear on the surface of the fluid in such a way as to produce an $\boldsymbol{E}$ that just balances the $\boldsymbol{v} \times \boldsymbol{B}$ [...] So here we have a conducting fluid failing to sweep the field in any sense [...]"

Since there is zero induced current, the net magnetic field is simply the applied field. So the magnetic field is unaffected by the moving fluid.

In his second example, he adds inside the fluid a longitudinal conducting plate that slows down the fluid in its neighborhood. There is still no electric current flowing to the tube. How he arrives at the following conclusion is not clear.

> "[...] [electric] currents do flow [in the fluid] because $\boldsymbol{v} \times \boldsymbol{B}$ is now non-uniform [...] The result is that [...] the field appears to be shifted *upstream* and it is evident that the sweeping concept [the frozen flux principle] is not reliable [...]"

The electromagnetic pump of Sect. 4.8.1 is his third example. Figure 4.6 shows how the applied current distorts the magnetic field lines *upstream*. Here, the electric current is applied, and not induced. The direction in which the lines are distorted is independent of σ. For given values of B and J, the field configuration is constant and independent of the value of σ.

The tubing through which the fluid flows is non-conducting. There is an externally applied magnetic field that points up, and an externally applied current that points into the paper. So the $\boldsymbol{J} \times \boldsymbol{B}$ force density on the fluid points right.

There is a Hall effect (Sect. 4.8), and charge carriers drift sideways, positive carriers flowing in the direction of the reader and negative carriers flowing in the opposite direction. This small current generates opposing charges on the tubing wall, and it soon stops.

The net magnetic field is the sum of the applied vertical field plus the magnetic field of the applied current. Figure 4.7 shows field lines for the net magnetic field.

THE MAGNETIC FIELD LINES ARE STATIONARY. THEY DO NOT FOLLOW THE MOVING CONDUCTOR.

THE FIELD LINES ARE DISTORTED UPSTREAM.

FOR GIVEN VALUES OF B AND J, THE MAGNETIC FIELD CONFIGURATION IS CONSTANT AND INDEPENDENT OF THE VALUE OF THE FLUID CONDUCTIVITY σ.

The direction in which the magnetic field lines are distorted is independent of σ.

Does the Newcomb criterion apply? Here, current flows between the electrodes, and $\boldsymbol{E} + (\boldsymbol{v} \times \boldsymbol{B}) \neq \boldsymbol{0}$. Now consider a closed curve C lying in a plane perpendicular to the paper. When C lies in the applied magnetic field, the

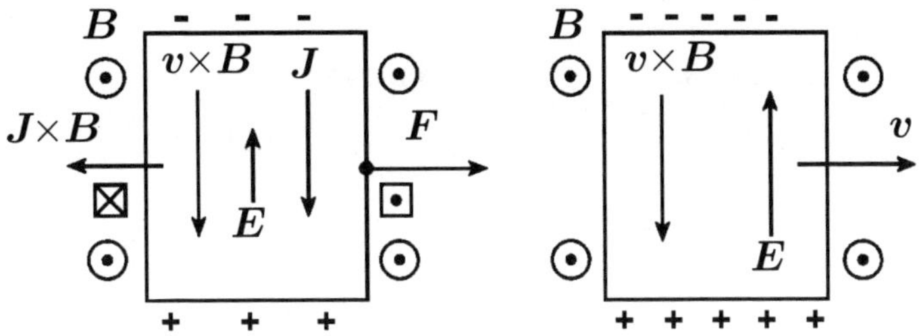

Fig. 8.1. Example 4. A rectangular copper plate moves to the right without rotating at the constant velocity $\boldsymbol{v}$ in a uniform magnetic field $\boldsymbol{B}$ that points out of the paper

magnetic flux Φ linking C has some non-zero value. But, as C moves away, out of the applied magnetic field, Φ tends to zero. So Φ is not constant, in agreement with the Newcomb criterion.

8.7.3 Example 4: Rigid Conductor

In this simple example, $\boldsymbol{B}_{\text{ind}} = \boldsymbol{0}$.

A rigid finite conducting object of arbitrary conductivity, shape, and size moves at a constant velocity $\boldsymbol{v}$ in a non-conducting fluid and in a uniform magnetic field $\boldsymbol{B}$. Figure 8.1 shows a square copper plate as an example. There is no external circuit.

The plate is first stationary in the magnetic field $\boldsymbol{B}$. a) An externally applied force $\boldsymbol{F}$ pulls the plate to the right at the constant velocity $\boldsymbol{v}$. The $\boldsymbol{v} \times \boldsymbol{B}$ field causes an induced current, of density $\boldsymbol{J}$, to flow downward, charging the top and bottom *surfaces*. The squares show the orientation of the magnetic field of the induced current. There is a braking force, of density $\boldsymbol{J} \times \boldsymbol{B}$, that opposes part of $\boldsymbol{F}$. b) Soon, the electrostatic surface charges have built up to their equilibrium value, $\boldsymbol{E} = -\boldsymbol{v} \times \boldsymbol{B}$, and $\boldsymbol{J} = \boldsymbol{0}$. There is then no magnetic braking force, $\boldsymbol{F} = \boldsymbol{0}$, and the plate continues moving to the right in the absence of any force. From then on, the copper plate has no effect on the magnetic field because there are no induced currents.

In reference frame S', in which the plate is at rest, a steady state is attained, in which $\boldsymbol{B}'$ is constant and, inside the plate,

$$\boldsymbol{\nabla}' \times \boldsymbol{J}' = \sigma' \boldsymbol{\nabla}' \times \boldsymbol{E}' = -\sigma' \frac{\partial \boldsymbol{B}'}{\partial t'} = \boldsymbol{0} \, . \tag{8.32}$$

From p. 98, note 2, $\boldsymbol{J}' = \boldsymbol{0}$ and then, in frame S also, $\boldsymbol{J} = \boldsymbol{0}$ (see Eq. 6.22). Again there is no induced current, and no induced magnetic field: $\boldsymbol{B}_{\text{ind}} = \boldsymbol{0}$.

From Eq. 7.10, there are no electrostatic *space* charges inside the copper plate because $\boldsymbol{v} \times \boldsymbol{B}$ is independent of the coordinates. So the electrostatic charges lie only at the surface.

Since $\boldsymbol{J} = \boldsymbol{0}$, it follows from Eq. 7.1 that, in this case,

$$\boldsymbol{E} + (\boldsymbol{v} \times \boldsymbol{B}) = \boldsymbol{0} \ : \tag{8.33}$$

the $\boldsymbol{E}$ of the *surface* charges cancels $\boldsymbol{v} \times \boldsymbol{B}$ exactly at every point *inside* the sheet, whatever the value of its conductivity σ. Both $\boldsymbol{E}$ and $\boldsymbol{v} \times \boldsymbol{B}$ are independent of σ. That is an important point.

There is no induced current and no induced magnetic field: the moving conductor does not disturb the magnetic field lines, and there is no magnetic braking force.

The magnetic field lines are stationary and unaffected, because there is no induced current. Field lines do *not* follow the moving conductor.

Since $\boldsymbol{E} + (\boldsymbol{v} \times \boldsymbol{B}) = \boldsymbol{0}$, the Newcomb criterion applies and $d\Phi/dt = 0$, as one can expect.

If the copper plate moves in a conducting fluid, then $\boldsymbol{E}$ does not completely cancel $\boldsymbol{v} \times \boldsymbol{B}$, and current flows both through the plate and through the fluid. Then there exists an induced magnetic field that adds to the applied field. The magnetic braking force density is $\boldsymbol{J} \times \boldsymbol{B}_{\text{net}}$.

8.7.4 Example 5: Rotating Cylinder

In this example also, $\boldsymbol{B}_{\text{ind}} = \boldsymbol{0}$.

Figure 8.2 shows a long conducting circular cylinder of radius R that rotates at a constant angular velocity ω in a uniform axial magnetic field $\boldsymbol{B}$. There is no external circuit.

As the cylinder starts rotating, the $\boldsymbol{v} \times \boldsymbol{B}$ field drives conduction electrons toward the axis, if ω and $\boldsymbol{B}$ are both positive as in the figure. The periphery becomes positive.

The $\boldsymbol{v} \times \boldsymbol{B}$ field also develops a negative uniform electrostatic volume charge density inside the cylinder, as in Eq. 7.10.

As shown in Sect. 8.5, everywhere inside the cylinder, the resulting $\boldsymbol{E}$ cancels $\boldsymbol{v} \times \boldsymbol{B}$, $\boldsymbol{J} = \boldsymbol{0}$, and $\boldsymbol{B}_{\text{ind}} = \boldsymbol{0}$.

As in Example 4, the rotating conducting cylinder has no effect on the magnetic field and there is no magnetic torque, whatever σ or the size of the cylinder.

Let us check the relation $\boldsymbol{E} + (\boldsymbol{v} \times \boldsymbol{B}) = \boldsymbol{0}$. From Eq. 7.10, the electrostatic space charge density $\widetilde{Q}_{\text{f}}$ is

$$\widetilde{Q}_{\text{f}} = -\epsilon_0 \boldsymbol{\nabla} \cdot (\boldsymbol{v} \times \boldsymbol{B}) = -\epsilon_0 \boldsymbol{\nabla} \cdot \left[\omega\rho\hat{\boldsymbol{\phi}} \times (B\hat{\boldsymbol{z}})\right] \tag{8.34}$$

$$= -2\epsilon_0 \omega B \ . \tag{8.35}$$

Thus $\widetilde{Q}_{\text{f}}$ is uniform and independent of σ.

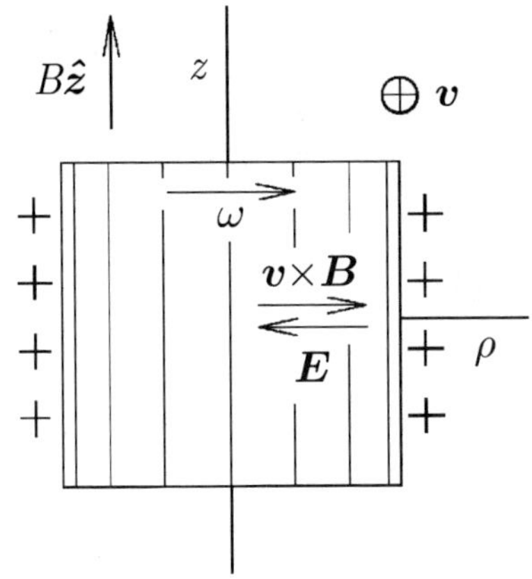

Fig. 8.2. Example 5. A conducting cylinder rotates at the constant angular velocity ω in a uniform axial $\boldsymbol{B}$

A surface charge of density $\widetilde{S}_f$ maintains the net charge on the cylinder equal to zero. Neglecting end effects,

$$2\pi R\widetilde{S}_f = -\pi R^2 \widetilde{Q}_f \, , \quad \widetilde{S}_f = -\frac{R\widetilde{Q}_f}{2} = +\epsilon_0 \omega BR \, . \tag{8.36}$$

The field of this surface charge extends only outside the cylinder, where it cancels the field of the space charge that is inside the cylinder. Thus, inside the cylinder, the electric field is only that of the space charge:

$$\boldsymbol{E} = \frac{\pi\rho^2 \widetilde{Q}_f}{2\pi\epsilon_0 \rho}\hat{\boldsymbol{\rho}} = -\omega\rho B\hat{\boldsymbol{\rho}} = -\boldsymbol{v} \times \boldsymbol{B} \, . \tag{8.37}$$

Therefore, neglecting end effects, the field $\boldsymbol{E}$ of the electrostatic *space* charge cancels the $\boldsymbol{v} \times \boldsymbol{B}$ field exactly at every point inside the cylinder, for any value of σ, and for any size of cylinder. Again, the vectors $\boldsymbol{E}$ and $\boldsymbol{v} \times \boldsymbol{B}$ are both independent of σ. The rotating conducting cylinder has no effect on the magnetic field because there are no induced currents. There is no magnetic braking torque.

As long as $(\omega\rho)^2 \ll c^2$, the magnetic field of the convection current that results from the rotation of the electrostatic volume and surface charges is negligible compared to the applied field, as in Sect. 7.2.3.

The magnetic field lines are stationary. They do not follow the moving conductor, whatever its conductivity.

Here, everywhere inside the cylinder, $\boldsymbol{E} + (\boldsymbol{v} \times \boldsymbol{B}) = \boldsymbol{0}$, so that $\boldsymbol{\nabla} \times [\boldsymbol{E} + (\boldsymbol{v} \times \boldsymbol{B})] = \boldsymbol{0}$. Also, $\boldsymbol{J} = \boldsymbol{0}$. Now the magnetic flux linking an arbitrary curve C is constant. So the Newcomb criterion applies.

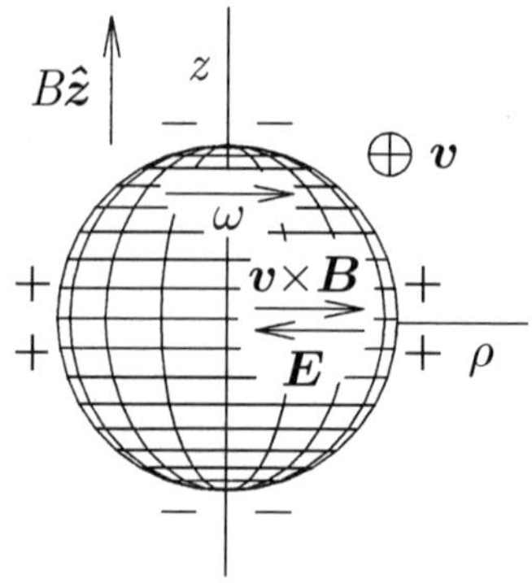

Fig. 8.3. Example 6. A conducting sphere rotates at a constant angular velocity ω in a uniform axial $\boldsymbol{B}$. This figure is the same as Fig. 7.3

8.7.5 Example 6: Rotating Solid Sphere

Again, $\boldsymbol{B}_{\text{ind}} = \boldsymbol{0}$.

Figure 8.3 shows a solid sphere of radius R and uniform conductivity σ that rotates at a constant angular velocity ω in a uniform axial magnetic field.

As the sphere starts rotating, the $\boldsymbol{v} \times \boldsymbol{B}$ field drives conduction electrons toward the axis, and the surface near the equator becomes positively charged, if ω and $\boldsymbol{B}$ are both positive as in the figure. The $\boldsymbol{v} \times \boldsymbol{B}$ field thus establishes a uniform electrostatic *space* charge density inside the sphere, as in Eq. 7.10. The resulting electrostatic field $\boldsymbol{E}$ cancels the $\boldsymbol{v}\times\boldsymbol{B}$ field everywhere inside the sphere, and current stops flowing outward. As in Example 5, there are then no induced currents, no magnetic torque, and $\boldsymbol{E} + (\boldsymbol{v} \times \boldsymbol{B}) = \boldsymbol{0}$: the sphere has no effect on the magnetic field. This fact was noted by J.J. Thomson (1893) over one century ago.

The charge space density $\widetilde{Q}_{\text{f}}$ inside the sphere is uniform as in Eq. 7.4, and the surface charge density is given by Eq. 7.74 (van Bladel, 1984):

$$\widetilde{S}_{\text{f}} = \epsilon_0 \omega B R \left(\tfrac{5}{2} \sin^2 \theta - 1 \right) \,. \tag{8.38}$$

The electric field $\boldsymbol{E}$ of these two charge distributions cancels the $\boldsymbol{v} \times \boldsymbol{B}$ field everywhere inside the sphere, $\boldsymbol{J} = \boldsymbol{0}$, and $\boldsymbol{B}_{\text{ind}} = \boldsymbol{0}$, whatever the conductivity.

It does not matter whether the source of the magnetic field rotates with the sphere or not (P. Lorrain, 1993a). The reason for this is as follows. By hypothesis, the *applied* magnetic field is uniform over the volume occupied by the sphere. It is the field of an appropriate coil. Now imagine that you

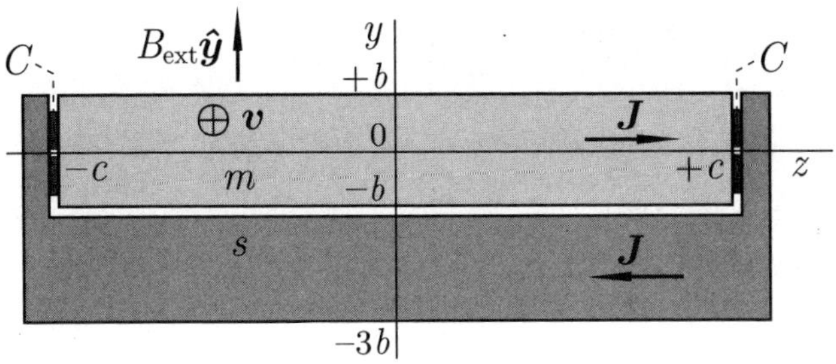

Fig. 8.4. Example 7. End view of a conducting plate m that moves into the paper at the constant velocity $\boldsymbol{v}$ in the direction of the x-axis. The stationary conducting plate s is connected to m through sliding contacts C. Both plates are infinitely long in the x-direction perpendicular to the paper. The induced currents, of density $\boldsymbol{J}$, point in the directions shown

install a $\boldsymbol{B}$-meter in that field and that you rotate the coil. The reading on the meter will of course not be affected. So, given a uniform magnetic field, a rotation of the source has no effect on the field.

More generally, rotating the source of an axisymmetric magnetic field about the axis of symmetry has no effect on the field.

Remember from Sect. 8.5 that, if a finite axisymmetric conducting body rotates as a solid about its axis, in a constant magnetic field that is symmetric around the same axis, then $\boldsymbol{J} = \boldsymbol{0}$ (in a steady-state situation). There are no induced currents and $\boldsymbol{E}$ cancels $\boldsymbol{v} \times \boldsymbol{B}$ exactly everywhere inside. The rotating axisymmetric conductor does not disturb the magnetic field, and there is no magnetic braking torque.

The magnetic field lines are stationary. They do not follow the moving conductor, whatever its conductivity.

The Newcomb criterion applies, as in the case of the rotating cylinder.

8.7.6 Example 7: Moving Plate

Once more, $\boldsymbol{v} \times \boldsymbol{B}_{\text{ind}} = \boldsymbol{0}$.

Figures 8.4 and 8.5 show two views of a conducting slab m of uniform conductivity σ_m and of thickness $2b$ that moves at a constant velocity $v\hat{\boldsymbol{x}}$. The slab is infinite in the x-direction and its width is $2c$ in the z-direction. It moves in a uniform externally applied magnetic field $B_{\text{ext}}\hat{\boldsymbol{y}}$. A similar stationary plate s, of conductivity σ_s lies underneath the moving one, is insulated from it, and connects the faces at $z = \pm c$ together through sliding contacts. We ignore corrections for the conductor edges, and we assume perfect sliding contacts.

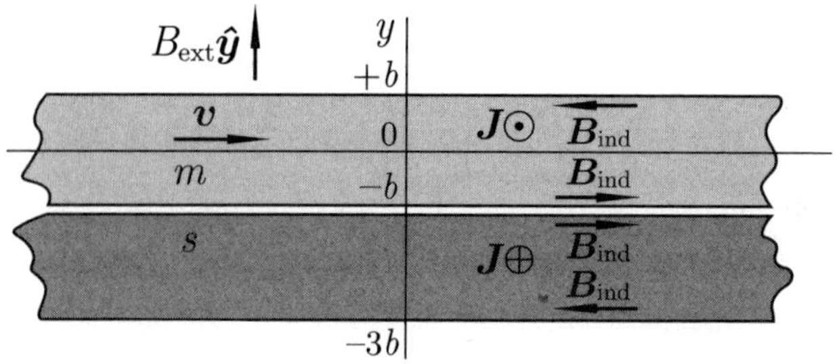

Fig. 8.5. Example 7. Side view of the two plates of Fig. 8.4 in the $z = 0$ plane. The upper plate m moves to the right at the velocity $v\hat{\boldsymbol{x}}$, while the lower plate s is stationary

1. First, set $\sigma_s = 0$: the stationary plate is an insulator. We then have Example 4: once the electrostatic surface charges are established, no current flows, and the magnetic field is undisturbed by the moving conducting plate.

In the z-direction, the resistances per meter of length measured in the x-direction are

$$R_m = \frac{c}{b\sigma_m}, \quad R_s = \frac{c}{b\sigma_s}. \tag{8.39}$$

2. Now set $\sigma_s = \infty$: the stationary plate short-circuits the moving plate, and $\boldsymbol{E}_s = \boldsymbol{E}_m = \boldsymbol{0}$. The current induced by the $\boldsymbol{v} \times \boldsymbol{B}$ field in the moving plate m flows in the $+z$-direction and its magnetic field $\boldsymbol{B}_{\text{ind}}$ is parallel to the x-axis. Then $\boldsymbol{v} \times \boldsymbol{B}_{\text{ind}} = \boldsymbol{0}$ and

$$\boldsymbol{v} \times \boldsymbol{B} = \boldsymbol{v} \times (\boldsymbol{B}_{\text{ext}} + \boldsymbol{B}_{\text{ind}}) = \boldsymbol{v} \times \boldsymbol{B}_{\text{ext}} = vB_{\text{ext}}\hat{\boldsymbol{z}} \tag{8.40}$$

is independent of the coordinates. Because of Eq. 7.10, there are no induced electrostatic space charges. Also, $\boldsymbol{E} = \boldsymbol{0}$, there are no electrostatic surface charges and

$$\boldsymbol{J}_m = \sigma_m(\boldsymbol{v} \times \boldsymbol{B}) = \sigma_m vB_{\text{ext}}\hat{\boldsymbol{z}} = -\boldsymbol{J}_s. \tag{8.41}$$

The vectors $\boldsymbol{E}_m$, $\boldsymbol{E}_s$, and $\boldsymbol{v} \times \boldsymbol{B}$ are all independent of the σ_m of the moving plate. Indeed, nothing happens to these vectors when σ_m increases indefinitely, whatever the size of the system, but $\boldsymbol{J}_m$ and $\boldsymbol{J}_s$ are both proportional to σ_m.

The equal and opposite currents in the two plates have a net zero induced magnetic field outside: the magnetic field line above and below is vertical. In the lower, stationary, plate s, current flows into the paper and rotates the magnetic field line in the clockwise direction. The upper plate m moves to the right at the velocity $\boldsymbol{v}$ and carries a current that flows out of the paper and rotates the field line in the counter-clockwise direction. The magnetic force density is $\boldsymbol{J} \times \boldsymbol{B}$. The magnetic force on m brakes the motion, while

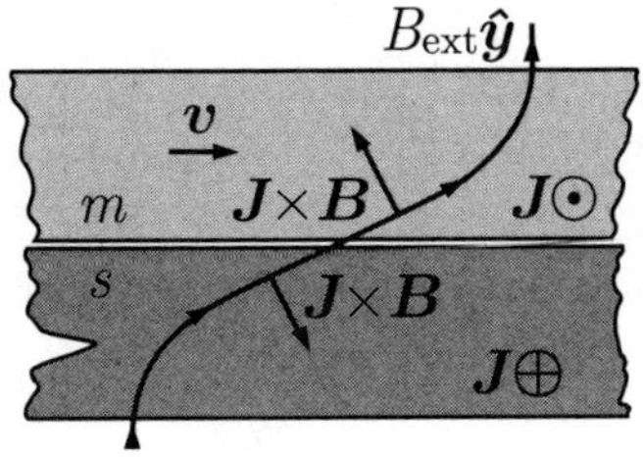

Fig. 8.6. Example 7. Magnetic field line for the conducting plates of Figs. 8.4 and 8.5, for $\mu_0\sigma v = 1$, $b = 1$, and $c \gg b$. The equal and opposite currents in the two plates have a net zero induced magnetic field outside: the magnetic field line above and below is vertical. In the lower, stationary, plate s, current flows into the paper and rotates the magnetic field line in the clockwise direction. The upper plate m moves to the right at the velocity $\boldsymbol{v}$ and carries a current that flows out of the paper and rotates the field line in the counter-clockwise direction

that on s points in the forward direction. The plates repel each other because they carry currents in opposite directions.

The induced magnetic field $\boldsymbol{B}_{\rm ind}$ is that of a pair of uniform sheets of current $2b$ meters thick and carrying $2bJ_m$ amperes in the plus and minus z-directions, per meter measured in the x-direction. Figure 8.6 shows a magnetic field line as observed in a stationary reference frame for the net magnetic field $\boldsymbol{B} = \boldsymbol{B}_{\rm ext} + \boldsymbol{B}_{\rm ind}$, with $\mu_0\sigma_m v = 1$, $\sigma_s = \infty$, $b = 1$, and $c \gg b$.

The magnetic field line is stationary, and its shape is independent of the time.

From Fig. 8.6, the magnetic force density $\boldsymbol{J} \times \boldsymbol{B}_{\rm net}$ points backward and upward in the moving plate, braking the motion, and downward and forward in the stationary plate. The magnetic force tends to separate the plates.

From Eq. 6.26, $\boldsymbol{B}' = \boldsymbol{B}$: an observer moving with the plate plots an identical field line, which is also independent of the time. This is a consequence of Special Relativity and has nothing to do with σ.

So, in this case, both $\boldsymbol{J}_m$ and $\boldsymbol{J}_s$ are independent of the coordinates, from Eq. 8.41. Also $\boldsymbol{E} = \boldsymbol{E}_s + \boldsymbol{E}_m = \boldsymbol{0}$ and is independent of the coordinates. So $\boldsymbol{\nabla} \times [\boldsymbol{E} + (\boldsymbol{v} \times \boldsymbol{B})] = \boldsymbol{0}$ and, from the Newcomb criterion, the flux Φ linked by the closed curve C that follows the conductor is constant.

3. Now set $\sigma_m = \sigma_s = \sigma$. Then

$$\boldsymbol{J}_m = \frac{\sigma v B_{\rm ext}\hat{\boldsymbol{z}}}{2} \,. \tag{8.42}$$

From Eq. 6.33,

$$\boldsymbol{E}_m = \frac{\boldsymbol{J}_m}{\sigma} - \boldsymbol{v} \times \boldsymbol{B}_{\rm ext} = -\frac{v B_{\rm ext}\hat{\boldsymbol{z}}}{2} \,, \tag{8.43}$$

$$\boldsymbol{E}_s = \frac{\boldsymbol{J}_s}{\sigma} = -\frac{\boldsymbol{J}_m}{\sigma} = -\frac{vB_{\text{ext}}\hat{\boldsymbol{z}}}{2} \,. \tag{8.44}$$

These electric fields result from electrostatic surface charges and are independent of σ. Observe how $\boldsymbol{E}_m$ and $\boldsymbol{v} \times \boldsymbol{B}$ of Eq. 8.2 have opposite signs, and how $\boldsymbol{E}_m$ cancels *half* of $\boldsymbol{v} \times \boldsymbol{B}$.

In the stationary plate s, $\boldsymbol{v} \times \boldsymbol{B} = \boldsymbol{0}$ and the return current density is

$$\boldsymbol{J}_s = \sigma \boldsymbol{E}_s = -\frac{\sigma v B_{\text{ext}}\hat{\boldsymbol{z}}}{2} = -\boldsymbol{J}_m \,. \tag{8.45}$$

Both $\boldsymbol{E}$ and $\boldsymbol{v}\times\boldsymbol{B}$ are again independent of σ, and $\boldsymbol{J}$ is again proportional to σ. The magnetic field line of Fig. 8.6 applies here also; the induced current and the induced magnetic field are both half as large, but the general shape of the line is the same.

The magnetic force density is $\boldsymbol{J} \times \boldsymbol{B}$. The magnetic force on m brakes the motion, while that on s points in the forward direction. The plates repel each other because they carry currents in opposite directions.

The magnetic field lines are stationary: they do *not* follow the moving conductor, whatever its conductivity.

8.7.7 Example 8: Rotating Fluid Sphere

Here, $\boldsymbol{v} \times \boldsymbol{B}_{\text{ind}} = \boldsymbol{0}$.

The conducting fluid inside a sphere of radius R has an azimuthal velocity $\boldsymbol{v}$ that is a function of ρ and z:

$$\boldsymbol{v} = \omega\rho\left(1 - K\frac{z^2}{R^2}\right)\hat{\boldsymbol{\phi}} \,. \tag{8.46}$$

If K is positive, the equator rotates faster than the poles. The sphere rotates in a uniform axial magnetic field $B_{\text{ext}}\hat{\boldsymbol{z}}$. It is immaterial whether the currents that generate the axial magnetic field lie inside or outside the rotating sphere, as noted in Example 4.

The field $\boldsymbol{v} \times \boldsymbol{B}_{\text{ext}}$ points everywhere in the radial direction, while $\boldsymbol{E}$ results from the presence of both space and surface electrostatic charges. Because of the axisymmetry, $\boldsymbol{E}$ has only radial and axial components. Then, from Eq. 6.33, $\boldsymbol{J}$ has only radial and axial components, and a current sheet has the shape of a toroid. So the induced magnetic field is azimuthal, like the velocity $\boldsymbol{v}$, and $\boldsymbol{v} \times \boldsymbol{B}_{\text{ind}} = \boldsymbol{0}$.

Under steady-state conditions, if σ is uniform,

$$\boldsymbol{v} \times \boldsymbol{B} = \boldsymbol{v} \times (B_{\text{ext}}\hat{\boldsymbol{z}} + B_{\text{ind}}\hat{\boldsymbol{\phi}}) = \boldsymbol{v} \times (B_{\text{ext}}\hat{\boldsymbol{z}}) \tag{8.47}$$

$$= \omega\rho B_{\text{ext}}\left(1 - K\frac{z^2}{R^2}\right)\hat{\boldsymbol{\rho}} \,, \tag{8.48}$$

$$\widetilde{Q}_{\text{f}} = -\epsilon_{\text{r}}\epsilon_0 \boldsymbol{\nabla} \cdot [\boldsymbol{v} \times (B_{\text{ext}}\hat{\boldsymbol{z}})] \tag{8.49}$$

$$= -2\epsilon_{\text{r}}\epsilon_0\omega B_{\text{ext}}\left(1 - K\frac{z^2}{R^2}\right) . \tag{8.50}$$

Here $\boldsymbol{v} \times \boldsymbol{B}$, $\widetilde{Q}_{\mathrm{f}}$, and $\boldsymbol{E}$, are all independent of the conductivity σ.

Also,

$$\nabla \times \boldsymbol{J} = \sigma[\nabla \times \boldsymbol{E} + \nabla \times (\boldsymbol{v} \times \boldsymbol{B})] \tag{8.51}$$

$$= \sigma \left[-\frac{\partial \boldsymbol{B}}{\partial t} + \nabla \times (\boldsymbol{v} \times \boldsymbol{B}) \right] \tag{8.52}$$

$$= \sigma \nabla \times (\boldsymbol{v} \times \boldsymbol{B}) = -2\sigma\omega B_{\mathrm{ext}} K \frac{\rho z}{R^2} \hat{\boldsymbol{\phi}} \, . \tag{8.53}$$

So the curl of $\boldsymbol{J}$ is not zero, in general, and, from Eq. 8.29, the flux Φ is not constant.

Say ω, B_{ext}, K, z are all positive. Then $\nabla \times \boldsymbol{J}$ points in the $-\hat{\boldsymbol{\phi}}$-direction, and current flows in closed loops in the direction that gives a magnetic field pointing also in the $-\hat{\boldsymbol{\phi}}$-direction. The magnetic field is that of a toroidal coil. So $\nabla \times \boldsymbol{J}$ and $\boldsymbol{B}_{\mathrm{ind}}$ point in the same direction.

With the first three above parameters positive, $\boldsymbol{B}_{\mathrm{ind}}$ points in the $-\hat{\boldsymbol{\phi}}$-direction in the upper hemisphere, and in the $+\hat{\boldsymbol{\phi}}$-direction in the lower hemisphere. A little thought shows that the rotation shifts a magnetic field line in the $+\phi$-direction in the lower hemisphere, and in the $-\phi$-direction in the upper hemisphere: magnetic field lines are distorted downstream. Increasing $\sigma\omega K$ increases the distortion. The magnetic field lines are stationary.

Now suppose that ω and B_{ext} are again positive, but that K lies between -1 and 0. Then every point in the sphere rotates in the positive direction about the z-axis, but the poles rotate faster than the equator. This changes the signs of $\nabla \times \boldsymbol{J}$ and of $B_{\mathrm{ind}}\hat{\boldsymbol{\phi}}$, and the magnetic field lines are distorted *upstream*. For a given angular velocity, they are stationary; they do not follow the moving conductor, whatever its conductivity.

Equation 8.53 shows that $\nabla \times \boldsymbol{J}$ is proportional to σ, which is uniform by hypothesis. Then $\boldsymbol{J}$ is proportional to σ. Then Eq. 6.33 shows that $\boldsymbol{E}+(\boldsymbol{v}\times\boldsymbol{B})$ is independent of σ.

The magnetic force density $\boldsymbol{J} \times (\boldsymbol{B}_{\mathrm{ext}} + \boldsymbol{B}_{\mathrm{ind}})$ has radial, azimuthal, and axial components.

Now consider a closed curve C that follows the conductor. Is the magnetic flux Φ linking C time-independent? The net magnetic flux density is $\boldsymbol{B}_{\mathrm{ext}}$, which is axial, plus $\boldsymbol{B}_{\mathrm{ind}}$, which is azimuthal. The velocity of a given point on C is azimuthal. Since it is difficult to visualize how C distorts in a general case, we investigate two simple cases.

1. The closed curve C moves with the conducting sphere and lies in a plane $z =$ constant. The surface $\mathcal{A}$ bounded by C lies in the same plane. The motion does not distort C, and $\Phi = \mathcal{A}B_{\mathrm{ext}}$ is constant.

2. The curve C is now a circle that lies originally in a radial plane through the axis of symmetry. The center of C is at $\rho = R/2$, $z = R/2$. Set $K > 0$ in Eq. 8.46. This makes the equator rotate faster than the poles.

Since the velocity of a point on C is purely azimuthal, the flux of the *azimuthal* $\boldsymbol{B}_{\mathrm{ind}}$ field through C is constant.

Is the flux of the *axial* $\boldsymbol{B}_{\text{ext}}$ field linked by C constant? The azimuthal velocity of a point on C decreases with z, as in Eq. 8.46. Seen from above, the area enclosed by C increases with time, and the flux of the axial $\boldsymbol{B}_{\text{ext}}$ through C also increases with time; it is not constant.

We conclude as above that the magnetic flux Φ linked by a closed curve C that moves with the fluid is *not* in general constant, whatever the value of σ.

8.7.8 Example 9: The Faraday Disk

In this Example, $\boldsymbol{v} \times \boldsymbol{B}_{\text{ind}} = \boldsymbol{0}$.

This last example concerns the Faraday disk, or disk dynamo (P. Lorrain et al., 1988; P. Lorrain, 1995, 1990) of Sect. 7.3. Disk D rotates at a constant angular velocity ω in an axial, constant, and uniform magnetic field $\boldsymbol{B}_{\text{ext}}$. There are contacts all around the disk, and all around the axle. Because of the $\boldsymbol{v}\times\boldsymbol{B}$ field in the disk, a current I flows in the direction shown in Fig. 7.1.

We ignore the magnetic field of the current I that flows in the external circuit. Then, inside the disk, the magnetic field has three components:

$$\boldsymbol{B}_{\text{net}} = \boldsymbol{B}_{\text{ext}} + \boldsymbol{B}_{\text{axle}} + \boldsymbol{B}_{\text{disk}} \; . \tag{8.54}$$

The disk current is radial. For a disk of thickness $2c$,

$$\boldsymbol{J} = \frac{I}{4\pi\rho c}\hat{\boldsymbol{\rho}} \; . \tag{8.55}$$

From Eq. 2.4 in cylindrical coordinates, assuming axisymmetry and disregarding edge effects,

$$-\frac{\partial B_{\text{disk},\,\phi}}{\partial z} = \mu_0 \frac{I}{4\pi\rho c} \; . \tag{8.56}$$

Setting $B_{\text{disk},\,\phi} = 0$ at $z = 0$ in the mid-plane,

$$\boldsymbol{B}_{\text{disk}} = -\frac{\mu_0 I}{4\pi\rho c} z\hat{\boldsymbol{\phi}} \; . \tag{8.57}$$

Above and below the disk, $\boldsymbol{J} = \boldsymbol{0}$ and, within these approximations, $\boldsymbol{B}_{\text{disk}}$ is independent of z.

The field $\boldsymbol{B}_{\text{axle}}$ is azimuthal, like the velocity $\boldsymbol{v}$. It adds to $\boldsymbol{B}_{\text{ext}}$, but does not contribute to the current, because $\boldsymbol{v} \times \boldsymbol{B}_{\text{axle}} = \boldsymbol{0}$. The value of $\boldsymbol{B}_{\text{axle}}$ depends on the connections to the axle. If the current enters at both ends, $\boldsymbol{B}_{\text{axle}}$ points in the $-\hat{\boldsymbol{\phi}}$-direction above the disk, and in the $+\hat{\boldsymbol{\phi}}$-direction below, like $\boldsymbol{B}_{\text{disk}}$.

With I flowing as in the figure, a magnetic field line has a negative ϕ-component above the midplane, and a positive ϕ-component below. The rotation of the disk thus distorts magnetic field lines downstream, as if they were dragged by the rotating disk. But the ϕ-component of the field is constant in Eq. 8.57, and the magnetic field lines are *stationary*. This result is independent of σ.

Here is another way of showing that the magnetic field lines in the disk dynamo are stationary, despite the rotation. Stop the rotation, open the circuit at the switch SW of Fig. 7.1, and apply at its terminals a voltage that restores the current. The current distribution and the magnetic field are the same as before.

The magnetic field lines are stationary. They do not follow the moving conductor, whatever its conductivity.

8.8 Summary

An induced current of course gives rise to an induced magnetic field. We are concerned here only with the *net* magnetic field, which is the vector sum of the applied and induced fields.

If a passive axisymmetric conductor rotates in a constant and axisymmetric magnetic field, then there are no induced currents, no induced magnetic field, and the rotating conductor has no effect on the applied magnetic field.

According to the Newcomb criterion, the magnetic flux Φ linking a closed curve C is constant if

$$\nabla \times [\boldsymbol{E} + (\boldsymbol{v} \times \boldsymbol{B})] = \boldsymbol{0} \,. \qquad (8.21)$$

This is a sufficient condition; the necessary condition is that either one of four integrals, that we deduced, is equal to zero.

The magnetic flux Φ' linked by a closed curve C' in S' is *not* constant, in general.

Assuming given, applied, constant magnetic fields and given constant velocities, magnetic field lines are either unaffected by a moving conductor, or distorted *downstream*, or distorted *upstream*. In all cases, the magnetic field lines are *stationary*, both in a fixed reference frame S and in the reference frame of the moving conductor S'.

The electrostatic space and surface charges play a fundamental role. In Examples 4 to 6, their $\boldsymbol{E}$ cancels $\boldsymbol{v} \times \boldsymbol{B}$ exactly at every point inside the conductor, there are no induced currents, no induced magnetic field, the moving conductor has no effect on the magnetic field lines, and there is no magnetic force.

In Examples 7 to 9, $\boldsymbol{E}$ cancels only part of $\boldsymbol{v} \times \boldsymbol{B}$. Then there are induced currents, an induced magnetic field, the moving conductor distorts the magnetic field lines, and there is a magnetic braking force.

Both $\boldsymbol{E}$ and $\boldsymbol{v} \times \boldsymbol{B}$ are independent of σ. A change in σ results in a corresponding change in $\boldsymbol{J}$.

The electric field $\boldsymbol{E}' = \boldsymbol{E} + (\boldsymbol{v} \times \boldsymbol{B})$ in the moving reference frame S' of the conductor is independent of σ.

9 Case Study: The Azimuthal Magnetic Field in the Earth's Core

Contents

The magnetic field in the core of a rotating body such as the Earth, or the planets, or the Sun, or the stars, originates in electric currents that flow both in the core and in the atmosphere. So the source of the magnetic field rotates with the body. Does it make sense to apply the results of Chapter 8 here? That problem will require some thought. We must of course distinguish between a fixed reference frame S and a frame S' that rotates with the body. But no! We have excluded rotating frames! So we have two problems: first, the source moves and, second, we must deal with rotation, and not with translation. As we shall see, the electric space charges in the rotating body play a crucial role.[1]

[1] See P. Lorrain, *Azimuthal magnetic fields in the Earth's core*, Physica Scripta **47**, 461–468 (1993a).

9.1 Introduction

We discussed briefly the Earth's magnetic field in Sects. 4.7 and 5.5.1. We are concerned here with the electric currents that are induced within the core of the Earth by its differential rotation in its own dipolar axisymmetric magnetic field, as in Sect. 4.7. By extension, our discussion applies, in general terms, to the rotation of the planets, of the Sun, and of stars. We consider here two particularly simple forms of differential rotation, with ω a function of z, and with ω a function of ρ.

Many authors have discussed the azimuthal magnetic field induced in this way in the Earth's core. For extensive reviews, see Gubbins and Roberts (1987), and Fearn et al. (1988). The discussions are no more than hand-waving, and rest on the assumption that magnetic field lines tend to follow a moving conductor. As we saw in the Examples of Chapter 8, that assumption is incorrect. To quote Moffatt (1978, pp. 66, 68),

> It is as if the $\boldsymbol{B}$ lines were gripped by the fluid and "cranked" around the z-axis in regions where ω is greatest [...] then B increases linearly with the time.

That is nonsense; the Earth has been spinning for quite a while!

As usual, we focus our attention on the *induced currents*, rather than on the magnetic field lines. Also, we take into account the field of the electrostatic space charges that almost invariably exist inside conductors that move in a magnetic field (Chapter 7 and Fig. 8.3).

We saw in Sect. 7.4 that a rotating *solid* conducting sphere does not disturb an externally applied axisymmetric magnetic field. In that case $\boldsymbol{J} = \boldsymbol{0}$ and $\boldsymbol{\nabla} \times \boldsymbol{B} = \boldsymbol{0}$. But the velocity $\boldsymbol{v}$ involved here is, in principle, $\omega\rho\hat{\boldsymbol{\phi}}$ plus the *convection* velocity within the Earth.

Since we are concerned with the rotation of a body in its own magnetic field, we must calculate $\boldsymbol{v} \times \boldsymbol{B}$, but for which $\boldsymbol{v}$, and for which $\boldsymbol{B}$? In order to answer these two questions we must distinguish clearly between the rotating reference frame S' of the Earth and the inertial frame S of a stationary observer who looks at the Earth, and we must decide which part of the magnetic field rotates, if any, and which part does not.

Since the liquid part of the Earth's core certainly does not rotate as a solid, we find the general condition for the generation of azimuthal magnetic fields by differential rotation, which condition is certainly satisfied within the core. We plot level lines for the azimuthal magnetic field, and find that it is much weaker than is generally believed.

We show that Ferraro's law of isorotation (Ferraro, 1937) applies only to a particular type of differential rotation.

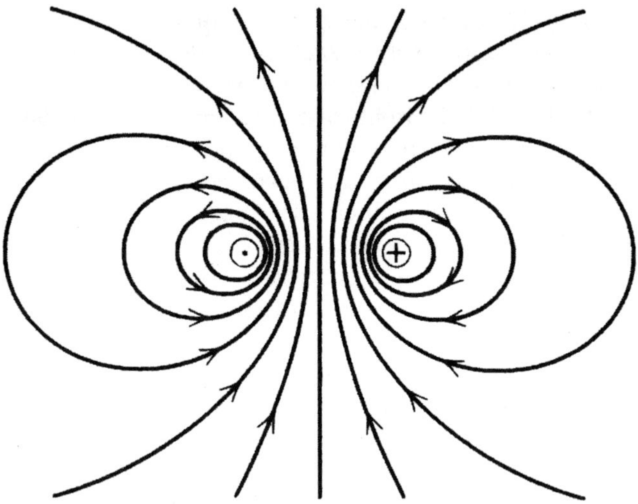

Fig. 9.1. Magnetic field lines for a current-carrying ring. The magnetic field is poloidal, while the current is toroidal

9.2 The Ratio B_T/B_P

We can estimate *very roughly* the ratio of the azimuthal magnetic field to the dipolar magnetic field in the Earth's core by using the magnetic Reynolds number of Sect. 6.6.

We use cylindrical coordinates ρ, ϕ, z.

First, we must define two new terms. In discussing fields in cylindrical coordinates, it is the custom to use the term *toroidal*, rather than azimuthal, for a field that has only a ϕ-component, and the term *poloidal* for a field that has no ϕ-component.

Remember: toroidal $\equiv$ azimuthal. If, for example,

$$\boldsymbol{B} = B_\rho \hat{\boldsymbol{\rho}} + B_z \hat{\boldsymbol{z}} + B_\phi \hat{\boldsymbol{\phi}} \,, \tag{9.1}$$

the sum of the first two terms is the *poloidal* component of $\boldsymbol{B}$, and the third term is its *toroidal* component:

$$\boldsymbol{B} = \boldsymbol{B}_\mathrm{P} + \boldsymbol{B}_\mathrm{T} \,. \tag{9.2}$$

Figure 9.1 shows *poloidal* magnetic field lines associated with a *toroidal* current distribution. The magnetic field lines of a straight current-carrying wire are toroidal.

A toroidal field is of the form $F\hat{\boldsymbol{\phi}}$, where F can be a function of ρ, ϕ, and z. A toroidal field line has a fixed ρ and a fixed z.

A poloidal field line, on the other hand, has a fixed ϕ. A poloidal field is of the form $G\hat{\boldsymbol{\rho}} + H\hat{\boldsymbol{z}}$, where G and H can be functions of ρ, ϕ, and z.

Observe that an axisymmetric toroidal *current* distribution has an axisymmetric poloidal *magnetic* field, and inversely.

From the Table facing the front cover, the curl of any toroidal vector field is poloidal, while the curl of an axisymmetric poloidal vector field is toroidal.

Let us now assume that all fields concerned are axisymmetric; they are functions of ρ and z only. We observe the Earth from an inertial reference frame S in which the center of the Earth is immobile.

Refer to Fig. 4.4, which shows a cross-section of the Earth. The *North magnetic* pole is near the *South geographic* pole. At the Equator, the magnetic field $\boldsymbol{B}_\mathrm{P}$ points *North.*

In addition, assume that the relative velocity in the core is not only axisymmetric, but also toroidal. These are rough approximations indeed because a) there is an angle of about 13 degrees between the axis of rotation of the Earth and the axis of symmetry of its poloidal field, and the two axes do not cross, and b) the relative velocity of a point in the convecting core is certainly not purely toroidal.

Recall now Eq. 7.1:

$$\boldsymbol{J} = \sigma[\boldsymbol{E} + (\boldsymbol{v} \times \boldsymbol{B})] = \sigma[\boldsymbol{E} + (\boldsymbol{v} \times \boldsymbol{B}_\mathrm{P})] \;, \tag{9.3}$$

since $\boldsymbol{v}$ is toroidal, like $\boldsymbol{B}_\mathrm{T}$.

Although the conductivity σ is certainly not uniform, let us nevertheless assume a uniform σ. Take the curl of both sides of the above equation:

$$\boldsymbol{\nabla} \times \boldsymbol{J} = \sigma\left[\boldsymbol{\nabla} \times \boldsymbol{E} + \boldsymbol{\nabla} \times (\boldsymbol{v} \times \boldsymbol{B}_\mathrm{P})\right] \tag{9.4}$$

$$= \sigma\left[-\frac{\partial \boldsymbol{B}}{\partial t} + \boldsymbol{\nabla} \times (\boldsymbol{v} \times \boldsymbol{B}_\mathrm{P})\right] . \tag{9.5}$$

Assume a steady state. Then $\partial \boldsymbol{B}/\partial t$ vanishes and

$$\boldsymbol{\nabla} \times \boldsymbol{J} = \sigma \boldsymbol{\nabla} \times (\boldsymbol{v} \times \boldsymbol{B}_\mathrm{P}) \,. \tag{9.6}$$

Now

$$\boldsymbol{v} = \omega\rho\hat{\boldsymbol{\phi}} + \boldsymbol{v}_\mathrm{rel} \;, \tag{9.7}$$

where $\boldsymbol{v}_\mathrm{rel}$ is the velocity of the conducting medium, relative to a solid Earth. Therefore

$$\boldsymbol{\nabla} \times \boldsymbol{J} = \sigma \boldsymbol{\nabla} \times \left[(\omega\rho\hat{\boldsymbol{\phi}} + \boldsymbol{v}_\mathrm{rel}) \times \boldsymbol{B}_\mathrm{P}\right] \tag{9.8}$$

$$= \sigma \boldsymbol{\nabla} \times (\omega\rho\hat{\boldsymbol{\phi}} \times \boldsymbol{B}_\mathrm{P}) + \sigma \boldsymbol{\nabla} \times (\boldsymbol{v}_\mathrm{rel} \times \boldsymbol{B}_\mathrm{P}) \tag{9.9}$$

$$= \sigma \boldsymbol{\nabla} \times (\boldsymbol{v}_\mathrm{rel} \times \boldsymbol{B}_\mathrm{P}) \;, \tag{9.10}$$

from Eqs. 8.13–8.17 for an axisymmetric $\boldsymbol{B}$. Finally,

$$\boldsymbol{\nabla} \times \boldsymbol{J} = \sigma \boldsymbol{\nabla} \times (\boldsymbol{v}_\mathrm{rel} \times \boldsymbol{B}_\mathrm{P}) \,. \tag{9.11}$$

Now, from Eq. 2.4,

$$\boldsymbol{J} = \frac{1}{\mu_0} \boldsymbol{\nabla} \times (\boldsymbol{B}_T + \boldsymbol{B}_P) , \tag{9.12}$$

so that

$$\boldsymbol{\nabla} \times [\boldsymbol{\nabla} \times (\boldsymbol{B}_T + \boldsymbol{B}_P)] = \mu_0 \sigma \boldsymbol{\nabla} \times (\boldsymbol{v}_{rel} \times \boldsymbol{B}_P) . \tag{9.13}$$

The RHS of Eq. 9.13 is toroidal, since $\boldsymbol{v}_{rel} \times \boldsymbol{B}_P$ is axisymmetric and poloidal. Similarly, in the same equation, $\boldsymbol{\nabla} \times (\boldsymbol{\nabla} \times \boldsymbol{B}_T)$ is toroidal and $\boldsymbol{\nabla} \times (\boldsymbol{\nabla} \times \boldsymbol{B}_P)$ poloidal. The latter quantity is therefore zero and Eq. 9.13 reduces to

$$\boldsymbol{\nabla} \times (\boldsymbol{\nabla} \times \boldsymbol{B}_T) = \mu_0 \sigma \boldsymbol{\nabla} \times (\boldsymbol{v}_{rel} \times \boldsymbol{B}_P) . \tag{9.14}$$

Let us estimate separately the orders of magnitude of typical values of both sides of this equation, with the methods of Appendix A. Let $\mathcal{L}$ be a characteristic length, say $R/6$, where R is the Earth's radius (Sect. A.8, Example 3).

Consider first the LHS of the above equation. With the terminology of Sect. A.6.3, $\boldsymbol{B}_T$ is a *transversal* field because its divergence is zero:

$$\boldsymbol{\nabla} \cdot \boldsymbol{B}_T = \frac{1}{\rho} \frac{\partial B_\phi}{\partial \phi} = 0 , \tag{9.15}$$

due to the azimuthal symmetry of the field. Then, according to Eq. A.87,

$$|\boldsymbol{\nabla} \times (\boldsymbol{\nabla} \times \boldsymbol{B}_T)|_{typical} \sim \frac{B_{T\,typical}}{\mathcal{L}^2} . \tag{9.16}$$

Consider now the RHS of Eq. 9.14. Since $\boldsymbol{v}_{rel}$ is toroidal by assumption and $\boldsymbol{B}_P$ is poloidal,

$$|\boldsymbol{v}_{rel} \times \boldsymbol{B}_P| = v_{rel} B_P \tag{9.17}$$

at all points in the field. Then, according to Eq. A.83,

$$\mu_0 \sigma |\boldsymbol{\nabla} \times (\boldsymbol{v}_{rel} \times \boldsymbol{B}_P)|_{typical} \lesssim \frac{\mu_0 \sigma |\boldsymbol{v}_{rel} \times \boldsymbol{B}_P|_{typical}}{\mathcal{L}} = \frac{\mu_0 \sigma (v_{rel} B_P)_{typical}}{\mathcal{L}} . \tag{9.18}$$

Now, assuming that

$$(v_{rel} B_P)_{typical} \sim v_{rel\,typical} B_{P\,typical} \tag{9.19}$$

(see Sect. A.8), it follows that

$$\mu_0 \sigma |\boldsymbol{\nabla} \times (\boldsymbol{v}_{rel} \times \boldsymbol{B}_P)|_{typical} \lesssim \frac{\mu_0 \sigma v_{rel\,typical} B_{P\,typical}}{\mathcal{L}} . \tag{9.20}$$

Equations 9.14, 9.16, and 9.20 imply that

$$\frac{B_{T\,typical}}{\mathcal{L}^2} \lesssim \frac{\mu_0 \sigma v_{rel\,typical} B_{P\,typical}}{\mathcal{L}} , \tag{9.21}$$

so that

$$\frac{B_{\mathrm{T\,typical}}}{B_{\mathrm{P\,typical}}} \lesssim \mu_0 \sigma v_{\mathrm{rel\,typical}} \mathcal{L} = R_{\mathrm{m}} , \qquad (9.22)$$

where R_{m} is the *magnetic Reynolds number* of Sect. 6.6.

The electric conductivity of the liquid part of the core is thought to be about 3×10^5 siemens per meter. Set $\mathcal{L} \approx 1$ megameter. As to the typical speed, set $v_{\mathrm{rel\,typical}} \approx 3 \times 10^{-4}$ meter/second, which is an interpolation of the western drift of the magnetic field at the Equator, down to the surface of the core (Stacey, 1992).

Then

$$\frac{B_{\mathrm{T\,typical}}}{B_{\mathrm{P\,typical}}} \lesssim R_{\mathrm{m}} \approx 100 , \qquad (9.23)$$

and Eq. 6.52 should apply. However, in Sect. 9.6.2 we find a value of about 6 for this ratio, and Eq. 6.52 is not too safe. The calculation of Sect. 9.6.3 does not lead to a value for R_{m}.

9.3 The Reference Frames S and S'

Refer to Sects. 6.1 and 6.2.

The fact that we observe the Earth's magnetic field from within its rotating frame is a source of confusion; we therefore distinguish between a non-rotating reference frame S and a rotating frame S' that is fixed to the Earth's mantle (P. Lorrain, 1990, 1993a). We consider that the Earth's center moves along a straight line at a constant velocity, and reference frame S follows the Earth without rotating. We perform our calculations in frame S, where physical phenomena are conceptually simpler.

The velocity of the liquid iron in the core has two components, one due to the rotation of the Earth, and a very much smaller one due to the relative motion of the core fluid. At the outer radius of the liquid part of the core, in the equatorial plane, the toroidal velocity due to the rotation is 255 meters/second. The speed of the relative motion of the core fluid is believed to be about 3×10^{-4} meter/second (Stacey, 1992) and the convection current density is about 5×10^{-16} ampere/meter^2, 14 orders of magnitude less than the conduction current.

We assume steady-state conditions in the rotating frame of reference S', a uniform, non-magnetic (see Sect. 4.7), conducting core, and we disregard electric currents in the mantle.

Figure 9.2 shows our coordinate system. With respect to a non-rotating reference frame S, we decompose the velocity $\boldsymbol{v}$ at a given point in the core into two parts:

$$\boldsymbol{v} = \boldsymbol{v}_\phi + \boldsymbol{v}_{\mathrm{o}} , \qquad (9.24)$$

where the subscript "o" stands for "other", and where, by definition, the azimuthal velocity $\boldsymbol{v}_\phi$ is axisymmetric,

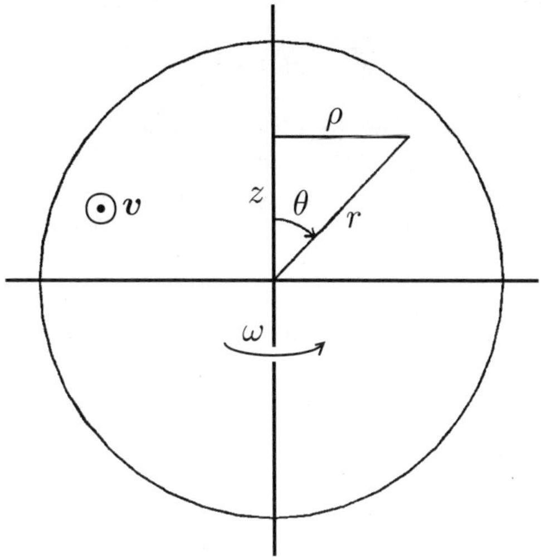

Fig. 9.2. Cylindrical coordinate system for the Earth's core. See Fig. 4.5

$$\boldsymbol{v}_\phi = v(\rho, z)\hat{\boldsymbol{\phi}} \,. \tag{9.25}$$

We disregard differential rotation until Sect. 9.6. Thus, for the moment, the core and the mantle rotate as a solid and

$$\boldsymbol{v} = \boldsymbol{v}_\phi = \omega\rho\hat{\boldsymbol{\phi}} \,, \tag{9.26}$$

where ω is the angular velocity of the Earth, 7.27×10^{-5} radian per second.

We are concerned here with azimuthal magnetic fields associated with the rotation. However, the induced current must result from an $\omega\rho\hat{\boldsymbol{\phi}} \times \boldsymbol{B}$ field, which disregards any azimuthal component of $\boldsymbol{B}$. It must also be emphasized that the $\boldsymbol{B}$ field involved here is the one observed at each point in space in the non-rotating reference frame S.

9.4 Does the Earth's Magnetic Field Rotate?

Before attempting to answer this question we must distinguish between the various components of the Earth's magnetic field inside the core, with respect to the non-rotating frame S.

We assume that the Earth's magnetic field consists of two parts: an axisymmetric component $\boldsymbol{B}_{\text{axi}}$ and the rest of the field $\boldsymbol{B}_{\text{o}}$, where the subscript "o" stands again for "other":

$$\boldsymbol{B} = \boldsymbol{B}_{\text{axi}} + \boldsymbol{B}_{\text{o}} \,. \tag{9.27}$$

The current distribution, of density $\boldsymbol{J}$, has two components:

$$\boldsymbol{J} = \boldsymbol{J}_{\text{axi}} + \boldsymbol{J}_{\text{o}} , \tag{9.28}$$

with $\boldsymbol{J}_{\text{axi}}$ associated with $\boldsymbol{B}_{\text{axi}}$, and $\boldsymbol{J}_{\text{o}}$ with $\boldsymbol{B}_{\text{o}}$.

To illustrate this decomposition of the magnetic field, imagine an observer equipped with a magnetometer, and located in the non-rotating reference frame S. The Earth rotates in front of him, or her. After correcting for solar and ionospheric effects, the magnetic field of the Earth, for that observer, is a periodic function with a period of 24 hours: a Fourier analysis shows a constant $\boldsymbol{B}$, plus harmonics. The constant component of the magnetic field is the axisymmetric $\boldsymbol{B}_{\text{axi}}$, and the rest is $\boldsymbol{B}_{\text{o}}$, at the position of the observer. For example, the poloidal component of the Earth's magnetic field gives a constant signal, corresponding to its axisymmetric component, plus a fluctuating signal corresponding to its transverse off-axis component.

Now the non-poloidal component of the Earth's magnetic field is prominent at the surface, despite the fact that the amplitudes of the quadrupole, octupole, etc. components decrease as $1/r^4$, $1/r^5$, etc., as in Sect. 4.6. We conclude that both $\boldsymbol{B}_{\text{axi}}$ and $\boldsymbol{B}_{\text{o}}$ are highly complex inside the core.

We disregard the toroidal, or azimuthal, or ϕ, component of the axisymmetric magnetic field because $\omega\rho\hat{\boldsymbol{\phi}} \times \boldsymbol{B}_\phi \equiv \boldsymbol{0}$, as noted in Sect. 9.3. We can therefore use the term $\boldsymbol{B}_{\text{axi}}$ for the poloidal component of the axisymmetric field. This axisymmetric poloidal magnetic field results from an axisymmetric toroidal current.

Our simulations of axisymmetric spherical self-excited dynamos show that such a toroidal current probably flows eastward in certain regions of the core, and westward in others, which makes $\boldsymbol{B}_{\text{axi}}$ badly contorted. We disregard the origin of the toroidal current that generates $\boldsymbol{B}_{\text{axi}}$.

9.4.1 Does the "*Other*" Magnetic Field Rotate?

Does the "other" magnetic field $\boldsymbol{B}_{\text{o}}$ of Eq. 9.27 rotate? For our observer in the non-rotating reference frame S, the field of a magnetic anomaly, say in Siberia, returns every 24 hours, and hence rotates with the Earth. The same applies to the transverse component of the dipolar magnetic field. Therefore, with respect to the non-rotating reference frame S, the field $\boldsymbol{B}_{\text{o}}$ rotates with the Earth.

9.4.2 Does the *Axisymmetric* Field $\boldsymbol{B}_{\text{axi}}$ Rotate?

Does the axisymmetric component of the Earth's magnetic field rotate? With respect to the non-rotating reference frame S, this axisymmetric component of magnetic field is constant. Does it rotate? Intuitively, it does not. Also, since it is impossible to measure the angular velocity of an axisymmetric magnetic field, it would not make sense to say that it rotates. However, let us look into this more closely.

A stationary copper ring of radius R carries a current I_1. That current is equal to the linear charge density λ_- of the conduction electrons, times their drift velocity:

$$I_1 = \lambda_- \, v_{\text{drift}} \, . \tag{9.29}$$

Now rotate the ring about its axis at a constant angular velocity ω. For a fixed observer, the new current I_2 is the superposition of two currents, one resulting from the rotation of the positive lattice, and one due to the drift of the conduction electrons:

$$I_2 = \lambda_+ \, \omega R + \lambda_- \, (\omega R + v_{\text{drift}}) \, . \tag{9.30}$$

For our fixed observer, the copper ring remains neutral, the charge densities λ_+ and λ_- are of the same magnitude, but of opposite signs, and

$$I_2 = \lambda_- \, v_{\text{drift}} = I_1 \, . \tag{9.31}$$

The rotation of the ring about its axis of symmetry therefore has no effect on the net current, and thus no effect on the magnetic field. We have assumed that $v^2 \ll c^2$, where v is the speed of a point within the conductor, and c is the speed of light.

It follows that, for non-relativistic speeds and for a fixed observer, the magnetic field of any toroidal current remains unaffected when the current distribution rotates about its axis of symmetry.

The axisymmetric component $\boldsymbol{B}_{\text{axi}}$ of the Earth's magnetic field therefore does *not* rotate with respect to the inertial reference frame S any more than if $\boldsymbol{B}_{\text{axi}}$ were generated by a current situated in a non-rotating coil outside the Earth.

9.4.3 Example: The One-piece Faraday Generator

The one-piece Faraday generator is a good illustration of the fact that an axisymmetric magnetic field does not rotate. Refer to Fig. 7.1. Remove the applied axial magnetic field, and replace the conducting disk by a disk that is both conducting and magnetized in the axial direction. Now rotate the disk. The device acts as a generator as if the magnetic field were supplied by an external, stationary coil. Faraday observed this phenomenon more than a century and a half ago, with a bar magnet rotating about its axis of symmetry (Faraday, 1852a; Corson, 1956; Crooks et al., 1978).

9.5 Solid Core

Assume again that the Earth's core rotates as a solid. As we saw in Sect. 9.3, the relative velocities are believed to be of the order of 10^{-6} times the azimuthal velocity $\omega\rho\hat{\boldsymbol{\phi}}$. Then Eq. 9.26 applies. Differential rotation will come later on, in Sect. 9.6.

With respect to the non-rotating reference frame S, the rotation of the Earth in its own axisymmetric field gives rise to a field $\boldsymbol{v} \times \boldsymbol{B} = \omega\rho\hat{\boldsymbol{\phi}} \times \boldsymbol{B}_{\text{axi}}$. The rest of the field $\boldsymbol{B}_\text{o}$ rotates with the Earth. As a first approximation, let us neglect $\boldsymbol{B}_\text{o}$.

Assuming thus that the conducting core rotates as a solid in its own axisymmetric magnetic field, we know from Sect. 8.5 that there is no induced current and thus no induced magnetic field. However recall Eq. 6.33 for the current density $\boldsymbol{J}$ inside a conductor of conductivity σ that moves at a velocity $\boldsymbol{v}$ in superposed $\boldsymbol{E}$ and $\boldsymbol{B}$ fields, all quantities being measured with respect to a stationary reference frame S. With the suffix "ind" standing for "induced",

$$\boldsymbol{J}_{\text{ind}} = \sigma(\boldsymbol{E}_{\text{ind}} + \omega\rho\hat{\boldsymbol{\phi}} \times \boldsymbol{B}_{\text{axi}}) \tag{9.32}$$

$$= \sigma\left(-\boldsymbol{\nabla} V_{\text{ind}} - \frac{\partial \boldsymbol{A}_{\text{ind}}}{\partial t} + \omega\rho\hat{\boldsymbol{\phi}} \times \boldsymbol{B}_{\text{axi}}\right) \tag{9.33}$$

$$= \sigma(-\boldsymbol{\nabla} V_{\text{ind}} + \omega\rho\hat{\boldsymbol{\phi}} \times \boldsymbol{B}_{\text{axi}}) \; . \tag{9.34}$$

Here all the terms are constant, by assumption.

Observe that the Earth's core acts as a one-piece Faraday disk (Sects. 7.3 and 9.4.3) that both supplies its own magnetic field and acts as its own load.

9.5.1 The Values of $\boldsymbol{v} \times \boldsymbol{B}$ and of $\widetilde{Q}$

Let us calculate some orders of magnitude for the Earth's core. Since there is no induced current,

$$\boldsymbol{E}_{\text{ind}} = -\boldsymbol{v} \times \boldsymbol{B} = -\omega\rho\hat{\boldsymbol{\phi}} \times \boldsymbol{B}_{\text{axi}} \; . \tag{9.35}$$

The magnitude of the poloidal component of the magnetic field *at the surface of the core*, at a radius of 3.5×10^6 meters, is generally taken to be about 5×10^{-4} tesla. This field presumably points toward the geographic *North*, as does the magnetic field at the Earth's surface. However, $\boldsymbol{B}_{\text{axi}}$ is certainly very much more complex than a poloidal field, and its average magnitude is certainly very much larger than that. Assuming nonetheless that value, Eq. 9.35 gives a radial electric field $\boldsymbol{v} \times \boldsymbol{B}$ of about $+0.13$ volt/meter at the surface of the core in the equatorial plane, where the toroidal velocity is about 250 meters/second.

What is the volume charge density? From Eq. 7.18,

$$\widetilde{Q} = -2\epsilon_0 \omega B \; . \tag{9.36}$$

Then, with a magnetic field of about 5×10^{-4} tesla that points up, and for an angular velocity $2\pi/(24 \times 3600) = 7.3 \times 10^{-5}$ radian per second,

$$\widetilde{Q} = +2 \times 8.85 \times 10^{-12} \times 7.3 \times 10^{-5} \times 5 \times 10^{-4} \tag{9.37}$$
$$= +6.46 \times 10^{-19} \text{ coulomb/meter}^3 . \tag{9.38}$$

This corresponds to a lack of about 4 electrons per cubic meter! There is also an equal and opposite small surface charge, as in Sect. 7.4.

Near the Earth's surface, at the Equator, $B \approx 3 \times 10^{-5}$ tesla, which is the best fit for the poloidal component, and $\widetilde{Q} \approx +4 \times 10^{-20}$ coulomb/meter^3, corresponding to a lack of two or three electrons per 10 cubic meters.

These slight electrostatic charges play an essential role in the Earth's dynamo. If there were no competing electrostatic field, the

$$\omega\rho\hat{\boldsymbol{\phi}} \times \boldsymbol{B}_{\text{axi}}$$

field would act alone in the core, and the ohmic power dissipation would be absurdly high, as the following calculation will show.

As we saw above, $vB \approx 0.13$ volt/meter. Since the conductivity of the core is about 3×10^5 siemens/meter, then, in the absence of the opposing electrostatic field, the current density at a radius of 3.5 megameters, at the surface of the liquid part of the core, would be 3.9×10^4 amperes/meter^2 and the ohmic power dissipation 5 kilowatts/meter^3. But our estimate of B is much too low, and the ohmic power dissipation is proportional to the square of vB, so the Joule losses would be larger than 5 kilowatts/meter^3 by orders of magnitude. Without the slight electrostatic volume charge, the Earth might melt, or even evaporate. Clearly, the $\omega\rho\hat{\boldsymbol{\phi}} \times \boldsymbol{B}_{\text{axi}}$ field cannot act alone.

9.6 Differential Rotation

The liquid part of the core cannot be expected to be uniform in temperature and in chemical composition, and hence in density. It is thus subjected to convection due to buoyancy, and to the centrifugal and Coriolis forces.

Assume that there is differential rotation: the toroidal velocity is again axisymmetric, but it is now a function of ρ and z, as in Eq. 9.25. Then

$$\boldsymbol{v}_\phi \times \boldsymbol{B}_{\text{axi}} = v\hat{\boldsymbol{\phi}} \times (B_{\text{axi},\,\rho}\,\hat{\boldsymbol{\rho}} + B_{\text{axi},\,z}\,\hat{\boldsymbol{z}}) \tag{9.39}$$
$$= v(B_{\text{axi},\,z}\,\hat{\boldsymbol{\rho}} - B_{\text{axi},\,\rho}\,\hat{\boldsymbol{z}}) \tag{9.40}$$

and, from Eq. 7.10, the induced electrostatic charge density is

$$\widetilde{Q}_{\text{ind}} = -\epsilon_0 \boldsymbol{\nabla} \cdot (\boldsymbol{v}_\phi \times \boldsymbol{B}_{\text{axi}}) \tag{9.41}$$
$$= -\epsilon_0 \left[\frac{1}{\rho}\frac{\partial}{\partial \rho}(\rho v B_{\text{axi},\,z}) - \frac{\partial}{\partial z}(v B_{\text{axi},\,\rho}) \right] . \tag{9.42}$$

Also,

$$\widetilde{Q}_{\text{ind}} = -\epsilon_0 \boldsymbol{\nabla} \cdot (\boldsymbol{v}_\phi \times \boldsymbol{B}_{\text{axi}}) \tag{9.43}$$

$$= -\epsilon_0 \left[\boldsymbol{B}_{\text{axi}} \cdot (\boldsymbol{\nabla} \times \boldsymbol{v}_\phi) - \boldsymbol{v}_\phi \cdot (\boldsymbol{\nabla} \times \boldsymbol{B}_{\text{axi}})\right] \tag{9.44}$$

$$= -\epsilon_0 \left(-B_\rho \frac{\partial v}{\partial z} + B_z \frac{v}{\rho} + B_z \frac{\partial v}{\partial \rho} - \mu_0 v J_\phi\right) . \tag{9.45}$$

From Eq. 9.34, under steady conditions and for a homogeneous core,

$$\boldsymbol{\nabla} \times \boldsymbol{J}_{\text{ind}} = \sigma \boldsymbol{\nabla} \times (-\boldsymbol{\nabla} V_{\text{ind}} + \boldsymbol{v}_\phi \times \boldsymbol{B}_{\text{axi}}) \tag{9.46}$$

$$= \sigma \boldsymbol{\nabla} \times (\boldsymbol{v}_\phi \times \boldsymbol{B}_{\text{axi}}) \tag{9.47}$$

$$= \sigma \left[\frac{\partial}{\partial \rho}(v B_{\text{axi},\,\rho}) + \frac{\partial}{\partial z}(v B_{\text{axi},\,z})\right] \hat{\boldsymbol{\phi}} . \tag{9.48}$$

Differential rotation as in Eq. 9.25 induces a poloidal current in the core, and thus a toroidal magnetic field, if the bracket in the above equation is *not* equal to zero. Since the magnetic field in the Earth's core is probably badly contorted, the bracket is certainly not equal to zero in the core.

It would be interesting to calculate the current density $\boldsymbol{J}_{\text{ind}}$ that is induced by the differential rotation, but that is more difficult than calculating its curl, for the following reason. When there is differential rotation, Eq. 9.34 becomes

$$\boldsymbol{J}_{\text{ind}} = \sigma(-\boldsymbol{\nabla} V_{\text{ind}} + v\hat{\boldsymbol{\phi}} \times \boldsymbol{B}_{\text{axi}}) , \tag{9.49}$$

where $\boldsymbol{v}$ is now a function of ρ and z, as in Eq. 9.25. So, to calculate $\boldsymbol{J}_{\text{ind}}$, we must calculate the potential field V_{ind} of the induced space charge density given by Eq. 9.45, which seems difficult. But, in taking the curl of $\boldsymbol{J}_{\text{ind}}$ we bypass V_{ind} because the curl of a gradient is identically equal to zero.

What is the configuration of the induced magnetic field? Since the curl of the current density is toroidal as in Eq. 9.48, the induced current is poloidal. In other words, the induced current flows in meridian planes. Its $\boldsymbol{B}$ is toroidal, or azimuthal, in the direction of $\boldsymbol{\nabla} \times \boldsymbol{J}$, and it has the sign of the term between brackets in Eq. 9.48.

Observe that the conductivity σ affects only the magnitude of J_{ind}, and not its distribution. That is because the induced magnetic field is toroidal, and thus has no effect on the $\boldsymbol{v}_\phi \times \boldsymbol{B}_{\text{axi}}$ field, which has no ϕ-component.

If the core rotates as a solid, as in Sect. 9.5, then $v = \omega\rho$ and the bracket in Eq. 9.48 is equal to $\rho \boldsymbol{\nabla} \cdot \boldsymbol{B}_{\text{axi}} \equiv 0$. (The fact that $\boldsymbol{B}_{\text{axi}}$ at a given point of frame S is the time-average of the $\boldsymbol{B}$ observed at that point implies that $\boldsymbol{\nabla} \cdot \boldsymbol{B}_{\text{axi}} \equiv 0$.)

9.6.1 Ferraro's Law of Isorotation

Various authors refer to Ferraro's law of isorotation (Ferraro, 1937), according to which differential rotation induces a toroidal magnetic field in the core, unless the *angular* velocity ω is constant over any surface traced out by a complete rotation of a magnetic field line around the axis of rotation.

We can show that this statement is correct, despite the fact that Ferraro's Physics is wrong, and that the law applies only to a specific type of differential rotation.

Ferraro starts his proof by writing that

$$\boldsymbol{J} = \sigma[\boldsymbol{E} + (\boldsymbol{v} \times \boldsymbol{B})] + \boldsymbol{i} \ , \tag{9.50}$$

where $\boldsymbol{J}$ is the current density and $\boldsymbol{i}$ is the current density associated with $\boldsymbol{B}$. The last term is redundant because $\boldsymbol{J}$ of course includes $\boldsymbol{i}$!

Ferraro sets

$$B_{\text{axi},\,\rho} = -\frac{1}{\rho}\frac{\partial U}{\partial z} \ , \qquad B_{\text{axi},\,z} = \frac{1}{\rho}\frac{\partial U}{\partial \rho} \ . \tag{9.51}$$

These equations define a specific class of axisymmetric poloidal field among those for which $\boldsymbol{\nabla}\cdot\boldsymbol{B}_{\text{axi}} = 0$. It is dubious that all axisymmetric poloidal fields satisfy these equations. The curl of $\boldsymbol{B}_{\text{axi}}$ is not equal to zero, for otherwise, from Eq. 2.4, the $\boldsymbol{J}$ required to generate $\boldsymbol{B}_{\text{axi}}$ would also be equal to zero.

A curve $U(\rho, z) = C$, where C is a constant, defines a magnetic field line. If the *angular* velocity is constant along such a line, then $v = \rho f(U)$, where f is a certain function. Substituting this relation, as well as Eq. 9.51 into Eq. 9.48 makes the bracket equal to zero, which confirms the law of isorotation, but only for that particular class of magnetic field that is defined by Eq. 9.51.

The general condition for the existence of an axisymmetric toroidal magnetic field induced by differential rotation is that the bracket in Eq. 9.48 be *not* equal to zero.

We now calculate the induced azimuthal field in the Earth's core, in two simple cases.

9.6.2 B_{T} with v_ϕ a Function of z

Assume that the angular velocity inside the core is a function of the z-coordinate alone, and set

$$\omega = \omega_0 \left(1 - K_z \frac{z^2}{R^2}\right) \ . \tag{9.52}$$

Then

$$v = \omega_0 \rho \left(1 - K_z \frac{z^2}{R^2}\right) \ , \tag{9.53}$$

with $K_z \approx 10^{-6}$ (Stacey, 1992). If K_z is positive, then the Equator rotates slightly faster than the poles. Differential rotation is certainly much more complex than that (Jault et al., 1988).

Assume also that $\boldsymbol{B}_{\text{axi}}$ is uniform and axial. That is again a simple-minded assumption, but the exercise will be instructive. Remember that the real $\boldsymbol{B}_{\text{axi}}$

is certainly highly complex. For clarity, we choose a magnetic field that is positive, and hence that points North, in the direction of the vector $\boldsymbol{\omega}$:

$$B_{\mathrm{axi},\,\rho} = 0\,, \quad B_{\mathrm{axi},\,z} = B_0\,, \tag{9.54}$$

where B_0 is positive. We set $B_0 = 5 \times 10^{-4}$ tesla, which is thought to be the value of B at the surface of the core, at the Equator (Stacey, 1992). From Eq. 9.48, with $\omega_0 = 7.27 \times 10^{-5}$ radian/second and $\sigma = 3 \times 10^5$ siemens/meter,

$$\nabla \times \boldsymbol{J}_{\mathrm{ind}} = \sigma B_0 \frac{\partial v}{\partial z}\hat{\boldsymbol{\phi}} = -2\sigma\omega_0 B_0 K_z \frac{\rho z}{R^2}\hat{\boldsymbol{\phi}} \tag{9.55}$$

$$= -2.18 \times 10^{-8}\frac{\rho z}{R^2}\hat{\boldsymbol{\phi}}\,. \tag{9.56}$$

The electrostatic space charge density is

$$\widetilde{Q}_{\mathrm{ind}} = -\epsilon_0 \nabla \cdot [\boldsymbol{v}_\phi \times (B_0 \hat{\boldsymbol{z}})] \tag{9.57}$$

$$= -\epsilon_0 \nabla \cdot \left[\omega_0 B_0 \rho \left(1 - K_z \frac{z^2}{R^2}\right)\hat{\boldsymbol{\rho}}\right] \tag{9.58}$$

$$= -2\epsilon_0 \omega_0 B_0 \left(1 - K_z \frac{z^2}{R^2}\right)\,. \tag{9.59}$$

With this geometry, $\nabla \times \boldsymbol{J}_{\mathrm{ind}}$ is independent of $\partial v/\partial \rho$, and $\widetilde{Q}_{\mathrm{ind}}$ is independent of ρ. Also, $\boldsymbol{E}$ depends not only on the volume charge density $\widetilde{Q}_{\mathrm{ind}}$, but also on surface charges that result from the rotation.

We can find an equation for the induced toroidal magnetic field B_{T} if we substitute $\nabla \times \boldsymbol{B}_{\mathrm{T}}/\mu_0$ for $\boldsymbol{J}_{\mathrm{ind}}$ in Eq. 9.56 and expand the left-hand side, remembering that $\boldsymbol{B}_{\mathrm{T}}$ has only a ϕ-component, and that derivatives with respect to ϕ are equal to zero. Then

$$\frac{\partial^2 B_{\mathrm{T}}}{\partial \rho^2} + \frac{\partial^2 B_{\mathrm{T}}}{\partial z^2} + \frac{1}{\rho}\frac{\partial B_{\mathrm{T}}}{\partial \rho} - \frac{B_{\mathrm{T}}}{\rho^2} = 2\mu_0 \sigma \omega_0 B_0 K_z \frac{\rho z}{R^2}\,. \tag{9.60}$$

Observe that B_{T} is proportional to the product $\sigma\omega_0 B_0 K_z$. For example, changing the sign of K_z changes the sign of B_{T}; doubling K_z doubles B_{T}.

We solved this equation by the finite-element method, setting

$$\omega_0 = 7.27 \times 10^{-5}\ \mathrm{second}^{-1}\,, \tag{9.61}$$

$$R = 3.5 \times 10^6\ \mathrm{meters}\,, \tag{9.62}$$

$$K_z = 10^{-6}\,, \tag{9.63}$$

$$\sigma = 3 \times 10^5\ \mathrm{siemens/meter}\,, \tag{9.64}$$

$$B_0 = 5 \times 10^{-4}\ \mathrm{tesla}\,. \tag{9.65}$$

This makes the factor in front of ρz on the RHS of Eq. 9.60 equal to 2.24×10^{-27}. The field B_{T} is zero at the surface of the core, in the equatorial plane,

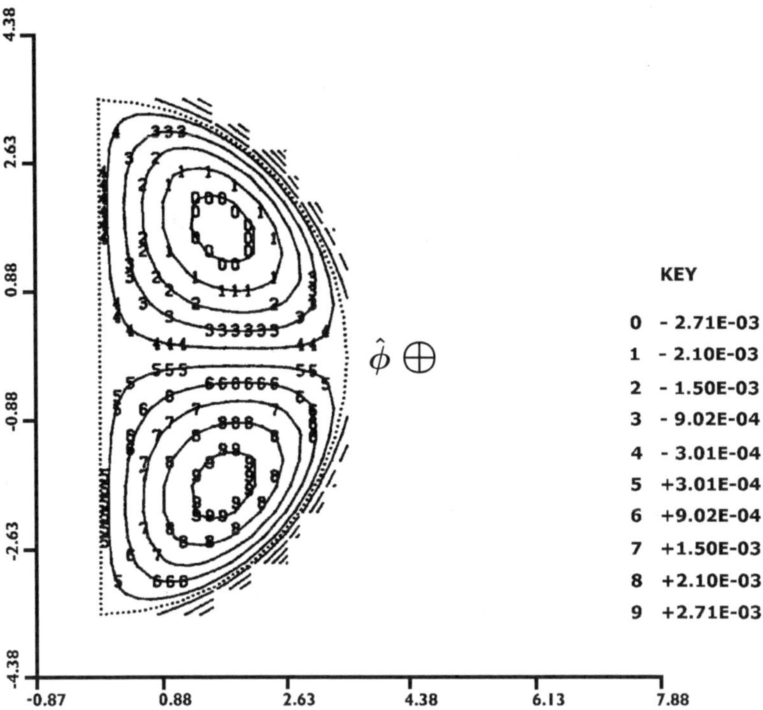

Fig. 9.3. Level lines for the azimuthal magnetic field induced in the Earth's core by differential rotation, with the angular velocity a function of the z-coordinate, as in Eq. 9.53. The figure shows only half of the core. The assumed axisymmetric magnetic field is uniform and points North. The core rotates in the direction of the unit vector $\hat{\phi}$. The axes are labeled in megameters, and the key is labeled in teslas

and on the axis of rotation, and it is antisymmetrical with respect to z. Figure 9.3 shows level lines of B_T for 150 triangles and 50×50 grid points. The key gives the values of B_T in teslas, the field being positive when it points into the paper. Dimensions are in megameters. We have assumed a uniform field that points North.

This field is similar to that of a pair of thick toroidal coils. The lines of $\boldsymbol{J}_{\text{ind}}$ are similar, clockwise in the Southern Hemisphere and counterclockwise in the Northern Hemisphere, on the assumption that $\boldsymbol{B}_{\text{axi}}$ points North inside the core, in this simple exercise.

We can calculate the ohmic power dissipation per cubic meter $\widetilde{P}$ as follows. From Eq. 9.56, setting $\rho z/R^2 \approx 1/4$, $\rho \approx R/2$, $z \approx R/2$,

$$|\boldsymbol{\nabla} \times \boldsymbol{J}_{\text{ind}}| \approx 5.5 \times 10^{-9} \text{ ampere/meter}^3 , \qquad (9.66)$$

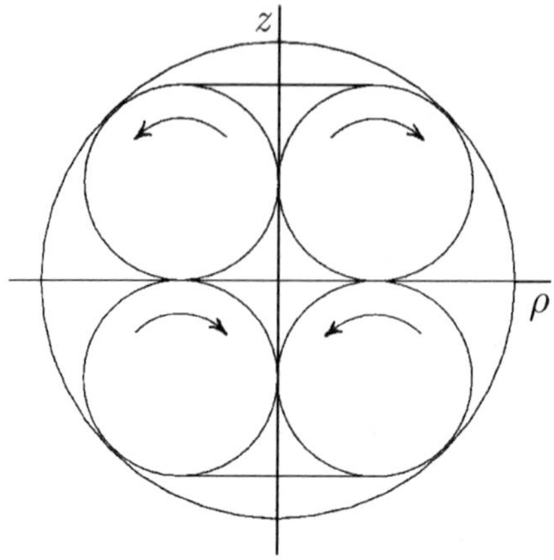

Fig. 9.4. Cross-section through a pair of superposed toroidal coils that simulate roughly the current distribution that generates the magnetic field of Fig. 9.3

$$J_{\text{ind}} \approx \frac{R}{2}|\boldsymbol{\nabla} \times \boldsymbol{J}_{\text{ind}}| \approx 10^{-2} \text{ ampere/meter}^2 , \tag{9.67}$$

$$\widetilde{P} = \frac{J_{\text{ind}}^2}{\sigma} \approx 3 \times 10^{-10} \text{ watt/meter}^3 . \tag{9.68}$$

This is roughly the average power density. For a sphere of radius 3×10^6 meters, the total power dissipation is about one gigawatt. Since the outward heat flow for the Earth is about 4000 gigawatts (De Bremaeker, 1985), $\widetilde{P}$ is negligible.

This finite-element solution yields a maximum value of 3.0×10^{-3} tesla for the toroidal magnetic flux density B_{T} that is induced by the differential rotation at $\rho = |z| = R/2$, and

$$\frac{B_{\text{T}}}{B_0} \approx 6 . \tag{9.69}$$

As a rough check of the above value of B_{T}, assume that the current distribution is that of a pair of superposed toroidal coils as in Fig. 9.4. The minor radius of the toroids is equal to $R/2.4$. Then the maximum magnitude of the azimuthal field, which occurs on the major radius, is

$$B_{\text{T}} \approx \mu_0 \frac{\pi (R/2.4)^2 J_{\text{ind}}}{2\pi(R/2.4)} \approx 8 \times 10^{-3} \text{ tesla} , \tag{9.70}$$

and

$$\frac{B_{\text{T}}}{B_0} \approx 16 . \tag{9.71}$$

The agreement with the finite-element calculation is satisfactory. The agreement with the magnetic Reynolds number of Sect. 9.2 is also satisfactory.

The finite-element value of 6 for the ratio toroidal/poloidal magnetic field is smaller than the values that are generally assumed by one or two orders of magnitude (Parker, 1979).

It is remarkable that a differential rotation of only one millionth of the toroidal velocity should generate a toroidal magnetic field $B_{\rm T}$ that is several times larger than the assumed uniform axial field B_0.

The magnetic force, of density $\boldsymbol{J}_{\rm ind} \times \boldsymbol{B}_{\rm axi}$, that results from the poloidal current in the core exerts a braking torque in certain regions, and an aiding torque in other regions. This provides mixing, and the Joule losses ensure a net braking effect.

Although, at first sight, it is impossible to deduce the azimuthal magnetic field inside the core from observations made at the surface, it is, in fact, possible to gain some information from seismological data (Tanimoto, 1989).

9.6.3 $B_{\rm T}$ with v_ϕ a Function of ρ

We now set

$$\boldsymbol{v} = \omega_0 \rho \left(1 - K_\rho \frac{\rho}{R}\right) \hat{\boldsymbol{\phi}} \,, \tag{9.72}$$

where R is again the radius of the core, and where the constant K_ρ can be either positive or negative, depending on whether the angular velocity near the axis of rotation is larger or smaller than at the periphery. If $K_\rho = 0$, then the core rotates as a solid.

According to Eq. 9.48, there is no induced current if the applied magnetic field is uniform and axial, and if the azimuthal velocity is only a function of the radial coordinate ρ. We therefore require a poloidal magnetic field, and we choose the simplest one, namely the field of a point magnetic dipole at the center of the core. Then, in spherical polar coordinates (P. Lorrain et al., 1988, p. 340),

$$B_{\rm axi,\, r} = \frac{2C}{r^3} \cos\theta \,, \quad B_{\rm axi,\, \theta} = \frac{C}{r^3} \sin\theta \,, \quad B_{\rm axi,\, \phi} = 0 \,, \tag{9.73}$$

$$C = \frac{\mu_0 m}{4\pi} \,, \tag{9.74}$$

where m is the magnetic dipole moment of the point dipole in amperes per square meter. In cylindrical coordinates,

$$B_{\rm axi,\, \rho} = 3C \frac{\rho z}{r^5} \,, \quad B_{\rm axi,\, z} = C \frac{2z^2 - \rho^2}{r^5} \,, \quad B_{\rm axi,\, \phi} = 0 \,. \tag{9.75}$$

Substituting $\boldsymbol{v}$ and $\boldsymbol{B}$ of Eqs. 9.72 and 9.75 into Eq. 9.48, and simplifying,

$$|\boldsymbol{\nabla} \times \boldsymbol{J}_{\rm ind}| = -3\sigma\omega C K_\rho \frac{\rho^2 z}{R(\rho^2 + z^2)^{5/2}} \tag{9.76}$$

$$= -3\sigma\omega \frac{\mu_0 m}{4\pi} K_\rho \frac{\rho^2 z}{R(\rho^2 + z^2)^{5/2}} \,. \tag{9.77}$$

The RHS is zero when $K_\rho = 0$, as expected, because the core then rotates as a solid.

We can calculate a numerical value for the constant C of Eq. 9.74 as follows. At the Equator, at the surface of the core,

$$r = \rho = 3.5 \times 10^6 \text{ meter} , \quad B_z = 5 \times 10^{-4} \text{ tesla} , \quad B_\rho = 0 .$$

Then, from Eqs. 9.74 and 9.75, $m = 2 \times 10^{23}$ amperes/meter^2. This is the dipole moment of a point dipole situated at the center of the Earth that would have a magnetic field equal to the poloidal field that is estimated to exist at the surface of the core, at the Equator.

Substituting the parameters of Eq. 9.65, except for K_z, and setting $K_\rho = 10^{-6}$ in Eq. 9.77, then substituting $\boldsymbol{\nabla} \times \boldsymbol{B}_\mathrm{T}/\mu_0$ for $\boldsymbol{J}_\mathrm{ind}$, and expanding the double curl as previously,

$$\frac{\partial^2 B_\mathrm{T}}{\partial \rho^2} + \frac{\partial^2 B_\mathrm{T}}{\partial z^2} + \frac{1}{\rho}\frac{\partial B_\mathrm{T}}{\partial \rho} - \frac{B_\mathrm{T}}{\rho^2} = 0.47 \frac{\rho^2 z}{(\rho^2 + z^2)^{5/2}} . \tag{9.78}$$

The value of B_T is now proportional to the product that appears on the right of Eq. 9.77, $\sigma \omega m K_\rho$.

Figure 9.5 shows level lines of the magnetic field B_T induced by the differential rotation defined by Eq. 9.72, calculated as in Sect. 9.6.2. The magnetic field is again similar to that of a toroidal coil, but now most of the field is located near the center of the core, as one would expect, since that is where $\boldsymbol{B}_\mathrm{axi}$ is maximum, with this simple model. The maximum B_T is about 3.0×10^{-2} tesla and it is situated at $\rho \approx |z| \approx 0.36$ megameter.

The magnetic force, of density $\boldsymbol{J}_\mathrm{ind} \times \boldsymbol{B}_\mathrm{axi}$, again provides mixing and a net braking torque, and the magnetic field extends either upstream or downstream, according to the value of K_ρ.

This result cannot be compared with the calculation based on the magnetic Reynolds number of Sect. 9.2 because, here, the magnitude of the poloidal field varies widely with ρ and z.

We can calculate the electric space charge density as follows:

$$\frac{\widetilde{Q}_\mathrm{f}}{\epsilon_0} = -\boldsymbol{\nabla} \cdot (\boldsymbol{v} \times \boldsymbol{B}) = -\boldsymbol{B} \cdot (\boldsymbol{\nabla} \times \boldsymbol{v}) + \boldsymbol{v} \cdot (\boldsymbol{\nabla} \times \boldsymbol{B}) \tag{9.79}$$

$$= -B_z \left(\frac{v}{\rho} + \frac{\partial v}{\partial \rho} \right) + B_\rho \frac{\partial v}{\partial z} - v \left(\frac{\partial B_z}{\partial \rho} - \frac{\partial B_\rho}{\partial z} \right) . \tag{9.80}$$

For the $\boldsymbol{v}$ of Eq. 9.72,

$$\widetilde{Q}_\mathrm{f} = -\epsilon_0 \omega F , \tag{9.81}$$

with

$$F = \rho \left(1 - K_\rho \frac{\rho}{R} \right) \left(\frac{\partial B_z}{\partial \rho} - \frac{\partial B_\rho}{\partial z} \right) + B_z \left(2 - 3K_\rho \frac{\rho}{R} \right) . \tag{9.82}$$

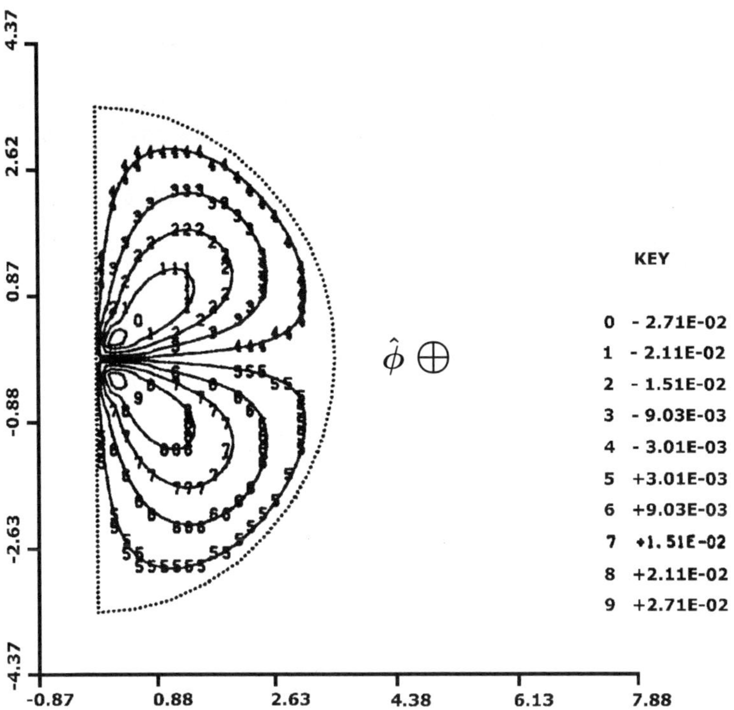

Fig. 9.5. Level diagram for B_T, as for Fig. 9.3, but with the differential rotation a function of ρ, as in Eq. 9.72. See the caption of Fig. 9.3

9.7 Summary

With respect to a non-rotating reference frame S', the non-axisymmetric component of the Earth's magnetic field rotates. But the axisymmetric component does not rotate, as if the source of the axisymmetric field were outside the Earth, and stationary. The rotation of the conducting core of the Earth in its own magnetic field generates both a $\boldsymbol{v} \times \boldsymbol{B}$ field and a competing electrostatic field $-\boldsymbol{\nabla} V$.

We assume that the core is uniform. If the core rotates as a solid, then the rotation does *not* induce a toroidal magnetic field because the induced electrostatic field $\boldsymbol{E}$ cancels the $\boldsymbol{v} \times \boldsymbol{B}$ field exactly. If $\boldsymbol{E}$ did not cancel the $\boldsymbol{v} \times \boldsymbol{B}$ field, then the ohmic power dissipation in the core would be absurdly high.

If there is differential rotation within the core, then it most probably induces a toroidal magnetic field. The general condition for the existence of such a field is that the bracket of Eq. 9.48 be *not* equal to zero.

Ferraro's law of isorotation is valid, despite the fact that his Physics is wrong, except that it applies only to one particular class of axisymmetric magnetic field.

Assuming that the axisymmetric magnetic field in the core is uniform and axial, with the azimuthal velocity a simple function of z, the induced toroidal magnetic field is only a few times larger than the axial field.

Assuming that the axisymmetric magnetic field is that of a point dipole at the center of the core, with the velocity a simple function of ρ, the azimuthal field is concentrated a short distance from the center of the core.

Part IV

Natural Dynamos

10 Case Study: The Disk Dynamo Model for Natural Dynamos

Contents

Gouy and Larmor were the first to propose, in the 1900's, that magnetic fields can originate in convecting conducting fluids (Sect. 1.7): given a seed magnetic field, fluid motion can give rise to a $v \times B$ field that generates an aiding magnetic field: there is positive feedback. Because of the complexity of the phenomena, Bullard (1955) suggested that a self-excited Faraday disk could be used as a model. Even though the rotating disk bears little resemblance to a convecting fluid, the model proves to be useful, if analyzed correctly.[1]

10.1 Introduction

A dynamo generates an electric current by moving part of an electric circuit in a magnetic field. If the output current of the dynamo generates the required magnetic field, then the dynamo is said to be *self-excited*. Commercial electric generators are partly self-excited because part of their output current is diverted to the windings that supply the magnetic field.

Some natural magnetic fields originate in magnetic materials such as magnetite in the Earth's crust, but all the others originate in dynamos in conducting, convecting fluids. There exist innumerable natural dynamos, in the

[1] P. Lorrain, *The disk-dynamo model for natural dynamos*, Physica Scripta **52**, 349–352 (1995).

liquid part of the Earth's core, in the solar plasma, in stars, in galaxies, and in Space.

Some dynamos are self-excited, for example sunspot dynamos (Chapter 13). But, in externally-excited dynamos, the fluid convects in a magnetic field that originates in one or more self-excited dynamos situated elsewhere. One example is the dynamo that generates the azimuthal magnetic field in the Earth (Chapter 9). In that case, the liquid part of the Earth's core convects and rotates in the Earth's dipolar magnetic field. There is no self-excitation.

One can also classify natural dynamos according to the nature of the observable output. For example, sunspot dynamos generate observable magnetic fields (Chapter 11), while others generate observable ion beams (Chapter 14).

Recall from Sect. 1.7 that it was the discovery of magnetic fields in sunspots by Bigelow and Hale that led Gouy and Larmor to postulate the self-generation of magnetic fields in the convecting solar plasma and in the Earth's core. The convection is driven by internal forces resulting, in particular, from thermal gradients and from rotation.

Given a seed magnetic field of outside origin, the $\boldsymbol{v} \times \boldsymbol{B}$ field of a self-excited dynamo generates, under appropriate conditions, an electric current that supplies an *aiding* magnetic field. As we shall see below, the system can become self-sustaining after the disappearance of the seed field. The energy stored in the magnetic field, as well as the ohmic, viscous, and other power losses, come from the forces that drive the convection.

In the case of a self-excited particle accelerator of Chapter 14, the observable output of the dynamo is an electric current, rather than a magnetic field. Then, given a seed electric current, the $\boldsymbol{v} \times \boldsymbol{B}$ field generates an *aiding* electric current.

A natural dynamo is never an isolated and closed system. For example, the magnetic field inside the liquid part of the Earth's core is known to be highly complex (Chapter 9). This requires a large number of self-excited and externally-excited dynamos, which are probably all more or less coupled together.

The classical model for natural self-excited dynamos is the disk dynamo. Even though natural dynamos are vastly more complex than that, the model is useful, when analyzed correctly.

The only reliable detailed papers on the self-excited *disk* dynamo are those of Knoepfel (1970) and of P. Lorrain (1995).

The only other detailed papers are by Bullard (1955, 1978), but his papers contain so many incredible errors that all his conclusions are totally invalid. In particular, his equations 3 and 4 are wrong on several counts. To give just one example, he confuses ω, the angular velocity of a rotating body, with the circular frequency ω (in $j\omega t = d/dt$) of an alternating current! But he does not use the factor j! Many authors have commented on the Bullard papers, without noticing the nonsense, but adding errors of their own!

Bullard wrote his first paper in 1955, and then, 23 years later, in 1978, repeated the nonsense on the occasion of a festschrift in his honor!

Another reference is the paper by Fearn et al. (1988), but their discussion is partly wrong, and purely hand-waving. Ingraham and Vulcan (1997) have investigated the magnetohydrodynamic equations for the disk dynamo.

The Cowling (1934, 1945, 1957, 1968, 1976) anti-dynamo "theorem", according to which a self-excited dynamo in a convecting, conducting fluid cannot be axisymmetric, has never been proven (Fearn et al., 1988; Alexeff, 1989; P. Lorrain, 1991; Ingraham, 1995, 1996). Contrary to the "theorem", the axis of symmetry of the magnetic field of Saturn coincides with its axis of rotation within experimental error. Also, the partly self-excited electromagnet of Kolm and Mawardi (1961) is axisymmetric. Unfortunately, most authors take the "theorem" for granted.

10.2 The Self-excited Disk Dynamo

Figure 10.1 shows a schematic diagram of a self-excited disk dynamo (P. Lorrain et al., 1988; P. Lorrain, 1995). We discussed this dynamo earlier in Sects. 7.3 and 8.7.8, but without considering self-excitation. The conducting disk D rotates at an angular velocity ω in the axial magnetic field $\boldsymbol{B} + \boldsymbol{B}_{\mathrm{s}}$, where $\boldsymbol{B}$ is the field of the solenoid S, and the seed magnetic field $\boldsymbol{B}_{\mathrm{s}}$ originates in some external source. The dynamo provides the field $\boldsymbol{B}$ by feeding the solenoid through the sliding contacts C_1 and C_2 on the axle and on the rim of the rotating disk.

With $\boldsymbol{B}$, $\boldsymbol{B}_{\mathrm{s}}$, and ω in the directions shown, current flows radially outward in the disk. The solenoid is connected to the contacts C_1 and C_2 in such a way that the current I then flows in the direction shown in the figure. Inverting both the direction of rotation and the connections to the solenoid has no effect on the polarity of the magnetic field. For a given configuration, changing the polarity of $\boldsymbol{B}_{\mathrm{s}}$ changes the direction of the current at the beginning, and thus of $\boldsymbol{B}$. So the induced magnetic field $\boldsymbol{B}$ has the same polarity as $\boldsymbol{B}_{\mathrm{s}}$.

10.2.1 The *Kinematic* Self-excited Disk Dynamo

Assume first that the angular velocity ω of disk D of Fig. 10.1 is given. We then have a *kinematic* self-excited disk dynamo. A feedback mechanism adjusts the power output of the motor M to keep its angular velocity constant, despite changes in the current I and in the magnetic field. The kinematic disk dynamo corresponds to an unrealistic natural dynamo in which neither the flow pattern of the conducting fluid, nor the magnitude of the flow velocity, would be affected by the magnetic field, and hence by the magnetic forces.

We assume axisymmetry, so that $\partial/\partial\phi = 0$. The axle has a radius a, and the disk a radius b. If the magnetic field $\boldsymbol{B} + \boldsymbol{B}_{\mathrm{s}}$ is axial and uniform over the surface of the disk, then the rotating disk generates an open-circuit voltage

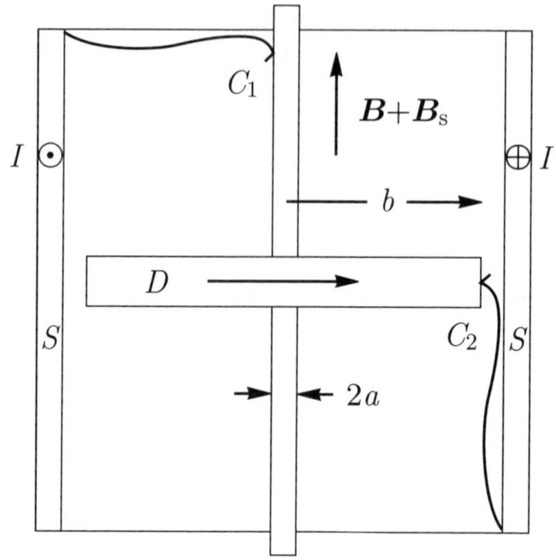

Fig. 10.1. Cross-section through a self-excited disk dynamo. The device is symmetrical about the vertical axis. At first, a motor rotates the disk D in the axial magnetic field $\boldsymbol{B}+\boldsymbol{B}_s$. Contacts C_1 on the axle and C_2 on the rim apply the voltage generated in the disk to the solenoid S that carries a current I. The magnetic field $\boldsymbol{B}$ is that of the solenoid, while $\boldsymbol{B}_s$ is a seed field that originates in an external source

$$\mathcal{V} = \omega(B + B_s)\frac{b^2 - a^2}{2} . \tag{10.1}$$

In Fig. 10.1 the rim is positive with respect to the axle if ω and the net magnetic field $\boldsymbol{B} + \boldsymbol{B}_s$ are both as shown. The above equation takes into account the field of the electrostatic space charge density $\widetilde{Q}$ of Chapter 7. If $\boldsymbol{B}$ increases, then the outward voltage $\mathcal{V}$ increases, which increases the current through the solenoid, and thus $\boldsymbol{B}$. So an increase in $\boldsymbol{B}$ leads to a further increase in $\boldsymbol{B}$: *this is a positive-feedback system.* Positive feedback is an essential characteristic of self-excited dynamos.

The magnetic field of the radial current through the disk is azimuthal.

The magnetic force, of density $\boldsymbol{J} \times (\boldsymbol{B} + \boldsymbol{B}_s)$, points in the direction opposite to the disk velocity and brakes the rotation.

Let the circuit resistance be R and the inductance L. In a natural dynamo R and L are distributed parameters because the currents are distributed over large volumes, but there are nonetheless Joule losses and an induced electric field $-\partial \boldsymbol{A}/\partial t$ throughout the circuit (Chapter 5).

From elementary circuit theory,

$$L\frac{dI}{dt} + RI = \mathcal{V} = \omega(B + B_s)\frac{b^2 - a^2}{2} \tag{10.2}$$

$$\approx \omega(B + B_s)\frac{b^2}{2} \quad (\text{with } b^2 \gg a^2) . \tag{10.3}$$

Assuming now a long solenoid (Sect. 4.4.1),

$$B = \mu_0 \widetilde{N} I , \tag{10.4}$$

where $\widetilde{N}$ is the number of turns per meter on the solenoid. If the length of the solenoid is not much larger than its diameter, then B is smaller by a factor $K < 1$. For simplicity, we set $K = 1$. Then

$$L\frac{dI}{dt} + \left(R - \omega\mu_0\widetilde{N}\frac{b^2}{2}\right) I = \omega B_s \frac{b^2}{2} , \tag{10.5}$$

$$\frac{dI}{dt} + \frac{R}{L}\left(1 - \frac{\omega}{\omega_0}\right) I = \omega B_s \frac{b^2}{2L} , \tag{10.6}$$

where

$$\omega_0 = \frac{2R}{\mu_0 \widetilde{N} b^2} \tag{10.7}$$

is positive. Inverting the connections to the solenoid would change the sign of ω_0, and ω would have to change sign for self-excitation. Then $\boldsymbol{B}$ would again have the same sign as $\boldsymbol{B}_s$.

Solving,

$$I = \frac{\omega B_s b^2}{2R(1 - \omega/\omega_0)} \left\{1 - \exp\left[-\frac{R}{L}\left(1 - \frac{\omega}{\omega_0}\right) t\right]\right\} . \tag{10.8}$$

We have assumed that $I = 0$ at $t = 0$. Hence, at $t = 0$, the solenoid has zero magnetic field and $\boldsymbol{B} = \boldsymbol{0}$.

Now start the dynamo with $0 \leq \omega/\omega_0 < 1$. The current I increases to the limiting value

$$I_\infty = \frac{\omega B_s b^2}{2R(1 - \omega/\omega_0)} . \tag{10.9}$$

If $\omega/\omega_0 > 1$, then the term between braces in Eq. 10.8 is negative. The argument of the exponential function is positive, and the current I builds up until it reaches a value I_1 at the time t_1. The current I_1 is positive, which means that it flows in the direction shown in the figure. Then, from the sign conventions shown, the induced magnetic field $\boldsymbol{B}$ has the same sign as the seed field $\boldsymbol{B}_s$.

At $t = t_1$ we turn the seed magnetic field $\boldsymbol{B}_s$ off. At that instant, the current I_1 is given by Eq. 10.8 with $t = t_1$. Of course $\boldsymbol{B}_s$ does not drop to zero instantaneously, because of the inductances, both in the circuit that generates $\boldsymbol{B}_s$, and in the circuit of the dynamo. Assume that this time constant is much shorter than t_1.

For $t > t_1$, with $\boldsymbol{B}_s = \boldsymbol{0}$ in Eq. 10.6,

$$\frac{dI}{dt} + \frac{R}{L}\left(1 - \frac{\omega}{\omega_0}\right) I = 0 \, . \tag{10.10}$$

Solving again, with $I = I_1$ at $t = t_1$,

$$I = I_1 \exp\left[-\frac{R}{L}\left(1 - \frac{\omega}{\omega_0}\right)(t - t_1)\right] \quad (t > t_1) \tag{10.11}$$

or, since B is proportional to I, as in Eq. 10.4,

$$B = B_1 \exp\left[-\frac{R}{L}\left(1 - \frac{\omega}{\omega_0}\right)(t - t_1)\right] \quad (t > t_1) \, , \tag{10.12}$$

where B_1 is the value of B at the time t_1. The factor R/L is positive.

These last two equations tell us what happens after the time t_1 when we turn $\boldsymbol{B}_s$ off.

a) If $\omega/\omega_0 < 0$, then I and B both decrease exponentially with time; the disk turns in the wrong direction for self-excitation.

b) If $\omega = 0$, the disk is stationary, and both I and B decrease exponentially with a time constant L/R, as expected.

c) If $0 < \omega/\omega_0 < 1$, then $\boldsymbol{B}$ again decreases exponentially with the time, but not as fast.

d) If $\omega/\omega_0 = 1$, the magnetic field is constant: $\boldsymbol{B} = \boldsymbol{B}_1$, as in Fig. 10.2. The magnetic field has whatever magnitude and polarity it had at the time t_1.

e) If $\omega/\omega_0 > 1$, then $\boldsymbol{B}$ increases exponentially with time.

In all cases the time constant is

$$\left| \frac{L/R}{1 - \omega/\omega_0} \right| \, .$$

Thus, according to the *kinematic* self-excited disk dynamo model, stability in a natural dynamo requires that some critical parameter, ω in the present case, have a specific value; otherwise, $\boldsymbol{B}$ can only increase or decrease exponentially with time, depending on the value of the critical parameter. The steady-state $\boldsymbol{B}$ can have any value; it is just $\boldsymbol{B}_1$, the magnetic field at the time when $\boldsymbol{B}_s$ was turned off.

The torque required to rotate the disk is

$$\Gamma = \frac{BIb^2}{2} = \frac{B^2 b^2}{2\mu_0 \widetilde{N}} \tag{10.13}$$

and the power P is ω times larger:

$$P = \frac{\omega BIb^2}{2} = \frac{\omega b^2}{2\mu_0 \widetilde{N}} B^2 \, , \tag{10.14}$$

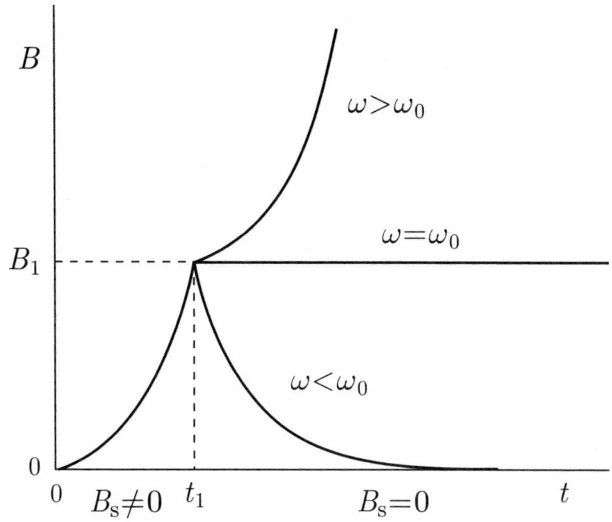

Fig. 10.2. The value of B as a function of the time for a *kinematic* self-excited disk dynamo. The seed field B_s is cut off at $t = t_1$. If the angular velocity ω is less than ω_0, B decreases exponentially. If $\omega = \omega_0$, B has the constant value B_1. But, if $\omega > \omega_0$, then B increases exponentially

from Eq. 10.4. Since the power input is proportional to B^2, an exponential increase in B requires a power input that increases with twice the exponent.

If the input mechanical power P drops to zero, then ω tends to zero, and both I and B decrease exponentially to zero with the time constant L/R, as if there were no rotating disk. Reactivating P will restart the dynamo if a seed magnetic field is available, and the new $\boldsymbol{B}$ will have the same polarity as the new seed field $\boldsymbol{B}_s$

The disk dynamo is not a good model for the magnetic forces that exist in natural dynamos and that we discussed in Sect. 6.7, but it illustrates well the "viscous" component, because the magnetic force density $\boldsymbol{J} \times \boldsymbol{B}$ at any given point on the disk points in the direction opposite to the velocity.

10.2.2 The *Dynamic* Self-excited Disk Dynamo

For the *dynamic* disk dynamo model we assume that the mechanical power P of Eq. 10.14 fed to the disk dynamo of Fig. 10.1 is constant. We assume that the seed field has been turned off. This makes the angular velocity ω of the disk depend on the magnetic flux density $\boldsymbol{B}$, and thus on the current I. From Eq. 10.13, the braking torque on the disk is proportional to B^2. The quantity P is now a constant and the angular velocity ω is proportional to P, but only for a given current I:

$$\omega = \frac{2P}{BIb^2} = \frac{2P}{\mu_0 \widetilde{N} I^2 b^2} . \tag{10.15}$$

We have again assumed that $b^2 \gg a^2$ in Fig. 10.1.

By conservation of energy,

$$P = I^2 R + \frac{d}{dt}\left(\frac{1}{2} L I^2\right) , \tag{10.16}$$

where the first term on the right is the ohmic power loss, and the second term is the rate of increase of magnetic energy (Sect. 4.10). We disregard friction, windage, and other losses. We also assume that the kinetic energy of the disk is negligible compared to the magnetic energy; otherwise the solution of Eq. 10.16 becomes difficult. The approximation is acceptable because the disk dynamo is clearly a crude model for dynamos in convecting, conducting fluids. Rearranging,

$$\frac{d(I^2)}{dt} + \frac{2R}{L} I^2 = \frac{2P}{L} . \tag{10.17}$$

Solving, with $I = I_1$ at $t = t_1$,

$$I^2 = \left(I_1^2 - I_\infty^2\right) \exp\left[-\frac{2R}{L}(t - t_1)\right] + I_\infty^2 \quad (t > t_1) , \tag{10.18}$$

where

$$I_\infty = \left(\frac{P}{R}\right)^{1/2} \tag{10.19}$$

is the current that flows through the circuit after an infinite time.

Recall that the kinematic disk dynamo has a steady-state solution for I only if $\omega = \omega_0$. However, the dynamic disk dynamo has a steady-state solution for I: once the transient first term of Eq. 10.18 has become negligible, $I = I_\infty$. The input power P is then equal to the ohmic power loss $I_\infty^2 R$, from Eq. 10.16.

From Eq. 10.18, at $t > t_1$,

$$B^2 = \left(B_1^2 - B_\infty^2\right) \exp\left[-\frac{2R}{L}(t - t_1)\right] + B_\infty^2 , \tag{10.20}$$

where, from Eq. 4.8,

$$B_1 = \mu_0 \widetilde{N} I_1 \quad \text{and} \quad B_\infty = \mu_0 \widetilde{N} I_\infty = \mu_0 \widetilde{N} \left(\frac{P}{R}\right)^{1/2} . \tag{10.21}$$

The steady-state solution, B_∞, is proportional to the square root of the input power P. There is no threshold power below which $B = 0$, as in Fig. 10.3.

The angular velocity of the disk after a time $t - t_1 \gg L/(2R)$ follows from Eqs. 10.15 and 10.19:

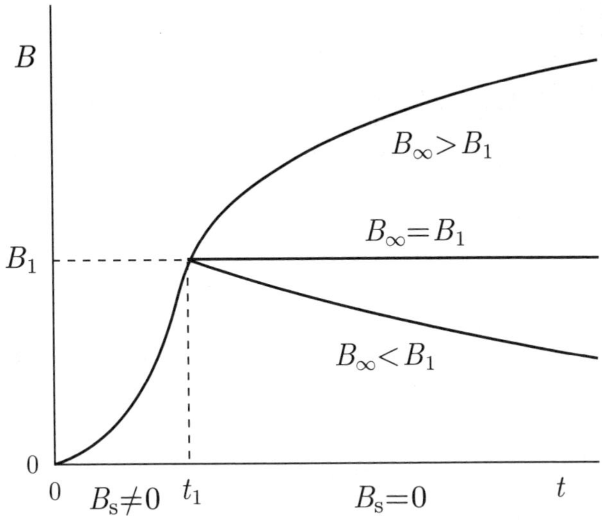

Fig. 10.3. The value of B as a function of time for the *dynamic* self-excited disk dynamo. The seed magnetic field B_s is again cut off at the time t_1. After a time $t > t_1$, if $B_\infty > B_1$, then B increases with time, but reaches the plateau B_∞. If $B_\infty < B_1$, B decreases with time and tends to B_∞

$$\omega_\infty = \frac{2P}{\mu_0 \tilde{N} I_\infty^2 b^2} = \frac{2R}{\mu_0 \tilde{N} b^2} = \omega_0 \, . \tag{10.22}$$

What if P is a function of the time? Then Eq. 10.17 is more difficult to solve, but we can easily guess what its solution is like. See Fig. 10.4. Assume that the applied torque varies erratically, both in sign and magnitude, by jumps. Then Eqs. 10.18 and 10.20 apply in the intervals, with the values of I_1 and B_1 readjusted, and the transient reappearing after each jump. Thus $B(t)$ drifts up and down, more smoothly than $P(t)$. As with the kinematic dynamo, if P drops to zero, then B decreases exponentially with a time constant L/R. The dynamo can later restart whenever P is reactivated, if a seed magnetic field is present. The new field has the polarity of the seed field. The kinematic dynamo exhibits the same behavior.

This is one explanation for the polarity reversals in the dipolar component of the Earth's magnetic field. Remember, however, that the disk dynamo model substitutes a simple rotational motion for an extremely complex flow pattern. Remember also that the magnetic field at the surface of the core results from the addition of a very large number of fields of pseudo-random magnitudes, phases, and orientations. The dipole component, there, is indiscernible. See Chapter 9.

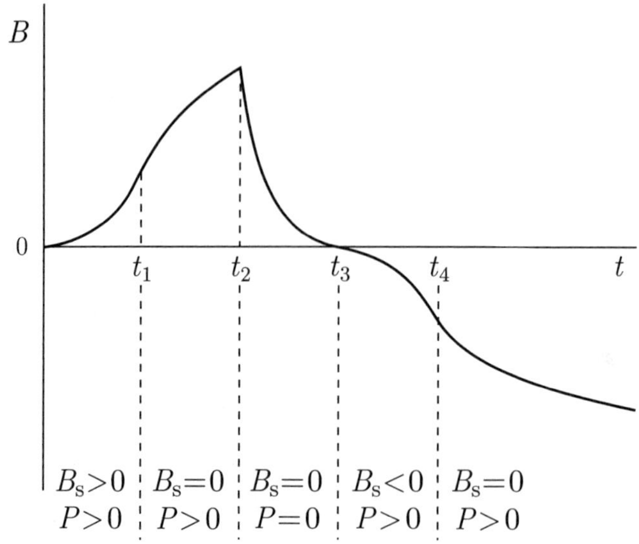

Fig. 10.4. Magnetic field as a function of time for a fluctuating power input P and a fluctuating seed field

10.3 A Laboratory-sized Dynamo?

Is it possible to build a laboratory model of such a dynamo? That would not be simple. The problem lies in Eq. 10.22 for ω_0. It is only after a time $t - t_1$ that is much larger than $L/(2R)$ that B_∞ reaches its maximum value $\mu_0 \widetilde{N}(P/R)$ as in Eq. 10.21 and that $\omega = \omega_\infty$.

Refer to Eq. 10.22. Try $R = 10$ ohms, $b = 0.25$ meter, $\widetilde{N} = 10^4$ turns per meter. Then

$$\omega_0 = \frac{20}{4\pi \times 10^{-7} \times 10^4 b^2} \tag{10.23}$$

$$= 2.55 \times 10^4 \text{ radians/second} \tag{10.24}$$

$$= 2.43 \times 10^5 \text{ revolutions/minute .} \tag{10.25}$$

Try

$$R = 10 \text{ ohms , } \omega = 1000 \text{ revolutions/minute , } b = 0.25 \text{ meter .} \tag{10.26}$$

Then

$$P = 1 \text{ kilowatt , } I = 10 \text{ amperes , and } B = 0.01 \text{ tesla ,} \tag{10.27}$$

which is reasonable. The power P dissipates in the solenoid. However, $\omega_\infty \approx 2.43 \times 10^5$ radians/second, which requires ultracentrifuge technology. The problem of ensuring contacts with the rotor would be difficult. The alternative is to operate at low temperature, which would reduce R, and thus ω_∞ and P, by orders of magnitude, for given values of B and I.

There is certainly no point in building such a dynamo to confirm the above simple theory.

10.4 Summary

The self-excited disk dynamo is a useful model for natural dynamos, even though it substitutes the rotation of a disk for a complex flow pattern.

In the *kinematic* self-excited disk dynamo the angular velocity of the disk is given; it therefore neglects the magnetic braking force, which is unrealistic because the flow pattern in a natural dynamo depends largely on magnetic forces. The kinematic self-excited disk dynamo has three modes of operation. In one mode the magnetic field is constant; that requires a specific angular velocity, corresponding to a specific, constant flow pattern. The magnetic field B can also either increase or decrease exponentially with time, and the power required to drive the disk is proportional to B^2.

The *dynamic* self-excited disk dynamo is more realistic: it operates with a given power input, and the angular velocity of the disk depends on the magnetic field. Then there exists both a steady-state solution and a transient solution. The time constant of the transient solution for B is equal to L/R, with L and R the inductance and the resistance of the circuit. The steady-state B is proportional to the square root of the mechanical power applied to the disk. There is no threshold power below which there is no magnetic field.

Both types of disk dynamo generate a magnetic field that has the polarity of the seed field, and both are subjected to polarity reversals if the power input is erratic.

11 Three Case Studies: Magnetic Flux Tubes, Flux Ropes, and Flux Coils

Contents

A magnetic flux tube in a convecting conducting fluid has an axial magnetic field and an azimuthal current, like a solenoid. For example, solar magnetic elements (Chapter 12) and sunspots (Chapter 13) are magnetic flux tubes that stand more or less vertically in the solar plasma. A flux rope is similar to a tube, except that its magnetic field lines are twisted: the rope carries an axial current, whose magnetic

field is azimuthal.[1] It may be that both tubes and ropes act as light guides, like optical fibers.

11.1 Introduction

From here on we shall be much concerned with the Sun. For an extensive and authoritative discussion of the Astrophysics of the Sun, see Zirin (1988). See also the *Special Historical Review on Macrophysics and the Sun* by Parker (1997).

The *photosphere* is the surface of the Sun, as observed in white light. Its temperature is about 5500 kelvins. High resolution observations show a pattern of bright patches, called *granules* separated by dark *lanes*. The granules have irregular shapes, and diameters of about 1000 kilometers. The hot plasma upwells in the granules, and downwells in the lanes, much as a boiling liquid, and the granules have a lifetime of 5 or 10 minutes: they appear and disappear without cease. There also exists a much larger-scale but similar phenomenon, called *supergranulation* whose cells have diameters of the order of 30 000 kilometers. We are concerned here rather with *pores* and with *magnetic elements* (Chapter 12). Pores are small bright spots on the photosphere that seem to be more or less vertical magnetic flux tubes.

As we shall see, downwelling in a convecting conducting fluid generates flux tubes or ropes.

One might say that a magnetic flux tube is a bundle of magnetic field lines; it is a cylindrical region inside which the axial magnetic field is much larger than the magnetic field outside. The magnetic field inside a solenoid is a flux tube, but we are only concerned here with flux tubes generated spontaneously within convecting conducting fluids.

A magnetic flux rope is a twisted flux tube, with helical field lines; it requires an axial electric current.

Flux tubes and ropes need not be straight, and their cross-sections can be neither circular, nor uniform along their lengths.

The magnetic elements observed on the Sun (Chapter 12) are presumably either magnetic flux tubes or ropes that stand more or less vertically. Sunspots (Chapter 13) are also either magnetic flux tubes or ropes. As we shall see, both occur in regions where there is downwelling in the solar plasma.

Magnetic flux tubes and ropes possibly also exist in the liquid part of planetary cores, if there is downwelling, as in the Earth's mantle (Bercovici et al., 1989; Schubert et al., 1990). The convection velocities in the Earth's core are orders of magnitude smaller than those in the solar plasma, but the conductivity is orders of magnitude larger.

---

[1] See P. Lorrain and N. Salingaros, *Local currents in magnetic flux tubes and flux ropes*, Amer. Jour. Phys. **61**, 811–817 (1993); P. Lorrain and S. Koutchmy, *Two dynamical models for solar spicules*, Solar Physics **165**, 115–137 (1996).

Many authors have discussed magnetic flux tubes and ropes, particularly Parker (1979) and Russell et al. (1990), who devote about 800 pages to the subject. However, except for P. Lorrain and Salingaros (1993), and for P. Lorrain and S. Koutchmy (1996), there is no mention in the literature of either the required local electric currents, nor of the local $\boldsymbol{v} \times \boldsymbol{B}$ fields that those currents require.

As we stressed in the Preface, we focus our attention on the electric currents. The usual procedure of simply discussing the shape of imaginary magnetic field lines is both pointless and senseless. See also Chapters 7 and 8.

In Sect. 11.5 we discuss the hypothesis that magnetic flux tubes and ropes in solar plasmas might act as light guides.

We assume rectilinear tubes, ropes, and coils. As usual, we use cylindrical coordinates ρ, ϕ, z, with the z-axis pointing up.

11.2 Convecting, Conducting Fluids

The following discussion will be useful for what follows.

Several fundamental equations apply to convecting conducting fluids.

For the moment, we assume a steady state, setting $\partial/\partial t = 0$, so that there is no displacement current, but, further on, we calculate time constants. The medium is non-magnetic. We use cylindrical coordinates, with both $\boldsymbol{v}$ and $\boldsymbol{B}$ unspecified, and we do not assume axisymmetry:

$$\boldsymbol{v} = v_\rho \hat{\boldsymbol{\rho}} + v_\phi \hat{\boldsymbol{\phi}} + v_z \hat{\boldsymbol{z}} \,, \quad \boldsymbol{B} = B_\rho \hat{\boldsymbol{\rho}} + B_\phi \hat{\boldsymbol{\phi}} + B_z \hat{\boldsymbol{z}} \,, \quad \frac{\partial}{\partial \phi} \neq 0 \,. \tag{11.1}$$

1. Let $\widetilde{m}$ be the mass per unit volume and $\boldsymbol{v}$ the velocity at a given point in the convecting fluid. From the continuity equation for the mass,

$$\boldsymbol{\nabla} \cdot (\widetilde{m}\boldsymbol{v}) = -\frac{\partial \widetilde{m}}{\partial t} = 0 \tag{11.2}$$

or, in cylindrical coordinates,

$$\frac{1}{\rho}\frac{\partial}{\partial \rho}(\rho \widetilde{m} v_\rho) + \frac{1}{\rho}\frac{\partial}{\partial \phi}(\widetilde{m} v_\phi) + \frac{\partial}{\partial z}(\widetilde{m} v_z) = 0 \,. \tag{11.3}$$

This equation is much more complex than it seems because $\widetilde{m}$, v_ρ, v_ϕ, and v_z are all functions, not only of the coordinates ρ and z, but also of $\boldsymbol{B}$, of $\boldsymbol{J}$, as well as functions of the forces that drive the convection.

2. From Ohm's law for moving conductor, Eq. 6.4, for a steady state,

$$\boldsymbol{J} = \sigma[-\boldsymbol{\nabla} V + (\boldsymbol{v} \times \boldsymbol{B})] \,. \tag{11.4}$$

All the variables are measured with respect to the same inertial reference frame. The conductivity σ is a tensor and is larger in the direction parallel to

$\boldsymbol{B}$ than in orthogonal directions (Sect. 3.2), but we disregard that distinction. The axial magnetic field reduces the azimuthal conductivity, here, by slowing down the azimuthal drift of the free electrons.[2]

The term $-\boldsymbol{\nabla} V$ is the electric field of the space charge $\widetilde{Q}_{\mathrm{f}}$ of Eq. 6.39 and of surface charges, if any.

There is a local electric current:

$$\begin{aligned} \boldsymbol{J} = \sigma &\left[-\frac{\partial V}{\partial \rho} + (v_\phi B_z - v_z B_\phi) \right] \hat{\boldsymbol{\rho}} \\ + \sigma &\left[-\frac{1}{\rho}\frac{\partial V}{\partial \phi} + (v_z B_\rho - v_\rho B_z) \right] \hat{\boldsymbol{\phi}} \\ + \sigma &\left[-\frac{\partial V}{\partial z} + (v_\rho B_\phi - v_\phi B_\rho) \right] \hat{\boldsymbol{z}} \,. \end{aligned} \tag{11.5}$$

The magnetic force, of density $\boldsymbol{J} \times \boldsymbol{B}$, brakes the convection.

3. From the Maxwell equation for the divergence of $\boldsymbol{B}$, Eq. 2.2,

$$\frac{1}{\rho}\frac{\partial}{\partial \rho}(\rho B_\rho) + \frac{1}{\rho}\frac{\partial B_\phi}{\partial \phi} + \frac{\partial B_z}{\partial z} = 0 \,. \tag{11.6}$$

4. From the Maxwell equation for the curl of $\boldsymbol{B}$, Eq. 2.4,

$$\mu_0 J_\rho = \frac{1}{\rho}\frac{\partial B_z}{\partial \phi} - \frac{\partial B_\phi}{\partial z} \,, \tag{11.7}$$

$$\mu_0 J_\phi = \frac{\partial B_\rho}{\partial z} - \frac{\partial B_z}{\partial \rho} \,, \tag{11.8}$$

$$\mu_0 J_z = \frac{1}{\rho}\left[\frac{\partial}{\partial \rho}(\rho B_\phi) - \frac{\partial B_\rho}{\partial \phi} \right] . \tag{11.9}$$

5. From the Maxwell equation for the divergence of $\boldsymbol{E}$, Eq. 2.1,

$$\frac{1}{\rho}\frac{\partial}{\partial \rho}(\rho E_\rho) + \frac{1}{\rho}\frac{\partial E_\phi}{\partial \phi} + \frac{\partial E_z}{\partial z} = \frac{\widetilde{Q}_{\mathrm{f}}}{\epsilon_0} \,. \tag{11.10}$$

6. From the Maxwell equation for the curl of $\boldsymbol{E}$, Eq. 2.3, with $\partial/\partial t = 0$,

$$\frac{1}{\rho}\frac{\partial E_z}{\partial \phi} - \frac{\partial E_\phi}{\partial z} = 0 \,, \tag{11.11}$$

$$\frac{\partial E_\rho}{\partial z} - \frac{\partial E_z}{\partial \rho} = 0 \,, \tag{11.12}$$

$$\frac{1}{\rho}\left[\frac{\partial}{\partial \rho}(\rho E_\phi) - \frac{\partial E_\rho}{\partial \phi} \right] = 0 \,. \tag{11.13}$$

[2] The paper by Oster (1968) on the electric conductivity in sunspots is not valid because it applies the results of Shkarofsky (1960), which concern *alternating* currents in plasmas! As a result his conductivities, in the presence of a steady magnetic field, are not real but complex. That is nonsense.

7. For a steady state, $\boldsymbol{\nabla} \cdot \boldsymbol{J} = -\partial \widetilde{Q}_{\rm f}/\partial t = 0$ and

$$\frac{1}{\rho}\frac{\partial}{\partial \rho}(\rho J_\rho) + \frac{1}{\rho}\frac{\partial J_\phi}{\partial \phi} + \frac{\partial J_z}{\partial z} = 0 \,. \tag{11.14}$$

8. Finally, according to Eq. 6.39, the $\boldsymbol{v} \times \boldsymbol{B}$ field sets up a non-zero electrostatic space charge of density

$$\widetilde{Q}_{\rm f} = -\epsilon_0 \boldsymbol{\nabla} \cdot (\boldsymbol{v} \times \boldsymbol{B}) \,. \tag{11.15}$$

Upon expanding, the term $v_z B_\phi/\rho$ is worrisome, but it is not infinite when ρ tends to zero because B_ϕ also tends to zero with ρ, and v_z is finite.

The space charge density $\widetilde{Q}_{\rm f}$ is a function of the coordinates and can be either positive or negative. The electrostatic potential V results, in principle, from $\widetilde{Q}_{\rm f}$ and from surface charges, but flux tubes, ropes, and coils presumably never have definite surfaces, and we may assume that V depends only on $\widetilde{Q}_{\rm f}$.

11.3 Magnetic Flux Tubes (MFT's)

We define a magnetic flux tube as a cylindrical region within a convecting conducting fluid, inside which there is an axial $\boldsymbol{B}$. The flux rope is a superposition of a flux tube and a flux coil (Sect. 11.6). Sunspots (Chapter 13) are the tops of magnetic flux tubes.

Figure 11.1 shows portions of two magnetic flux tubes, and Fig. 11.2 shows a cross-section of a region, say in the solar photosphere, where there is downwelling. We assume axisymmetry, here, so that $\partial/\partial\phi = 0$.

There is downwelling, and thus a converging flow at the surface, in and around a growing magnetic flux tube, say a sunspot. On the contrary, there is upwelling, and a diverging flow, when a flux tube decays.

In Fig. 11.2, both v_ρ and v_z are negative, and we set $v_\phi = 0$. So

$$v_\rho < 0 \,, \quad v_\phi = 0 \,, \quad v_z < 0 \,. \tag{11.16}$$

The magnetic field lines have roughly the same configuration as the streamlines of Fig. 11.2, except that they can point either up or down, and except that they form a smaller angle with the vertical than the downflowing $\boldsymbol{v}$ (see below):

$$\frac{B_\rho}{B_z} < \frac{v_\rho}{v_z} \,. \tag{11.17}$$

The components B_ρ and B_z have the same sign. We set $B_\phi = 0$, which excludes any axial current: $J_z = 0$. So

$$\boldsymbol{B} = B_\rho \hat{\boldsymbol{\rho}} + B_z \hat{\boldsymbol{z}} \,. \tag{11.18}$$

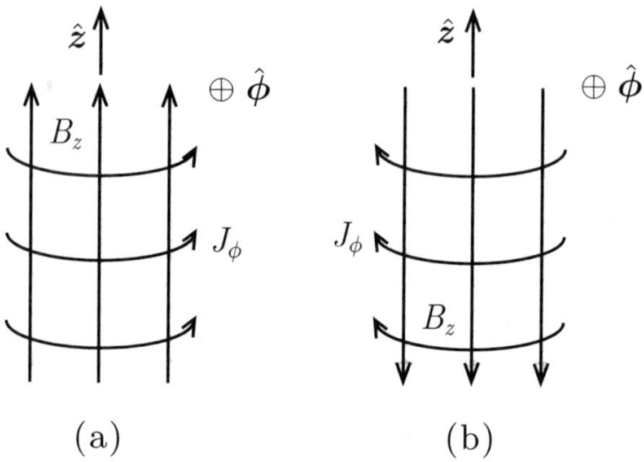

Fig. 11.1. Portions of two magnetic flux tubes. The magnetic field outside is negligible. According to the Maxwell equation for the curl of $\boldsymbol{B}$, the axial magnetic field B_z cannot exist without a corresponding local azimuthal current, of density J_ϕ. a) With B_z positive, J_ϕ is positive. b) With B_z negative, J_ϕ is negative. So B_z and J_ϕ have the same sign

Then

$$\boldsymbol{v} \times \boldsymbol{B} = (v_\rho \hat{\boldsymbol{\rho}} + v_z \hat{\boldsymbol{z}}) \times (B_\rho \hat{\boldsymbol{\rho}} + B_z \hat{\boldsymbol{z}}) = (v_z B_\rho - v_\rho B_z)\hat{\boldsymbol{\phi}} \,, \tag{11.19}$$

and $\boldsymbol{v} \times \boldsymbol{B}$ has only a ϕ-component. Then its divergence is zero, the electric space charge density Q_f is zero, from Eq. 7.10, and $V = 0$, assuming zero surface charges.

Downwelling in and around sunspots have long been reported by many authors (Zirin and Wang, 1989, 1991; Wang and Zirin, 1992; Brekke et al., 1990) and several groups have devised methods for observing flows in and around sunspots below the photosphere (Duvall et al., 1993; Duvall, 1995; Duvall et al., 1996; Kosovichev, 1996; Lindsey et al., 1996; Braun, 1995; Bogdan and Braun, 1995; Braun et al., 1996; Sun et al., 1997; Bogdan et al., 1998).

Kosovichev (1996), Lindsey et al. (1996), Braun (1995), and Braun et al. (1996) observed upwelling and outflows around decaying flux tubes.

There are two types of radial force exerted on the solar plasma near magnetic flux tubes: the convective force associated with either downwelling or upwelling, and the magnetic force that we discuss here. These two forces point in opposite directions, the magnetic force braking the convection.

Assuming zero surface charges, the azimuthal electric current density is now

$$\boldsymbol{J} = \sigma(\boldsymbol{v} \times \boldsymbol{B}) = \sigma(v_z B_\rho - v_\rho B_z)\hat{\boldsymbol{\phi}} \tag{11.20}$$

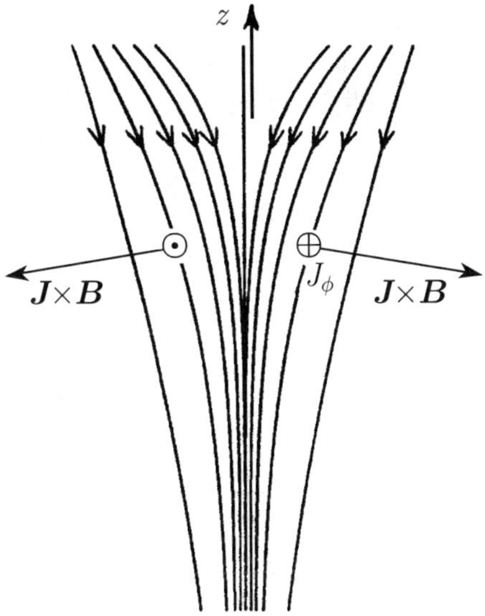

Fig. 11.2. Streamlines in a convecting conducting plasma in a region where there is downwelling. The plasma flows downward and inward, the density and the pressure both increasing with depth. Magnetic field lines have roughly the same shape, but form smaller angles with the axis of symmetry, whether the field points up or down. With $\boldsymbol{B}$ pointing up, the magnetic force density $\boldsymbol{J} \times \boldsymbol{B}$ points outward. With $\boldsymbol{B}$ pointing down, $\boldsymbol{J}$ changes sign, and the magnetic force density again points outward

$$= \sigma v_z B_z \left(\frac{B_\rho}{B_z} - \frac{v_\rho}{v_z} \right) \hat{\boldsymbol{\phi}} \, . \tag{11.21}$$

All the terms of this equation are interrelated, and functions of the coordinates ρ and z. Here, J_ϕ, B_z, and B_ρ all have the same sign, while v_ρ and v_z are both negative.

Assume a B_z that is positive. If

$$\frac{B_\rho}{B_z} < \frac{v_\rho}{v_z} \tag{11.22}$$

(as in Eq. 11.17), then there is positive feedback because the induced current points in the positive direction of $\hat{\boldsymbol{\phi}}$ (recall that v_z and v_ρ are both negative), and this provides a magnetic field that also points up. If $\boldsymbol{B}$ points down instead of up, then B_ρ and B_z are both negative, and there is again positive feedback: the magnetic flux tube is a self-excited dynamo. The parenthesis in Eq. 11.21 is negative, like v_z, and J_ϕ has the sign of B_z.

The total magnetic field at first grows exponentially with time, and eventually reaches a stable value as in the dynamic disk dynamo of Sect. 10.2.2.

This is just the principle of operation of the partly self-excited electromagnet of Kolm and Mawardi (1961).

A magnetic flux tube can exist only in regions of a convecting fluid where the velocity has a negative radial component, in cylindrical coordinates, all along the tube.

Various authors, for example Moffatt (1978), discuss the shapes of flux tubes that form closed loops, or that are linked to other closed loops, or that are knotted! Such shapes are absurd because the fluid would then have to flow inward all along the contorted tube to generate the required local current.

The azimuthal electric current results from the rotation of positively charged particles in the direction of $\boldsymbol{v} \times \boldsymbol{B}$, and of negative particles in the opposite direction. The electrons carry most of the current because of their lower mass and higher velocity. The plasma remains neutral. The ions and electrons move inward with the plasma as in Fig. 11.2, and moreover rotate about the axis of symmetry. The magnetic force density $\boldsymbol{J}_\phi \times \boldsymbol{B}$, points radially outward, whatever the sign of B_z: the magnetic flux tube supports a field of either polarity (Hénoux and Somov, 1987; Heyvaerts and Hagyard, 1991; P. Lorrain, 1991, 1993b, 1995).

The angular momentum of the orbiting particles is conserved because there is no externally applied torque. So their azimuthal velocity increases as they move inward.

As z increases, or as we look at shallower and shallower depths, the flux tube flares more and more, and the parenthesis in Eq. 11.21, and J, become smaller and smaller, while B_ρ becomes larger and larger until, at the top of the flux tube, the parenthesis is equal to zero, and

$$\frac{B_\rho}{B_z} = \frac{v_\rho}{v_z} \; . \tag{11.23}$$

Above that level, $\boldsymbol{v} \times \boldsymbol{B}$ points in the $-\hat{\boldsymbol{\phi}}$-direction, there is negative feedback, there is no self-excitation, and $J_\phi = 0$. Observations show that flux tubes indeed flare near the top (Solanki and Schmidt, 1993; Lites et al., 1993).

The components of $\boldsymbol{v}$ and of $\boldsymbol{B}$ are unknown except that, at a given point, for a steady state,

$$\boldsymbol{\nabla} \cdot \boldsymbol{B} = 0 \, , \quad \text{or} \quad \frac{B_\rho}{\rho} + \frac{\partial B_\rho}{\partial \rho} + \frac{\partial B_z}{\partial z} = 0 \, , \tag{11.24}$$

$$\boldsymbol{\nabla} \cdot \boldsymbol{v} = 0 \, , \quad \text{or} \quad \frac{v_\rho}{\rho} + \frac{\partial v_\rho}{\partial \rho} + \frac{\partial v_z}{\partial z} = 0 \, , \tag{11.25}$$

$$\boldsymbol{\nabla} \times \boldsymbol{B} = \mu_0 \boldsymbol{J} \, , \quad \text{or} \quad \frac{\partial B_\rho}{\partial z} - \frac{\partial B_z}{\partial \rho} = \mu_0 J_\phi \; . \tag{11.26}$$

Near the axis of symmetry, both B_ρ and ρ tend to zero, but the current density $\boldsymbol{J}$ remains azimuthal.

Let us check the induction equation for this dynamo:

$$\frac{\partial \boldsymbol{B}}{\partial t} = \boldsymbol{\nabla} \times (\boldsymbol{v} \times \boldsymbol{B}) - \frac{\boldsymbol{\nabla} \times (\boldsymbol{\nabla} \times \boldsymbol{B})}{\mu_0 \sigma} \tag{11.27}$$

$$= \boldsymbol{\nabla} \times (\boldsymbol{v} \times \boldsymbol{B}) - \frac{\boldsymbol{\nabla} \times \boldsymbol{J}}{\sigma} . \tag{11.28}$$

For a steady state, the LHS is zero. Then, if the conductivity σ is uniform, and from Eq. 11.20, the induction equation reduces to the identity $\mathbf{0} \equiv \mathbf{0}$.

11.3.1 The Magnetic Force Density in MFT's

Let us look at the magnetic force density in a magnetic flux tube. From Eq. 11.21,

$$\widetilde{\boldsymbol{F}} = \boldsymbol{J} \times \boldsymbol{B} = \sigma v_z B_z \left[\frac{B_\rho}{B_z} - \frac{v_\rho}{v_z} \right] \hat{\boldsymbol{\phi}} \times (B_\rho \hat{\boldsymbol{\rho}} + B_z \hat{\boldsymbol{z}}) \tag{11.29}$$

$$= \sigma v_z B_z \left[\frac{B_\rho}{B_z} - \frac{v_\rho}{v_z} \right] (-B_\rho \hat{\boldsymbol{z}} + B_z \hat{\boldsymbol{\rho}}) . \tag{11.30}$$

Now remember, from Eq. 11.22, that v_z and the bracket are both negative, so their product is positive. Moreover, the two B components are of the same sign. Say B_ρ and B_z are both positive: the magnetic field points up. Then the z-component of the magnetic force is negative, while the ρ-component is positive. If the magnetic field points down, both B_ρ and B_z are negative, and Fig. 11.2 again applies. So the force density is perpendicular to both $\boldsymbol{J}$ and $\boldsymbol{B}$, and it points outward and downward, whether $\boldsymbol{B}$ points up or down.

The downward component of the magnetic force adds to the gravitational force, increasing the effective value of g and depressing the surface. It anchors the flux tube, opposing the force of buoyancy. This explains the Wilson depression of sunspots, as we shall see in Sect. 13.9.

What happens, eventually, to the fluid that downflows? At deeper and deeper levels, $\boldsymbol{B}$ and $\boldsymbol{v}$ become more and more vertical, B_ρ tends to zero, the downward component of $\widetilde{\boldsymbol{F}}$ decreases and, eventually, the plasma returns to the surface.

11.3.2 The Gas Pressure Inside MFT's

The radial component of the magnetic force points outward because J_ϕ and B_z are both positive, as in Fig. 11.2. This is in agreement with the rule that parallel magnetic field lines repel laterally. Thus the magnetic force points outward and brakes the inward motion of the fluid. This applies, whether B_z points up or down.

Positive and negative particles circle around the tube axis in opposite directions to give the azimuthal current, and the magnetic force density points radially outward, whatever the sign of B_z.

A magnetic flux tube must therefore expand radially and disappear, unless there is an opposing force. Now, since the mean-free-path of the circling charged particles is finite, the magnetic force on the particles also pulls the ambient gas particles outward. Thus, under static conditions, the flux tube is partly evacuated: the gas pressure is lower inside the tube than outside. If p is the gas pressure, $\boldsymbol{\nabla} p$ points outward. So the pressure gradient opposes the magnetic force:

$$\boldsymbol{J} \times \boldsymbol{B} = \boldsymbol{\nabla} p \, . \tag{11.31}$$

Since the parenthesis in Eq. 11.21 is negative, like v_z, then J_ϕ has the sign of B_z. Also,

$$\boldsymbol{\nabla} p = \frac{\partial p}{\partial \rho} \hat{\boldsymbol{\rho}} = J_\phi B_z \hat{\boldsymbol{\rho}} \, . \tag{11.32}$$

Substituting the value of $\boldsymbol{J}_\phi$ from Eq. 11.8,

$$\boldsymbol{\nabla} p = -\frac{1}{\mu_0} \left(\frac{\partial B_z}{\partial \rho} - \frac{\partial B_\rho}{\partial z} \right) B_z \hat{\boldsymbol{\rho}} \, . \tag{11.33}$$

Let us limit ourselves to lower levels on the magnetic flux tube, where both B_ρ and $\partial / \partial z$ are nearly zero. Then

$$\boldsymbol{\nabla} p \approx -\frac{1}{\mu_0} \frac{\partial B_z}{\partial \rho} B_z \hat{\boldsymbol{\rho}} = -\frac{\partial}{\partial \rho} \frac{B_z^2}{2\mu_0} \hat{\boldsymbol{\rho}} \tag{11.34}$$

and

$$\frac{\partial}{\partial \rho} \left(\frac{B_z^2}{2\mu_0} + p \right) = 0 \, . \tag{11.35}$$

The sum of the magnetic pressure, $B_z^2/(2\mu_0)$ (Sect. 4.9), and of the gas pressure p is independent of ρ, within our approximations.

Observe that a magnetic flux tube cannot exist in a vacuum, except inside a solenoid. Even then there is an outward magnetic force on the wire, and coils have been known to explode at high magnetic fields.

With increasing ρ, B_z decreases, as inside the winding of a thick solenoid, while p increases until, outside the tube, $B_z = 0$ and $p = p_{\text{outside}}$. Thus, at any radial position ρ inside the flux tube,

$$\frac{B_z^2}{2\mu_0} + p = p_{\text{outside}} \, . \tag{11.36}$$

Near the top and bottom ends of the flux tube, the magnetic field lines fan out, as at the ends of a solenoid. From Eq. 11.30, the $\hat{\boldsymbol{\rho}}$-component of $\widetilde{\boldsymbol{F}}$ is positive, while the $\hat{\boldsymbol{z}}$-component is negative. So the magnetic force tends to expand the flux tube radially and compress it axially.

Since the gas pressure cannot be negative, at ρ,

$$\frac{B_z^2}{2\mu_0} \leq p_{\text{outside}} \, . \tag{11.37}$$

For example, at a depth of 36 megameters in the Sun, the gas pressure is 3×10^{10} pascals (Bahcall and Pinsonneau, 1992) and, inside the flux tube of a sunspot, at that depth,

$$B \leq 300 \text{ teslas} . \tag{11.38}$$

Whatever be the radial dependence of the azimuthal current density J_ϕ, the axial component of the magnetic field is maximum on the axis, so the gas pressure is minimum there: B is a decreasing function and p an increasing function of the radial position ρ.

The azimuthal current heats the gas and reduces its density, for a given pressure.

In the Sun, the ambient pressure p_{ext} increases with depth, while the tube radius decreases. The flux tube is thus conical, with the smaller radius below, where the pressure is higher. The radial component v_ρ of the plasma velocity remains negative all along the tube.

11.3.3 Φ, B_z, and J_ϕ

Assuming that, in a tube of radius b, except near the ends, the axial magnetic flux

$$\Phi = \int_0^b B_z 2\pi\rho \, d\rho \tag{11.39}$$

is independent of depth, then $d\Phi/dz = 0$ and B_z increases with depth, since b decreases with depth.

At the radius ρ,

$$B_z = \mu_0 \int_\rho^b J_\phi \, d\rho' , \tag{11.40}$$

as in a thick solenoid, and

$$\Phi = 2\pi\mu_0 \int_0^b \int_\rho^b J_\phi \, d\rho' \, \rho \, d\rho . \tag{11.41}$$

Assume, for simplicity, that J_ϕ is independent of ρ. Then

$$B_z = \mu_0 J_\phi (b - \rho) \tag{11.42}$$

is maximum on the axis and zero at the periphery and beyond, and

$$\Phi = \frac{2\pi\mu_0 J_\phi b^3}{6} , \tag{11.43}$$

or

$$J_\phi = \frac{3\Phi}{\pi\mu_0 b^3} \tag{11.44}$$

and, from Eq. 11.42 at $\rho = 0$,

$$B_{\text{axis}} = \frac{3\Phi}{\pi b^2} , \tag{11.45}$$

or three times larger, for a given Φ, than if B_z were uniform.

11.3.4 The Magnetic Energy in MFT's

From Eq. 4.29, the magnetic energy per meter for a magnetic flux tube of outer radius b is

$$\widetilde{\mathcal{E}} = \int_0^b \frac{B_z^2 + B_\rho^2}{2\mu_0} 2\pi\rho \, d\rho \, . \tag{11.46}$$

Of course, real magnetic flux tubes in conducting fluids probably never have well-defined outer radii, their cross-sections are never perfectly circular, and their "radius" is a function of the axial coordinate z, with $B_\rho \neq 0$, as above.

11.3.5 Power Dissipation in MFT's

The local current heats the gas at a rate J^2/σ watts/meter^3.

Since, by hypothesis, $J_z = 0$, and $J_\rho = 0$, the power dissipation per meter of length is

$$\widetilde{P} = \int_0^b \frac{J_\phi^2}{\sigma} 2\pi\rho \, d\rho \, . \tag{11.47}$$

The conductivity σ is minimum on the axis, where the magnetic field is largest.

This power comes from the convection energy of the fluid: part of the ordered kinetic energy of the fluid becomes thermal energy.

11.3.6 The Resistance per Meter in MFT's

Now assume that the tube radius is independent of z. Then

$$B_\rho = 0 \, , \quad \frac{\partial B_z}{\partial z} = 0 \, , \quad J_\rho = 0 \, , \tag{11.48}$$

from Eq. 11.7, and since V and B_ϕ are both zero.

The azimuthal current per meter of length is

$$\widetilde{I}_\phi = \int_0^b J_\phi \, d\rho \, . \tag{11.49}$$

We can define a resistance per meter $\widetilde{R}$ for a tube of uniform radius b, independent of z, by setting

$$\widetilde{P} = \widetilde{I}_\phi^2 \widetilde{R} \, , \quad \text{or} \quad \widetilde{R} = \frac{\widetilde{P}}{\widetilde{I}_\phi^2} \, . \tag{11.50}$$

As a rough approximation, assume that both J_ϕ and σ are independent of z out to a radius b, and that J_ϕ is zero outside. This is the situation inside a thick wire-wound solenoid. This approximation is not as poor as it seems, because we have set, above, $B_\rho = 0$. Then, from Eq. 11.20, our approximation makes $v_\rho B_z$ independent of z, and B_z is maximum on the axis, where v_ρ is zero. However, at the periphery, B_z tends to zero, and v_ρ does not tend to infinity. The magnetic force points outward and brakes the inward motion of the fluid.

With this approximation, the azimuthal current per meter is independent of z:

$$\tilde{I}_\phi = J_\phi b \tag{11.51}$$

and

$$B_z = \mu_0 \int_\rho^b J_\phi \, d\rho = \mu_0 J_\phi (b - \rho) \, . \tag{11.52}$$

The dissipated power per meter and the resistance per meter are then

$$\widetilde{P} = \pi b^2 \frac{J_\phi^2}{\sigma} = \pi \frac{\tilde{I}_\phi^2}{\sigma} \, , \quad \widetilde{R} = \frac{\pi}{\sigma} \, . \tag{11.53}$$

11.3.7 The Inductance per Meter in MFT's

We again assume a tube whose radius is independent of z.

Magnetic flux tubes are inductive, like solenoids. We can calculate the inductance per meter of length $\widetilde{L}$ from the relation

$$\widetilde{\mathcal{E}} = \frac{1}{2} \widetilde{L} \tilde{I}^2 \, , \tag{11.54}$$

where $\widetilde{\mathcal{E}}$ is the electric energy per meter. Thus

$$\widetilde{L} = \frac{2\widetilde{\mathcal{E}}}{\tilde{I}^2} \, . \tag{11.55}$$

With the above approximation, and with $B_\rho = 0$,

$$\widetilde{\mathcal{E}} = \frac{\pi \mu_0 J_\phi^2 b^4}{12} \, , \quad \widetilde{L} = \frac{\pi \mu_0 b^2}{6} \, . \tag{11.56}$$

11.3.8 The Time Constant of an MFT

We now return to the time derivative of the vector potential $\boldsymbol{A}$ in Eq. 6.34, that we have neglected until now. This time derivative is an electric field that tends to oppose changes in $\boldsymbol{B}$, and thus changes in $\boldsymbol{J}$, according to Lenz's law, as in Sect. 5.2.1.

In a uniform flux tube, both $\boldsymbol{J}$ and $\boldsymbol{A}$ are independent of z and are azimuthal, as in a solenoid. At the radius ρ,

$$A = \frac{1}{2\pi\rho} \int_0^{\rho} B_z 2\pi\rho \, d\rho \, \hat{\phi} \,. \tag{11.57}$$

This relation follows from the general rule that the line integral of $\boldsymbol{A}$ around a closed curve is equal to the enclosed magnetic flux (Sect. 4.5). If J_ϕ decreases, then the $-\partial \boldsymbol{A}/\partial t$ field points in the direction of $\boldsymbol{J}$, and thus opposes the decrease.

Imagine now that the flow of plasma starts abruptly, in a weak seed field of outside origin. Assume that the radius of the flux tube does not change with time. Then both J and B increase with a time constant

$$T = \frac{\widetilde{L}}{\widetilde{R}} \,, \tag{11.58}$$

like the current in an inductive circuit of inductance $\widetilde{L}$ and resistance $\widetilde{R}$. The final, or steady-state, value is a function of the mechanical power input (Chapter 10). Inversely, if the flow stopped abruptly, J and B would both decrease by a factor of e in a time constant T, and the magnetic energy by a factor of e^2.

With the above approximations,

$$T = \frac{\widetilde{L}}{\widetilde{R}} = \frac{\mu_0 \sigma b^2}{6} \,. \tag{11.59}$$

Since this time constant is proportional to the square of the tube radius b, it can be long for large-scale phenomena such as those in the Sun or in Space.

11.3.9 Fluctuating MFT's

Of course, our assumption that $\partial/\partial t = 0$ in a magnetic flux tube is a gross approximation. As we shall see, flux tube fluctuations have important effects. See Fig. 11.3.

Chromospheric Heating

The solar chromosphere lies above the photosphere. It owes its name to the fact that, during a total eclipse, with the photosphere well hidden, the Sun is surrounded by a pink ring, due to H_α radiation, which is predominant. Spicules (Chapter 14) penetrate the chromosphere.

The temperature of the photosphere of the Sun is a few thousand degrees, while the temperature of the atmosphere is a few million degrees. How can a body at a few thousand degrees heat another one to millions of degrees? This is a major unsolved problem of solar Physics. Astrophysicists have long discussed this problem. See, for example, Ionson (1978, 1982, 1983) related to coronal loops, a paper by Holweg (1987) related to MHD surface waves,

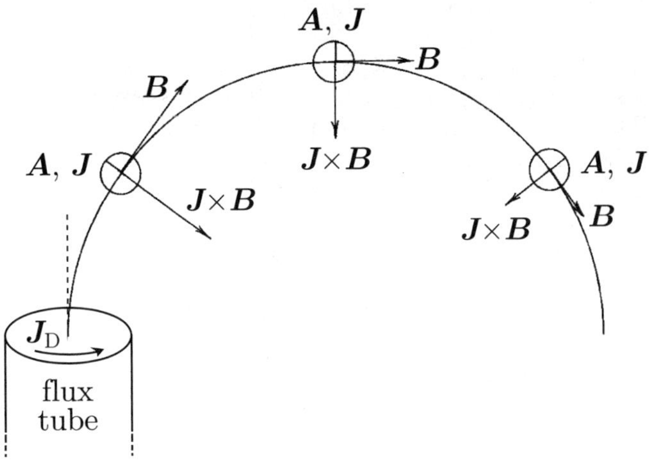

Fig. 11.3. Imagine a vertical magnetic flux tube in the plane of the paper at the lower left-hand corner of the figure. Its current density $\boldsymbol{J}_{\mathrm{D}}$ is in the direction shown. A magnetic field line emerges from the top of the flux tube. Of course, a field line of a magnetic flux tube is not an arc of a circle, but the arc shown is sufficient for our purposes. The field is symmetrical about the axis of the tube. We assume that $\boldsymbol{B}$ points up in the tube. The vector $\boldsymbol{A}$ points in the direction shown for the following reason. According to Eq. 4.13, the line integral of $\boldsymbol{A}$ over a closed curve is equal to the linking magnetic flux. Imagine a circle, perpendicular to the paper, and centered on the axis of symmetry of the tube. By hypothesis, $\boldsymbol{B}$ points up, and the flux through the circle is positive. Then $\boldsymbol{A}$ points in the $+\phi$-direction. The current density $\boldsymbol{J}$ is one that appears when $\boldsymbol{B}$ is not constant

and also by Davila (1987) related to Alfvén waves. See also Goodman (1995, 1996). Here is what seems to be a partial explanation.

See Fig. 11.3. Assuming that the driving current $\boldsymbol{J}_{\mathrm{D}}$ in the flux tube stops suddenly, which is equivalent to assuming that the photospheric plasma freezes at $t = 0$, the magnitude of $\boldsymbol{A}$ decreases exponentially with time, but its orientation does not change. Cutting off $\boldsymbol{J}_{\mathrm{D}}$ has no immediate effect on $\boldsymbol{A}$ or on $\boldsymbol{B}$ but, from then on, their magnitudes decrease exponentially, while their orientations also do not change. After $\boldsymbol{J}_{\mathrm{D}}$ is cut off, the vectors $\boldsymbol{E} = -\partial \boldsymbol{A}/\partial t$ and $\boldsymbol{J} = \sigma \boldsymbol{E}$ appear, and point in the same direction as $\boldsymbol{A}$. The power dissipation per cubic meter is

$$W = EJ = \sigma E^2 = \sigma \left| \frac{\partial \boldsymbol{A}}{\partial t} \right|^2 . \qquad (11.60)$$

So a fluctuating flux tube, in the photosphere, heats the atmosphere above. However, this phenomenon is probably not sufficient to explain the large temperature difference observed.

Photospheric Seismic Oscillations

Refer again to Fig. 11.3. Observe how the magnetic force, of density $\boldsymbol{J} \times \boldsymbol{B}$, compresses the gas, and thus exerts a vertical pressure on the photosphere. This might be a partial explanation for the vertical seismic oscillations observed in the photosphere.

See the papers by Goodman (1995, 1996) discussing the effects of fluctuating magnetic elements on the photosphere. He supplies numerical values: $J_{\max} = 1.33 \times 10^{-3}$ ampere/meter^2 at a radial position of 0.7 megameter from the center of a magnetic element (Chapter 12), where $B = 1.8 \times 10^{-2}$ tesla and the hydrogen atom density is 7×10^{19}/meter^3. The magnetic force density is 2.4×10^{-5} newton/meter^3, while the gravitational force density is 3.2×10^{-5} newton/meter^3.

Horizontal Magnetic Forces

Refer once more to Fig. 11.3. Assume that $\boldsymbol{B}$ decreases in the flux tube. Then $\partial \boldsymbol{A}/\partial t$ points out of the paper, opposite to $\boldsymbol{A}$. Then both $-\partial \boldsymbol{A}/\partial t$ and $\boldsymbol{J} = (-\partial \boldsymbol{A}/\partial t)/\sigma$ point into the paper, and the magnetic force density $\boldsymbol{J} \times \boldsymbol{B}$ near the magnetic flux tube, and near the surface, points away from the tube.

So a decreasing magnetic field causes the field lines to move outward, and pushes the plasma outward.

If now the tube magnetic field points down and decreases, the vectors $\boldsymbol{B}$, $\boldsymbol{A}$, and $\boldsymbol{J}$ all change sign, and the magnetic force density again points outward. So, for either polarity of the flux tube, a decreasing $\boldsymbol{B}$ pushes the plasma outward.

Conversely, if $\boldsymbol{B}$ increases, whatever its polarity, the magnetic force points inward.

The magnetic force has both ρ and z-components. With $\boldsymbol{B}$ constant, there is no induced current, and no magnetic force.

Of course, there is a horizontal force at the surface due to upwelling or downwelling and, in a specific case, the plasma does not necessarily move in the direction of the magnetic force. The magnetic and convective forces point in opposite directions, the magnetic force braking the convective force.

11.4 Magnetic Flux Ropes (MFR's)

Figures 11.4 and 11.5 show the $\boldsymbol{B}$ and $\boldsymbol{J}$ vectors in two magnetic flux ropes.

The axial magnetic field of a flux rope requires an azimuthal electric current, of density J_ϕ, that is generated as in a flux tube, by an inward radial plasma velocity v_ρ. The magnetic force associated with this field again points outward. The mechanism that generates this axial magnetic field is a self-excited dynamo as above.

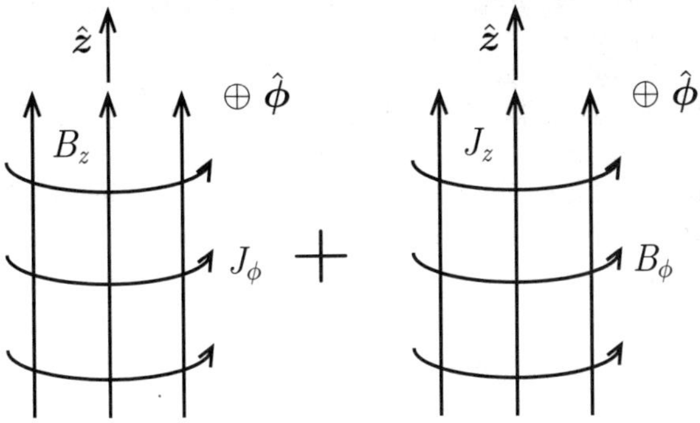

Fig. 11.4. A flux rope has both a B_z and a B_ϕ. Here, both B_z and B_ϕ are positive. The net $\boldsymbol{B}$ is helical: the magnetic field lines are twisted like the thread of a right-hand screw. The net $\boldsymbol{J}$, not shown, is also helical and twisted in the same direction

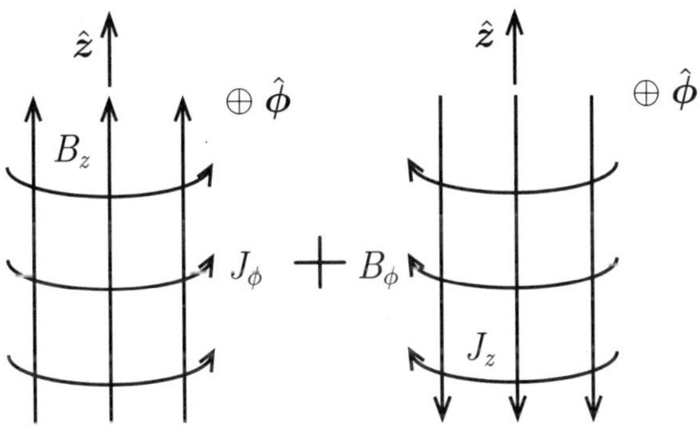

Fig. 11.5. A flux rope with B_z positive, while B_ϕ is negative. The net $\boldsymbol{B}$ is twisted like the thread of a left-hand screw. The net $\boldsymbol{J}$ is also helical and twisted in the same direction. See Fig. 11.1

The azimuthal component of the magnetic field of a flux rope can only result from an axial current. The axial current can have either polarity. If J_z points in the general direction of $\boldsymbol{B}$, then the field lines of the net magnetic field have the shape of a right-hand screw.

So a magnetic flux rope is a magnetic flux tube, plus an axial current of density J_z and its associated azimuthal magnetic field B_ϕ. It is thus the superposition of two structures, a flux tube and what one might call a "flux coil", in which the current is axial and the magnetic field azimuthal. The two components of $\boldsymbol{B}$ are independent of each other. Flux coils are interesting in themselves because they are charged particle accelerators, as in Chapter 14. See Sect. 11.6 below.

The magnetic force associated with the axial magnetic field in a flux rope points radially outward, as in a flux tube. But the radial magnetic force associated with the azimuthal magnetic field points inward. So, in a flux rope, the magnetic force density $\boldsymbol{J} \times \boldsymbol{B}$ is radial and has both positive and negative components. If the net magnetic force points inward, then the gas pressure is higher inside the rope than outside.

11.4.1 The Axial Current in an MFR

The axial magnetic field, if it is predominant, channels low-energy charged particles moving in the axial direction, even if the rope is not straight: a particle of mass m and charge Q would follow helical trajectories of radius $mv_\perp/(BQ)$ around magnetic field lines. Electrons are better channeled than ions because of their much smaller mass. The velocity component perpendicular to $\boldsymbol{B}$ should not be large for efficient channeling.

A particle of charge Q in the axial beam gains energy at the rate

$$\frac{d\mathcal{E}}{dz} = Q|\boldsymbol{v} \times \boldsymbol{B}| = Qv_\rho B_\phi \ , \tag{11.61}$$

whatever the polarity of the beam particles.

The axial beam is thus unusual. It is the superposition of two particle beams, a beam of positive particles moving in the direction of $\boldsymbol{v} \times \boldsymbol{B}$, and a beam of negative particles moving in the opposite direction. Both the particle energy and the current density are functions of ρ.

An electron or ion beam is subjected to both focusing and defocusing forces. The focusing force is exerted by the magnetic field: parallel currents attract. The defocusing force results from the mutual electrostatic repulsion of the particles. The *net* defocusing force is equal to the electrostatic repulsive force, multiplied by the factor $1 - (v^2/c^2)$, where v is the axial speed of the beam particles and c is the speed of light. At non-relativistic speeds, and in a vacuum, that factor is approximately equal to unity, and only the defocusing electrostatic force matters.

Imagine a beam of ions propagating in a low-pressure gas; the same argument applies to a beam of electrons. The ions ionize the ambient gas, and thus form further ions and electrons. Now the average energy of these new particles is only of the order of a few electron-volts and the new ions drift away from the beam by electrostatic repulsion, while the electrons stay trapped inside. The electrostatic charge of the ion beam is thus canceled, and we are left only with the magnetic focusing force. This phenomenon is called *self-pinching*.

Self-pinching has its limits because of scattering: every time a beam particle ionizes a gas particle, it loses energy and acquires a small transverse momentum. Eventually, the beam disappears. For a given energy, electrons ionize much less than ions.

Since electrons are much better channeled by the axial magnetic field and lose less energy by ionization, magnetic flux ropes seem to be more likely to carry electron beams, rather than ion beams.

Since there are electric currents in both magnetic flux tubes and flux ropes, both structures are dissipative. For the same reason, they are inductive. So each structure has its own time constant.

In a flux rope, we set

$$\frac{\partial}{\partial t} = 0 \,, \quad \frac{\partial}{\partial \phi} = 0 \,, \quad v_\phi = 0 \,, \tag{11.62}$$

as for flux tubes, but we also set

$$B_\rho \neq 0 \,, \quad B_\phi \neq 0 \,, \quad B_z \neq 0 \,. \tag{11.63}$$

Then, from Eqs. 11.7 to 11.9,

$$J_\rho = \sigma \left(-\frac{\partial V}{\partial \rho} - v_z B_\phi \right) \,, \tag{11.64}$$

$$J_\phi = \sigma (v_z B_\rho - v_\rho B_z) \,, \tag{11.65}$$

$$J_z = \sigma \left(-\frac{\partial V}{\partial z} + v_\rho B_\phi \right) \,. \tag{11.66}$$

Equation 11.65 is the same as Eq. 11.21, and there is positive feedback for J_ϕ if the parenthesis has the same sign as J_ϕ, which has the same sign as B_z, as in Fig. 11.5. If the rope radius is independent of z, then $B_\rho = 0$, and v_ρ must be negative for positive feedback, as for a flux tube.

The flux tube part of a flux rope is a self-excited dynamo. The injection of the axial particle beam presumably occurs in a flux coil, as in Sect. 11.6 and in Chapter 14.

Equation 11.15 applies, the electric space charge density $\widetilde{Q}_{\rm f} \neq 0$ and the electric potential $V \neq 0$.

In Eq. 11.66, $v_\rho B_\phi$ is the z-component of $\boldsymbol{v} \times \boldsymbol{B}$. That component sweeps positive charges either up or down, depending on its sign, and negative charges in the opposite direction. The term $-\partial V/\partial z$ is the resulting electric field, which partly opposes the vertical component of $\boldsymbol{v} \times \boldsymbol{B}$.

Say J_z and B_ϕ are both positive as in Fig. 11.4, above the flux coil. As in a magnetic flux tube, v_ρ is negative. Then the term $v_\rho B_\phi$ is negative and opposes J_z. Thus the negative radial velocity that provides positive feedback for the axial magnetic field provides negative feedback for J_z and B_ϕ in a flux rope.

A positive J_z can be generated locally if

$$\left| \frac{\partial V}{\partial z} \right| < |v_\rho B_\phi| \,. \tag{11.67}$$

Inside the rope, at the radius ρ, B_ϕ depends on the axial current inside that radius:

$$B_\phi = \frac{\mu_0}{2\pi\rho} \int_0^\rho 2\pi\rho J_z \, d\rho \,. \tag{11.68}$$

If J_z is independent of ρ, then B_ϕ decreases linearly with ρ, as in a current-carrying wire at low frequency. If J_z has the same sign as B_z, the magnetic field lines have the shape of a right-hand screw.

Outside the rope, the magnetic field lines are mostly azimuthal, because of the axial current, with a slight axial component from the return flux of the axial field.

If the axial electric current flows in the same direction as B_z, then the magnetic field lines of the net $\boldsymbol{B}$ have the shape of a right-hand screw, inside the rope. If J_z were uniform, then B_ϕ inside the rope would increase linearly with ρ, as in a current-carrying wire at low frequency.

11.4.2 Magnetic Forces on MFR's

In a magnetic flux rope of uniform radius, the magnetic force density is purely radial:

$$\widetilde{\boldsymbol{F}} = \boldsymbol{J} \times \boldsymbol{B} = (J_\phi \hat{\boldsymbol{\phi}} + J_z \hat{\boldsymbol{z}}) \times (B_\phi \hat{\boldsymbol{\phi}} + B_z \hat{\boldsymbol{z}}) \tag{11.69}$$

$$= (J_\phi B_z - J_z B_\phi) \hat{\boldsymbol{\rho}} \,. \tag{11.70}$$

The first term on the right of this latter equation points outward, J_ϕ and B_z having the same sign, as in a flux tube. The second term points inward because J_z and B_ϕ have the same sign as in Figs. 11.4 and 11.5: parallel currents attract. This force pinches the axial current.

The net outward force on a magnetic flux rope can thus be either positive or negative, and it need not have the same sign for all values of ρ and of z. Thus the gas pressure inside a rope can be either higher or lower than ambient.

Of course the vector products $\boldsymbol{J}_z \times \boldsymbol{B}_z$ and $\boldsymbol{J}_\phi \times \boldsymbol{B}_\phi$, are both identically equal to zero.

We can rewrite Eq. 11.70 in terms of $\boldsymbol{B}$ alone by substituting the current densities of Eqs. 11.8 and 11.9:

$$\widetilde{\boldsymbol{F}} = -\left[\frac{\partial}{\partial\rho} \left(\frac{B_\phi^2 + B_z^2}{2\mu_0} \right) + \frac{B_\phi^2}{\mu_0 \rho} \right] \hat{\boldsymbol{\rho}} \,, \tag{11.71}$$

where the term between parentheses is the magnetic energy density (Sect. 4.10). This is the general expression for the magnetic force density in an axisymmetric magnetic field with ϕ and z-components, and with $B_\rho = 0$.

Often, in the literature, this magnetic force density is set equal to the negative gradient of the magnetic energy, without the second term in the bracket. This leads to a pinching force in a Z-pinch that is too low by a factor of 2 (Miller, 1982; Salingaros, 1992).

11.4.3 Power Dissipation and Time Constants in MFR's

The values of $\widetilde{P}$, $\widetilde{R}$, $\tilde{I}$, $\widetilde{\mathcal{E}}$, $\widetilde{L}$, and T that we found in Sect. 11.3 also apply to the flux tube part of a flux rope.

For the flux coil (Sect. 11.6) part, it is easy to calculate the resistance $\widetilde{R}$ and the power dissipation $\widetilde{P}$ per meter, but the azimuthal magnetic field does not lend itself to a simple analysis. The reason is that the magnetic field of a linear current decreases only as $1/\rho$ in the radial direction for an infinitely long current, which makes the magnetic energy, the inductance, and the time constant all infinite. Real flux ropes are of course not infinitely long and straight, and these quantities could be computed only if one knew the actual field configuration at the ends. So we cannot calculate here the time constant for the flux coil part of the field; we only know that it is not zero.

If both the inward radial flow of conducting plasma and the axial current stopped suddenly and simultaneously, then the pitch of the magnetic field lines would change with time, and both fields would eventually disappear.

11.5 Are Tubes and Ropes Light Guides?

Could it be that solar magnetic flux tubes guide light like gigantic optical fibers? There is evidence to that effect.

Pores seem to be more or less vertical magnetic flux tubes, and they are bright, possibly because the light that one sees originates deep in the photosphere, where the temperature is higher than at the surface. Magnetic elements are bright, possibly for the same reason.

What about sunspots (Chapter 13)? They are also, apparently, more or less magnetic flux tubes, but they are dark, not bright! Now there do not exist isolated sunspots: a given sunspot is always accompanied by one or more spots of the *opposite polarity*. Say sunspot S has a companion S' of the opposite polarity. The two spots presumably share a common magnetic flux tube, below the photosphere. Often, a spot S is accompanied, not by one, but by several satellite spots, all of the polarity opposite that of S. Then the flux tube of S is apparently split into several smaller tubes, again below the photosphere. If the above is correct, and if flux tubes are really light guides, then, looking *down* S is in fact looking *up* S', at the atmosphere above the photosphere, which is much less bright.

This section is special in that, to date, we have not been able to show that flux tubes are light guides. Guidance requires that the phase speed of a light wave be an increasing function of the radius, and the calculations below leads to the conclusion that it is independent of the radius. Further work is required.

If you have not read the previous sections of this chapter, you would be well advised to read Sect. 11.3 before continuing; you need not read the subsections.

11.5.1 Electromagnetic Waves

An electromagnetic wave has two components, an electric field and a magnetic field. The electric field is $\boldsymbol{E}$, as usual, but, when dealing with electromagnetic waves, there are good reasons to use $\boldsymbol{H}$ for the magnetic field, instead of $\boldsymbol{B}$, with $\boldsymbol{H} = \boldsymbol{B}/\mu_0$ in a non-magnetic medium. In a plane wave, the $\boldsymbol{E}$ and $\boldsymbol{H}$ vectors are mutually orthogonal, and perpendicular to the direction of propagation. The vector $\boldsymbol{E} \times \boldsymbol{H}$ points in the direction of propagation and gives the transmitted power per square meter. In a gas, the relative permittivity $\epsilon_r = 1$.

The Maxwell equations 2.1 to 2.4 are now

$$\boldsymbol{\nabla} \cdot \boldsymbol{E} = \frac{\widetilde{Q}}{\epsilon_0}\,, \tag{11.72}$$

$$\boldsymbol{\nabla} \cdot \boldsymbol{H} = 0\,, \tag{11.73}$$

$$\boldsymbol{\nabla} \times \boldsymbol{E} + \mu_0 \frac{\partial \boldsymbol{H}}{\partial t} = \boldsymbol{0}\,, \tag{11.74}$$

$$\boldsymbol{\nabla} \times \boldsymbol{H} = \boldsymbol{J}\,. \tag{11.75}$$

Electromagnetic waves can have any frequency, but we are only concerned here with waves in the visible region, with frequencies of the order of 6×10^{14} hertz.

11.5.2 Guided Electromagnetic Waves

Hollow metallic tubes for guiding electromagnetic waves, known simply as *waveguides*, have been in use for well over a half century. They have cross-sections of a few centimeters by a few centimeters. Propagation inside the tube occurs by reflection at the walls, and a ray follows a zig-zag path. Pairs of wires are also waveguides.

Some optical fibers guide captive light waves by total reflection at the boundary between the core and a cladding. The index of refraction of a medium is $n = c/v_{ph}$, where v_{ph} is the *phase* speed of a plane wave in the medium of propagation. The core has an index of refraction $n_1 = c/v_{ph1}$, while the cladding has a lower index of refraction $n_2 = c/v_{ph2}$, and hence a higher phase speed. In that type of optical fiber a light ray also follows a zig-zag path.

The phase speed is the speed of propagation of a crest, or of a trough. That speed is often larger than c. It is the *group* speed v_g, or the speed of propagation of a group, or of a signal, that cannot be larger than c. The difference between the two speeds is easy to observe in the case of ripples at the surface of a pond of water.

Graded-index optical fibers have an index of refraction that decreases gradually with ρ, while the phase speed increases. Then the reflection is gradual, as in a mirage, and a given light ray oscillates from one side of the fiber to the other, following a trajectory that resembles a sine wave.

Mono-mode optical fibers carry a captive electromagnetic wave, without reflection. They bear no resemblance with flux tubes and ropes.

11.5.3 Electromagnetic Waves in Plasmas

In the absence of a magnetic field, at low temperatures, and in the absence of collisions, the phase speed of an electromagnetic wave in a plasma is (P. Lorrain et al., 1988, pp. 542–546)

$$v_{\mathrm{ph}} = \frac{c}{(1 - \omega_{\mathrm{pl}}^2/\omega^2)^{1/2}}\,, \tag{11.76}$$

where

$$\omega_{\mathrm{pl}} = \left(\frac{\widetilde{N} e^2}{\epsilon_0 m_{\mathrm{e}}}\right)^{1/2} \tag{11.77}$$

is the *plasma circular frequency*, $\widetilde{N}$ is the electron density, e is the absolute value of the charge of the electron, and m_{e} is the mass of the electron. So

$$v_{\mathrm{ph}} \approx \frac{c}{(1 - 80.6\widetilde{N}/f^2)^{1/2}}\,. \tag{11.78}$$

For $80.6\widetilde{N} > f^2$, the denominator is imaginary and there is attenuation without propagation.

In plasmas, $v_{\mathrm{g}} v_{\mathrm{ph}} = c^2$.

11.5.4 Guiding Light in Magnetic Flux Tubes and Ropes

A cylindrical structure such as optical fiber can guide an electromagnetic wave on two conditions: first, the attenuation must be low and, second, the radial gradient of phase speed must be positive: the phase speed must increase with the radius ρ.

Since magnetic flux tubes are largely evacuated, as we saw in Sect. 11.3.2, maybe they can satisfy the low-attenuation requirement. Radiation transfer within the Sun and within stars is a major branch of astrophysics, because that is how the nuclear energy generated near the center reaches the surface. However, we are concerned here only with superficial effects, because the radial extent of a pore, of a solar magnetic element, or of a sunspot, is totally negligible compared to the solar radius.

The intensity of a wave is the power transferred in watts per square meter. If we consider a plane wave of a given wavelength, we may write that, over a distance dz, a wave of intensity I_0 is attenuated to an intensity I according to the equation

$$\frac{I}{I_0} = \exp(-\kappa\, dz) = \exp(-d\tau)\,, \tag{11.79}$$

where κ is the *absorption coefficient* and $d\tau$ is the *optical thickness* over the distance dz, with $\kappa = d\tau/dz$. Over an optical thickness equal to unity the intensity decreases by a factor of e. The absorption coefficient κ is a function of wavelength. In the case of the solar atmosphere, the reference wavelength is usually 500 nanometers, or 0.5 micrometer, which is in the visible spectrum. Tabulated values of τ are $\tau_{0.5}$.

At the level of the photosphere, the temperature is about 6000 kelvins, the pressure 1.2×10^4 pascals, and the density 3×10^{-4} kilogram/meter^3. At an altitude of 200 kilometers, the gas pressure is 2.61×10^3 pascals, and the electron density 3.65×10^{18}/meter^3.

At a depth of 36 megameters below the photosphere, the temperature is 2.8×10^5 kelvins, the pressure 3×10^{10} pascals, or about 3×10^5 times the pressure at the surface of the Earth, and the density about 1 kilogram/meter^3 (Bahcall and Pinsonneau, 1992), about the same as that of the Earth's atmosphere. The density is about the same, but the absorption is much lower because the Sun consists almost exclusively of atomic hydrogen; there are no molecules, not even hydrogen molecules, to excite.

If the plasma density inside a flux tube were the same as that of the solar atmosphere at an altitude of 200 kilometers, where $\kappa \approx 4.9 \times 10^{-7}$, then the intensity of 500-nanometer light would decrease by a factor of $e = 2.718$ over a distance of $1/\kappa$, or about 2 megameters, in the flux tube.

Sixty kilometers below the photosphere, the gas pressure is about 1.71×10^4 pascals, and the electron density 7.04×10^{20}/meter^3. For that electron density and at $\lambda = 500$ nanometers, the index of refraction is nearly the same as that of a vacuum: $n \approx 1 - 8 \times 10^{-8}$.

That would seem to indicate that the attenuation of light in solar magnetic flux tubes is negligible.

What about the positive radial gradient of the phase speed? The phase speed depends on the electron density $\widetilde{N}$, which is the same as the proton density, here.

We showed in Sect. 11.3.2 that the sum of the gas pressure p, plus the magnetic pressure $B^2/(2\mu_0)$, at any radius ρ within the flux tube, some distance below the surface, is a constant and equal to the external pressure p_{ext}. Since B is a decreasing function of ρ, then the gas pressure p is an increasing function of ρ.

We now wish to know how the phase speed v_{ph} varies with the radius inside the magnetic flux tube. According to Eq. 11.78, at a given optical frequency f, v_{ph} increases with increasing $\widetilde{N}$. We may assume that $\widetilde{N}$ increases with the pressure, and hence that it increases with ρ.

If the electron (or proton) density in the flux tube changes from 3.65×10^{18}/meter^3, as at an altitude of 200 kilometers, to 7.04×10^{20}/meter^3 at the periphery of the flux tube, as at a depth of 60 kilometers, then, from Eq. 11.78, the ratio of the phase speeds is

$$\frac{v_{\text{periphery}}}{v_{\text{axis}}} \approx \frac{c/(1 - 8 \times 10^{-8})}{c/(1 - 4 \times 10^{-10})} \approx 1 + 10^{-7} . \tag{11.80}$$

So, according to this calculation, the magnetic flux tube is not a light guide. However, light guidance is such an attractive hypothesis that it is worthwhile investigating it further. The following notes might be useful.

1. At 6000 kelvins, the thermal speed of an electron is

$$\left(\frac{3kT}{m_{\text{e}}}\right)^{1/2} \approx 5.2 \times 10^5 \text{ meters/second} , \tag{11.81}$$

which is about three orders of magnitude smaller than the phase speed. Then one may consider the plasma to be cold.

2. The cyclotron frequency is $eB_{\text{ext}}/m_{\text{e}}$. For $B_{\text{ext}} \approx 0.1$ tesla, this frequency is approximately 2×10^{10} radians/second, which is very much less than the circular frequency of the wave, $2\pi f \approx 3.8 \times 10^{15}$ radians/second. That would seem to indicate that the phase speed is independent of the tube's magnetic field. As we have seen, this field is maximum on the axis and zero at the periphery.

3. The mean free path of a hydrogen atom is

$$l = \frac{1}{\sigma_{\text{c}} N_{\text{H}}} , \tag{11.82}$$

where N_{H} is the number of hydrogen atoms per unit volume and σ_{c} is the collision cross-section:

$$\sigma_{\text{c}} = \pi(2r)^2 \approx \pi(10^{-10} \text{ meter})^2 = \pi \times 10^{-20} \text{ meter}^2 . \tag{11.83}$$

Here $N_{\text{H}} < 10^{21}$ per meter^3 (see above). Then

$$l > \frac{1}{\pi \times 10} \approx 0.03 \text{ meter} . \tag{11.84}$$

Now the collision frequency is equal to the thermal speed divided by the mean free path, which is greater than 0.03 meter. The frequency of collisions is smaller than 2×10^7 per second. Since this frequency is orders of magnitude smaller than the light's frequency, collisions do not affect its propagation.

4. We assume that the conductivity σ is a constant independent of $\boldsymbol{E}$ and $\boldsymbol{B}$. Then, for a given $\boldsymbol{v}$, the current density $\boldsymbol{J}$ is a linear function of these two fields. In fact, σ is a function of $\boldsymbol{E}$ and $\boldsymbol{B}$, and $\boldsymbol{J}$ is a non-linear function of these two fields. This may be worth exploring.

5. Other non-linear phenomena might affect the propagation of light in solar flux tubes. However, the ponderomotive force discussed in F.F. Chen (1984, pp. 305 ff.) seems to be irrelevant.

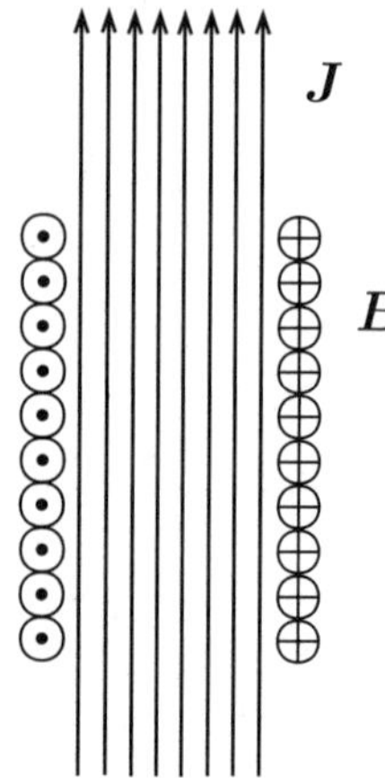

Fig. 11.6. Flux coil associated with a particle beam. The magnetic flux density in the coil is $\boldsymbol{B}$, and the current density in the beam is $\boldsymbol{J}$. The magnetic force density vector $\boldsymbol{J} \times \boldsymbol{B}$ points inward, in the direction of the axis of symmetry

11.6 Magnetic Flux Coils As Particle Accelerators

Astrophysicists have long pondered over the way a plasma can accelerate charged particles spontaneously. See, for example, Fichtel and McDonald (1967) and Tandberg-Hanssen (1974). Here we propose a plausible type of particle accelerator that exploits a $\boldsymbol{v} \times \boldsymbol{B}$ field in a convecting plasma. We return to it in Chapter 14.

Figure 11.6 shows a magnetic field configuration that might be called a magnetic flux *coil*. The magnetic field near a current-carrying wire is a flux coil.

As we shall see below, such a structure might exist in a convecting conducting fluid, and it might act as a particle accelerator. Of course, in a real situation, both $\boldsymbol{J}$ and $\boldsymbol{B}$ would be complex functions of the coordinates.

This hypothetical $\boldsymbol{v} \times \boldsymbol{B}$ accelerator could generate a charged particle beam. It starts spontaneously, as soon as a seed electric current appears. The seed current must originate in some outside source. The power source for the accelerator comes from the thermal forces that drive the convection in the photosphere. It is doubtful that such accelerators could exist in laboratory-sized plasmas.

According to this model, a region where the radial component of the plasma velocity v_ρ is positive, in cylindrical coordinates, can be the site of a self-excited particle accelerator that generates an axial beam of either sign, depending on the sign of the seed current. See Fig. 14.4. Given a seed current directed, say, upward in the figure, its magnetic field points in the positive direction of ϕ. A positive radial plasma velocity then provides a $\boldsymbol{v} \times \boldsymbol{B}$ field whose axial component also points upward, like the seed field, and that adds a current of the same sign as the seed current.

If the beam current points instead in the negative direction of the z-axis, then its $\boldsymbol{B}$ and $\boldsymbol{v} \times \boldsymbol{B}$ vectors both change sign, and the induced current again points in the same direction as the seed current. The induced current, of density $\sigma(\boldsymbol{v} \times \boldsymbol{B})$, therefore amplifies any axial current. This is positive feedback, which is the essential property of self-excited dynamos. With time, the magnitude of the axial current does not tend to infinity, but rather reaches an asymptotic value that is a function of the power that drives the convection (Sect. 10.2.2).

We assume an arbitrary plasma velocity:

$$\boldsymbol{v} = v_\rho \hat{\boldsymbol{\rho}} + v_\phi \hat{\boldsymbol{\phi}} + v_z \hat{\boldsymbol{z}} \, . \tag{11.85}$$

The current density is given by

$$\boldsymbol{J} = \sigma[-\boldsymbol{\nabla} V + (\boldsymbol{v} \times \boldsymbol{B})] \, . \tag{11.86}$$

If we assume that $\boldsymbol{J}$ has only an axial component, then the radial component of $\boldsymbol{\nabla} V$ exactly cancels the radial component of $\boldsymbol{v} \times \boldsymbol{B}$ and we are left with the simple relation

$$J_z = \sigma v_\rho B_\phi \, . \tag{11.87}$$

Since the three variables on the right are functions of ρ, then J_z is also a function of ρ. The azimuthal and axial components of the particle velocity, v_ϕ and v_z, have no effect on the axial current.

Now J_z and B_ϕ have the same sign, and the DC plasma conductivity is positive. So Eq. 11.87 requires that v_ρ be positive. Thus flux coils might exist in convecting plasmas in regions where the velocity has a positive radial component, for example where there is upwelling. The flux coil is a self-excited dynamo (Chapter 10).

With time, J_z increases, σ increases because of the increasing B_ϕ, but v_ρ decreases because of the braking action of the magnetic force. Eventually, J_z presumably becomes stable as in the self-excited Faraday disk (Sect. 10.2.2).

The electrostatic space charge density is not zero because, in this case, $\boldsymbol{\nabla} \cdot (\boldsymbol{v} \times \boldsymbol{B}) \neq 0$ (Eq. 7.10).

The azimuthal magnetic field pinches positive particles moving upward in the beam. A negative particle moves downward and is similarly pinched.

A particle of charge Q in the axial beam gains an energy $d\mathcal{E}_{\text{particle}}$ at the rate

$$\frac{d\mathcal{E}_{\text{particle}}}{dz} = Q|\boldsymbol{v} \times \boldsymbol{B}| = Q v_\rho B_\phi = \frac{Q}{\sigma} J_z \, . \tag{11.88}$$

The particle energy is a function of the radial coordinate ρ, like the axial current density J_z and σ. The equation applies, whatever the polarity of the accelerated particles or of $\boldsymbol{J}$.

If J_z were uniform, then the energy gain would be proportional to $1/\sigma$, and hence would decrease with increasing ρ because B_ϕ would be proportional to ρ.

The axial beam is like that of Sect. 11.4.1: it is the superposition of two particle beams, a beam of positive particles moving along z in the direction of $\boldsymbol{v} \times \boldsymbol{B}$, and a beam of negative particles moving in the opposite direction. Both the current density and the particle energies are functions of ρ.

The flux coil is an inductive circuit, and it has a time constant. If the convection stopped suddenly, then, over one time constant, the current density and the magnetic field would both decrease by a factor of e.

The time constant T is given by

$$T = \frac{\widetilde{L}}{\widetilde{R}} , \quad \widetilde{L} = 2\frac{\widetilde{\mathcal{E}}_{\text{mag}}}{I^2} , \quad \widetilde{R} = \frac{\widetilde{P}}{I^2} , \tag{11.89}$$

where $\widetilde{L}$ and $\widetilde{R}$ are, respectively, the inductance and the resistance per meter of length, $\widetilde{\mathcal{E}}_{\text{mag}}$ is the magnetic energy stored per meter of length, $\widetilde{P}$ is the ohmic power dissipation per meter, and I is the axial current.

For a flux coil of radius b,

$$I = \int_0^b J_z 2\pi\rho \, d\rho . \tag{11.90}$$

If J_z were uniform, then we would have that $I = \pi\rho^2 J_z$. Also,

$$\widetilde{P} = \int_0^b \frac{J_z^2}{\sigma} 2\pi\rho \, d\rho , \tag{11.91}$$

where σ is a function of ρ.

To calculate the inductance per meter $\widetilde{L}$, we require $\widetilde{\mathcal{E}}_{\text{mag}}$, and

$$\widetilde{\mathcal{E}}_{\text{mag}} = \int_\infty \frac{B^2}{2\mu_0} dv = \frac{1}{2} I^2 \widetilde{L} , \tag{11.92}$$

where I is the current.

If $d/dz = 0$, then the flux coil is infinitely long, B_ϕ decreases only as $1/\rho$, and the stored magnetic energy is infinite.

In a flux coil, the axial current density is given by Eq. 11.87 and the magnetic force density is

$$\widetilde{\boldsymbol{F}} = \boldsymbol{J} \times \boldsymbol{B} = -J_z B_\phi \hat{\boldsymbol{\rho}} = -\sigma v_\rho B_\phi^2 \hat{\boldsymbol{\rho}} . \tag{11.93}$$

The magnetic force points inward and is counteracted by a pressure gradient. The pressure is higher inside than outside: the pressure gradient brakes the inward motion of the gas.

As usual in self-excited dynamos, the power required to establish the magnetic field and the ohmic power dissipation are both supplied by the forces that drive the convection.

We discussed above, in Sect. 11.4.1, how an axial current might be channeled in a flux rope. The same applies to the axial current in a flux coil.

11.7 Summary

A magnetic flux tube (MFT) is a bundle of magnetic field lines, as in a solenoid. We discuss here flux tubes that develop spontaneously as self-excited dynamos within convecting, conducting fluids, for example in the solar plasma. The medium must have a negative radial velocity, in cylindrical coordinates, typically in a region where there is downwelling. A flux tube carries a local azimuthal current, and is partly evacuated by the outward magnetic force, which points in the same direction as the pressure gradient. A flux tube possesses magnetic energy and is both inductive and dissipative. It thus has a time constant: if the medium freezed suddenly, the magnitude of the magnetic field would decrease by a factor of e in one time constant.

A magnetic flux rope (MFR) is a magnetic flux tube that carries an axial current. The magnetic field of the axial current is azimuthal, and the net field is helical. The magnetic force on a flux rope has two components, one that points outward, as in a flux tube, and one that points inward. The gas pressure gradient is equal to the net magnetic force density. Flux ropes are dissipative and inductive, like flux tubes.

Magnetic flux tubes and ropes possibly act as light guides. a) Magnetic elements are bright, compared to the surroundings, because one sees the solar plasma below the photosphere, where the temperature is higher. b) A sunspot of a given polarity is always accompanied by one or more spots of the opposite polarity. So, if sunspot flux tubes are light guides, then, when one looks down into a sunspot, one is actually looking up the twin spot, at the solar atmosphere, which is much less bright than the photosphere.

A magnetic flux coil has azimuthal magnetic field lines, associated with an axial current. The magnetic field near a current-carrying wire is a flux coil. We show that a region where the radial plasma velocity is positive can act as an accelerator that generates an axial beam of positive particles one way, and a beam of negative particles the other way. We apply this concept in Chapter 14. The magnetic force points inward and is counteracted by the pressure gradient, which points inward: the gas pressure is higher inside than outside.

12 Case Study: Solar Magnetic Elements

Contents

Magnetic elements are small bright spots on the solar photosphere. They exhibit magnetic fields, with *B* pointing either up or down, and they are apparently more or less vertical flux tubes. But how are the required electric currents generated? And why are they bright?[1]

12.1 Introduction

The present chapter, as well as Chapters 13 and 14, are extensions of Chapter 11.

We propose a self-excited dynamo model for magnetic elements: an element is a magnetic flux tube that stands more or less vertically and that results from the presence of azimuthal electric currents, as in Chapter 11.

[1] See P. Lorrain and S. Koutchmy, *Two dynamical models for solar spicules*, Solar Physics **165**, 115–137 (1996); and P. Lorrain and S. Koutchmy, *Chromospheric heating by electric currents induced by fluctuating magnetic elements*, Solar Physics **178**, 39–42 (1998).

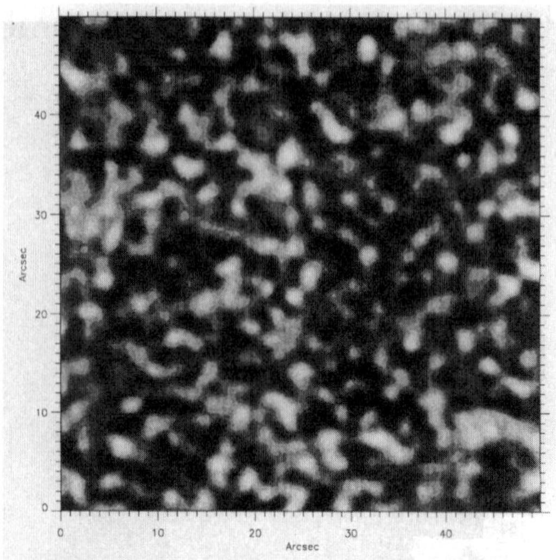

Fig. 12.1. Magnetic elements near disk center. The field of view covers 50×50 arcseconds^2, and one arcsecond corresponds to about 720 kilometers at the surface of the Sun. These observations were made at the focus of the Vacuum Tower Telescope at the National Solar Observatory at Sacramento Peak, New Mexico by Serge Koutchmy. The maximum measured B over this area is about 0.07 tesla

There is some repetition here, so as to make it possible to read the present chapter and Chapter 11 independently. Ideally, you should read the first three sections of Chapter 11 before proceeding.

For background information on the Sun, see the classic book *Astrophysics of the Sun* by Zirin (1988). See also "Reflections on macrophysics and the Sun" by Parker (1997).

The *photosphere* is the surface of the Sun that we see in white light. It is a plasma, composed largely of hydrogen, of mass density 3×10^{-4} kilogram/meter^3, at a temperature of about 6000 kelvins and a pressure of about 10^4 pascals. The free electron density is about 6×10^{19}/meter^3 (Vernazza et al., 1981).

Magnetic elements are small bright regions on the photosphere that exhibit vertical magnetic fields of roughly 0.15 tesla. They are about 150 kilometers in diameter. See Fig. 12.1. Magnetic elements come in pairs with opposite polarities. The axial magnetic field requires an azimuthal electric current, and we have estimated (P. Lorrain and S. Koutchmy, 1993) that the current density is about 1.6 amperes/meter^2 (see Sect. 12.3.2 below).

The inward radial velocity at the surface of the photosphere is of the order of 300 meters/second.

With the flow geometry of Fig. 11.2, the inward radial component of velocity, and thus the self-excited dynamo effect, eventually disappears with increasing depth and, at still greater depths, the radial velocity becomes positive and the plasma returns to the surface (Chapter 14). Thus the vertical extent of the current distribution, and its radius as a function of depth, depend on the flow pattern.

As a first approximation, we assume that the azimuthal electric current density $\boldsymbol{J}$ inside the element is uniform.

Dara et al. (1990) have reported that magnetic elements of opposite polarity attract, while elements of the same polarity repel. This is in agreement with our model: magnetic elements should attract and repel, like parallel and anti-parallel solenoids and bar magnets.

This model accounts for the existence of magnetic elements, for their stability, for the existence of neighboring fields of opposite polarity, and for their mutual attraction and repulsion.

We also propose a more realistic, but more elaborate, model in which the current distribution is not uniform (Sect. 12.3.9).

It is difficult to measure the net vertical flow inside a magnetic element, but there is little doubt that there exists a radial convergent flow pattern in the region surrounding an element.

Finally, the power dissipation associated with the azimuthal currents accounts for only a small part of the observed radiation (S. Koutchmy, 1991). In Sect. 11.5 we discussed the hypothesis that the element might act as a light guide. If this is true, then, when looking at the element, one would see the hot plasma well below the photosphere.

12.2 Local Currents in Magnetic Elements

As has been stressed repeatedly by Alfvén (1975, 1981), it is pointless to discuss the shapes of imaginary magnetic field lines, while disregarding the required electric currents. So, as usual, we focus on the *electric currents* that generate the observed magnetic field.

Our argument runs as follows.

a) There exist patches of vertical magnetic field in the Sun's photosphere that are presumably the top ends of magnetic flux tubes.

b) Then there exist local azimuthal electric currents of density $\boldsymbol{J}$, according to the Maxwell equation

$$\boldsymbol{\nabla} \times \boldsymbol{B} = \mu_0 \boldsymbol{J} , \tag{12.1}$$

with

$$\boldsymbol{J} = \sigma(\boldsymbol{E} + \boldsymbol{v} \times \boldsymbol{B}) = \sigma \left(-\boldsymbol{\nabla} V - \frac{\partial \boldsymbol{A}}{\partial t} + \boldsymbol{v} \times \boldsymbol{B} \right) . \tag{12.2}$$

In writing these equations we assume that the displacement current density $\partial \boldsymbol{D}/\partial t$ is negligible compared to the conduction current density $\boldsymbol{J}$, a condition that applies to low-frequency phenomena in the Sun's plasma (Sect. 2.4). The medium is non-magnetic.

The $-\boldsymbol{\nabla} V$ term concerns electrostatic charges that may reside at the surface, or within the magnetic element with a volume density (Eq. 7.10)

$$\widetilde{Q}_{\mathrm{f}} = -\epsilon_0 \boldsymbol{\nabla} \cdot (\boldsymbol{v} \times \boldsymbol{B}) \tag{12.3}$$

in a homogeneous medium.

c) We seek an $\boldsymbol{E} + \boldsymbol{v} \times \boldsymbol{B}$ field that will generate $\boldsymbol{J}$.

As we saw in Chapter 11, regions where the plasma has a negative radial velocity, in cylindrical coordinates, generate an appropriate $\boldsymbol{J}$. See Fig. 11.2. The gas density increases, and the inward radial velocity decreases with increasing depth. We assume an axisymmetric geometry, but that is not essential; all that our model requires is an inward radial flow. It seems impossible to imagine a flow pattern that would be qualitatively different from that of Fig. 11.2.

As we shall see in Sect. 13.6.1, the current density in sunspots appears to be uniform all the way from the axis of symmetry to the outer radius. Nonetheless, we propose a more general and more realistic model in Sect. 12.3.9, where J is not uniform.

Assuming that J is uniform, we find the stored magnetic energy, the dissipated power, the time constant, etc.

12.3 A Simple Model

Equations 2.4 and 12.6 are the only fundamental equations that we require for the simple model. As a second approximation, we propose a more elaborate model that requires more fundamental equations. Further approximations would require still more.

With the simple model, the azimuthal current density J is uniform out to the outer radius b. Current flows in the azimuthal direction, and thus in a direction perpendicular to $\boldsymbol{B}$, which makes the conductivity σ smaller than in the absence of a magnetic field (Cambel, 1963). We disregard oscillations and waves.

As to the radius b and the length l of the current distribution, we perform our *theoretical* calculations for $l \gg 2b$, for simplicity. Nonetheless, in our *numerical* calculations, we set $l = 2b$.

We disregard gravity until Sect. 12.3.8. Finally we set, in cylindrical coordinates,

$$\frac{\partial}{\partial t} = 0\,, \qquad \frac{\partial}{\partial z} = 0\,, \tag{12.4}$$

$$v_\phi = 0\,, \quad B_\rho = 0\,, \quad B_\phi = 0\,. \tag{12.5}$$

Our assumption of a uniform electric current density J makes some sense for the following reason. If Eqs. 12.4 and 12.5 apply, there is zero electrostatic

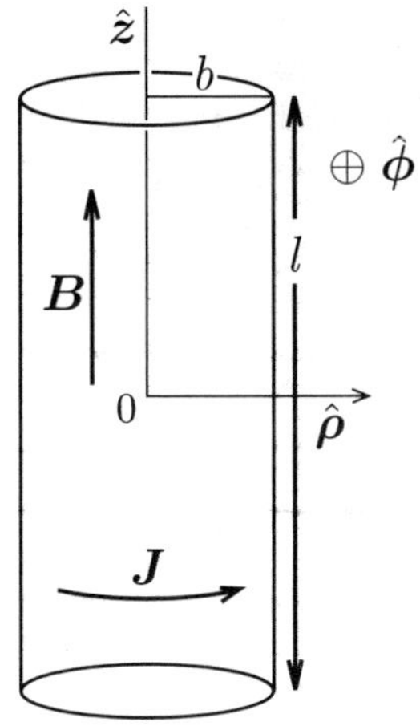

Fig. 12.2. The current distribution for the model. The magnetic field is axial, and the current distribution is azimuthal

space charge (Chapter 7), zero electrostatic surface charges, because $\boldsymbol{v} \times \boldsymbol{B}$ is purely azimuthal, $\boldsymbol{\nabla} V = \boldsymbol{0}$, and, with the sign conventions of Fig. 12.2,

$$\boldsymbol{J} = \sigma(\boldsymbol{v} \times \boldsymbol{B}) \ . \tag{12.6}$$

Then

$$J = -\sigma v_\rho B_z \ , \tag{12.7}$$

where σ is the plasma conductivity, v_ρ is the radial velocity of the plasma, and B_z is the axial magnetic flux density. Now a conductivity σ is positive by definition and, as we saw in Sect. 12.2, v_ρ is negative, so that, in Eq. 12.7, J and B_z have the same sign. With increasing ρ, B_z decreases, σ increases, and $|v_\rho|$ increases as in Fig. 11.2. It thus makes sense, as a first approximation, to set the product of the three quantities on the right of Eq. 12.7 equal to a constant.

As we shall see in Sect. 12.3.6, the time constant of a magnetic element is hardly affected by the particular current distribution chosen.

The calculations that follow are a first approximation: no magnetic element has a perfectly circular cross-section, the radius b of the current distribution is certainly a function of the depth, the current density J is not really uniform, and the conductivity σ is a complex function of position because it depends on the gas density and on B, both of which are functions of the radial position (Sect. 12.3.4) and of the depth.

12.3.1 The Magnetic Flux Density *B*

We can calculate the field of a magnetic element as follows. At the radius ρ inside an azimuthal current distribution of outer radius b, length l, and current density J, in a plane passing through the center,

$$B = \frac{1}{2}\int_\rho^b \frac{\mu_0 (J\,d\rho) l}{[\rho^2 + (l^2/4)]^{1/2}} \tag{12.8}$$

$$= \frac{\mu_0 J l}{2} \ln\left\{\frac{b + [b^2 + (l^2/4)]^{1/2}}{\rho + [\rho^2 + (l^2/4)]^{1/2}}\right\} \tag{12.9}$$

$$\approx \mu_0 J (b - \rho) \qquad (\text{if } l \gg 2b)\ . \tag{12.10}$$

This last expression also applies inside a long azimuthal current distribution of uniform current density J, at a distance of a few times b from the ends. At radii $\rho > b$, $B \approx 0$ if $l \gg 2b$. If this latter condition is not satisfied, then the field outside is weaker than inside, and it points in the opposite direction.

The magnetic flux inside a magnetic element whose uniform current density extends from the axis of symmetry to the radius b is

$$\Phi \approx \int_0^b \mu_0 J (b - \rho) 2\pi\rho\, d\rho = \pi\mu_0 J \frac{b^3}{3}\ , \tag{12.11}$$

which is three times less than if the field were uniform and equal to the field on the axis.

12.3.2 The Current Density *J* and the Conductivity σ

Set the radius b of the magnetic element equal to 75 kilometers, and B on the axis equal to 0.15 tesla. For simplicity, we use Eq. 12.10, despite the fact that we shall set $l = 2b$ below. That is a good enough approximation at this stage. Then

$$J \approx \frac{B}{\mu_0 b} \approx 1.6 \text{ amperes/meter}^2 \tag{12.12}$$

(P. Lorrain and S. Koutchmy, 1993). The free electrons then drift at a speed of 0.17 meter/second.

In our simple model, this current density applies throughout the volume of the magnetic element.

How does the above drift speed compare with the thermal speed of the electrons? The mass of an electron is 9.11×10^{-31} kilogram, the Boltzmann constant k is 1.38×10^{-23} joule/kelvin, and the temperature T is about 6000 kelvins. Then

$$\tfrac{1}{2} m v_{\rm th}^2 = \tfrac{3}{2} kT\ , \quad v_{\rm th} = 5.22 \times 10^5 \text{ meters/second}\ , \tag{12.13}$$

which is much larger than the above v_ρ.[2]

We can now deduce the conductivity σ in the direction of $\boldsymbol{J}$, or in the azimuthal direction, which is perpendicular to the magnetic field lines. In Eq. 12.7 we can set $v_\rho \approx 300$ meters/second (Simon et al., 1988; Alissandrakis et al., 1991), and $B \approx 0.075$ tesla, one half of a maximum value of 0.15 tesla. Then

$$vB \approx 22.5 \text{ volts/meter} , \quad \sigma \approx 0.07 \text{ siemens/meter} . \tag{12.14}$$

To calculate the conductivity in the absence of a magnetic field we proceed as follows. Observations are made at an altitude of about 300 kilometers. According to Vernazza et al. (1976), at that altitude, the ratio of the number density of free electrons to the number density of gas atoms is given by

$$\frac{n_\mathrm{e}}{n_\mathrm{g}} \approx 9.5 \times 10^{-5} , \quad \log \frac{n_\mathrm{e}}{n_\mathrm{g}} \approx -4.02 . \tag{12.15}$$

Then, according to Kopecký and Soytürk (1971),

$$\log \sigma \approx 1.26 , \quad \sigma \approx 18.2 \text{ siemens/meter} . \tag{12.16}$$

In selecting the above value of the ratio $n_\mathrm{e}/n_\mathrm{g}$ we have accorded more weight to the former reference.

The orthogonal conductivity of Eq. 12.16 is smaller than the normal conductivity by a factor of 260.

Note that the above orthogonal conductivity is *nine* orders of magnitude less than that of copper at room temperature, which is 5.8×10^7 siemens/meter.

The Reynolds number

$$R_\mathrm{m} \approx \mu_0 \sigma v_\mathrm{typical} \mathcal{L} \tag{12.17}$$

$$\approx (4\pi \times 10^{-7}) \times 0.07 \times 300 \times 75 \times 10^3 \approx 2 . \tag{12.18}$$

How does this compare with generally accepted values for the electric conductivity? In the *absence* of a magnetic field the electrical conductivity of the photosphere at a height of about 100 kilometers is roughly 2300 siemens/meter (Kopecký and Soytürk, 1971), and the orthogonal conductivity that we calculated above is four and a half orders of magnitude smaller, which is not a bad agreement (Cambel, 1963), in view of the fact that our calculation is only a first approximation.

Observe that a plasma conductivity of 2.3×10^3 siemens/meter in the absence of a magnetic field is *four and a half* orders of magnitude *smaller* than the conductivity of copper at room temperature, which is 5.8×10^7

[2] In copper at room temperature, the drift speed of the conduction electrons is typically about 10^{-4} meter/second, while their thermal speed is larger by *nine* orders of magnitude.

siemens/meter. The orthogonal conductivity of the plasma is even smaller by a few further orders of magnitude. It is therefore highly inappropriate to set the electrical conductivity of the photospheric plasma equal to infinity, as some authors do.

12.3.3 The Stored Magnetic Energy

The energy stored in the magnetic field over a length l is

$$\mathcal{E} = \int_0^b \frac{B^2}{2\mu_0} l 2\pi\rho \, d\rho \tag{12.19}$$

or, from Eq. 12.10,

$$\mathcal{E} = \mu_0 \pi J^2 b^4 l/12 = 4 \times 10^{18} \text{ joules} , \tag{12.20}$$

with $J = 1.6$ amperes/meter^2 as above, $l = 2b$, and $b = 7.5 \times 10^4$ meters. This is equivalent to the energy supplied by 130 one-gigawatt nuclear reactors operating for one year.

12.3.4 The Gas Pressure Inside the Element

We showed in Sect. 11.3.2 that, inside a flux tube, the gas pressure plus the magnetic pressure is equal to the external gas pressure. Here is a somewhat different proof.

The magnetic force tends to spread out the current distribution in the radial direction:

$$\widetilde{\boldsymbol{F}}_{\text{mag}} = \boldsymbol{J} \times \boldsymbol{B} = \mu_0 J^2 (b - \rho) \hat{\boldsymbol{\rho}} . \tag{12.21}$$

We now set the outward magnetic force density equal to the gradient of the gas pressure p:

$$\widetilde{\boldsymbol{F}}_{\text{mag}} = \boldsymbol{\nabla} p = \frac{dp}{d\rho} \hat{\boldsymbol{\rho}} . \tag{12.22}$$

The two vectors point outward.

We have disregarded the acceleration of the fluid. With this approximation, at the radius ρ,

$$dp = |\widetilde{\boldsymbol{F}}_{\text{mag}}| \, d\rho = \mu_0 J^2 (b - \rho) \, d\rho . \tag{12.23}$$

Integrating, and calling the gas pressure outside the element p_{out}, we have that, inside the element,

$$p_{\text{out}} - p = \mu_0 \frac{J^2 (b - \rho)^2}{2} = \frac{B^2}{2\mu_0} . \tag{12.24}$$

Substituting the above values of J and b,

$$p_{\text{axis}} = p_{\text{out}} - \mu_0 \frac{J^2 b^2}{2} = p_{\text{out}} - 9 \times 10^3 \text{ pascals} . \tag{12.25}$$

The outside gas pressure is at least 9×10^3 pascals, which corresponds to an altitude of about +50 kilometers or less, where the temperature is about 5500 kelvins or higher (Allen, 1973).

12.3.5 The Dissipated Power

The *dissipated* power P_{d} results from the ohmic losses of J^2/σ watts/meter^3. Setting $l = 2b$, from Eq. 12.12, with $J \approx 1.6$ amperes/meter^2,

$$P_{\text{d}} = (\pi b^2)(2b)\frac{J^2}{\sigma} \approx 10^{17} \text{ watts} . \tag{12.26}$$

This power is supplied by the forces that drive the convecting plasma against the magnetic forces. Generating that amount of power on Earth would require 10^8 one-gigawatt nuclear reactors for each magnetic element!

Since magnetic elements fluctuate, their magnetic fields in the chromosphere above also fluctuate (P. Lorrain and S. Koutchmy, 1998).

Let us now calculate an approximate value for the *radiated* power P_{r} from the observed temperature of a typical magnetic element. The power radiated by a black body maintained at a temperature T is equal to $\mathcal{S}T^4$ watts/meter^2, where $\mathcal{S} = 5.68 \times 10^{-8}$ is the Stefan-Boltzmann constant. Set the temperature T_{p} of the photosphere equal to 5500 kelvins, as above, and that of an element, T_{e}, 1000 degrees higher (S. Koutchmy and Stellmacher, 1978), or 6500 kelvins. The element radiates a power P_{rt} from the top surface and a power P_{rs} from the sides; we disregard the power radiated by the bottom surface because the photosphere is hotter there. Then

$$P_{\text{r}} = P_{\text{rt}} + P_{\text{rs}} \tag{12.27}$$

$$= \pi b^2 \mathcal{S} T_{\text{e}}^4 + (2\pi b)(2b)\, \mathcal{S}\, \frac{T_{\text{e}}^4 - T_{\text{p}}^4}{2} \tag{12.28}$$

$$= \pi b^2 \mathcal{S} \left(3T_{\text{e}}^4 - 2T_{\text{p}}^4\right) = 3.5 \times 10^{18} \text{ watts} . \tag{12.29}$$

The factor of $\frac{1}{2}$ in the calculation of P_{rs} is meant to account roughly for the fact that the temperature of the photosphere, T_{p}, increases with depth.

This observed radiated power P_{r} is 35 times larger than the dissipated power P_{d} that we calculated above. The discrepancy can be explained, but probably only in part, as follows.

a) The above calculation, and our assumption that the current distribution is cylindrical and uniform, are both crude.

b) The flux tube is partly evacuated, and thus somewhat transparent, so that one sees deeper into the photosphere, where the temperature is higher.

c) Because of the presence of the axial magnetic field, hot electrons cannot easily cross the side walls, but they enter the flux tube freely through the lower end, which is hotter.

d) Observations suggest a more complex geometry for the magnetic field, which might be a magnetic flux rope with both axial and azimuthal magnetic fields, rather than a flux tube with only an axial field. If that is correct, then there exists an axial electric current, which increases the dissipation (P. Lorrain, 1993b).

e) If a flux tube acts as a light guide (Sect. 11.5), then, when looking at a magnetic element, one sees the plasma well below the photosphere, where the temperature is higher.

Stenflo (1989) postulated the presence of a hot cylindrical "wall" enclosing the element, which would heat the gas inside. However, it is difficult to imagine a mechanism that would limit the electric current in that way, because that would require a zero radial plasma velocity inside the element. See Sect. 12.2. Hirayama (1992) pursued this proposal and found, as we do, that the ohmic losses account for only a small percentage of the observed radiated power.

12.3.6 The Time Constant

One striking characteristic of magnetic elements is that they are quite stable, despite the fact that they are violently buffeted by the turbulent solar atmosphere. Let us calculate a rough value for the time constant associated with the stored magnetic energy.

Imagine that the inward radial plasma velocity that drives the dynamo stops abruptly at the time $t = 0$. The magnetic energy $\mathcal{E}$ dissipates slowly through ohmic losses and

$$\frac{d\mathcal{E}}{dt} = -P_\mathrm{d} \, . \tag{12.30}$$

Assume that Eq. 12.26 continues to apply and that b remains constant after the flow has stopped. We can calculate a time constant for the magnetic energy as follows.

From Eqs. 12.20 and 12.26,

$$P_\mathrm{d} = \frac{12}{\mu_0 \sigma b^2} \mathcal{E} \, . \tag{12.31}$$

The dissipated power P_d, which is equal to the power that drives the dynamo, is therefore proportional to the stored magnetic energy, and thus to B^2, from Eq. 12.19, where B is the magnetic flux density at any given point. In other words, B at a given point is proportional to the square root of the power that drives the convection. The self-excited disk dynamo (Chapter 10) follows the same rule.

Thus

$$P_{\mathrm{d}} = -\frac{d\mathcal{E}}{dt} = \frac{12}{\mu_0 \sigma b^2}\mathcal{E} \tag{12.32}$$

and

$$\mathcal{E} = \mathcal{E}_0 \exp\left(-\frac{12}{\mu_0 \sigma b^2} t\right) , \tag{12.33}$$

where $\mathcal{E}_0$ is the stored magnetic energy at $t = 0$. If the system is left to itself with zero radial velocity, and under the gross assumption that the outer radius of the magnetic element b remains constant, then the magnetic energy $\mathcal{E}$ decreases exponentially with a time constant

$$T_{\mathcal{E}} = \frac{\mu_0 \sigma b^2}{12} \text{ seconds} . \tag{12.34}$$

We now require the time constant for B, which is measurable, contrary to $\mathcal{E}$. Refer to Eq. 12.19: if B decreases everywhere by a factor of e, then $\mathcal{E}$ decreases by a factor of e^2. It follows that

$$B = B_0 \exp\left(-\frac{6}{\mu_0 \sigma b^2} t\right) , \tag{12.35}$$

and the time constant for B is

$$T_B = \frac{\mu_0 \sigma b^2}{6} \text{ seconds} . \tag{12.36}$$

These time constants depend on the conductivity σ, which itself depends on the local value of B. They also depend on the radius b of the magnetic element, but they are independent of its length l, as one would expect with our simple model. This rough calculation shows that, if the convection stopped suddenly, then the magnetic field would decrease to about 30% of its original value in a time T_B. Conversely, if the convection started suddenly, the field would increase to about 70% of its final value in the same time T_B. This calculation fixes a rough lower bound for the real time constant. With the above values of σ and b, $T_B \approx 80$ seconds. Since the observed lifetimes of magnetic elements are rather of the order of 10^5 seconds, their stability probably comes mostly from the stability of the converging flow, despite the buffeting.

We can also calculate the time constant for I, or for B, by defining an equivalent resistance R and an equivalent inductance L by means of the equations

$$R = \frac{P_{\mathrm{d}}}{I^2} , \quad \mathcal{E} = \frac{1}{2} L I^2 , \quad I = blJ = 2b^2 J . \tag{12.37}$$

Then the time constant T_B, or T_I, is equal to L/R, as for a simple electric circuit.

These calculated time constants are fairly independent of the current distribution chosen. We have assumed a uniform current density J, but a current density that is proportional to the radius leads to the same time constant, and a current density that is inversely proportional to the radius, within radii a and b, gives only slightly different time constants.

12.3.7 Magnetic Element "Dipoles"

Occasionally, one sees opposing magnetic fields immediately next to a magnetic element. This might mean that the magnetic element is both short and slanting, and that one sees the magnetic field of the opposite polarity that emerges from its lower end.

Dara et al. (1990) have reported that magnetic elements of opposite polarity attract, while elements of the same polarity repel. This is in agreement with our model: magnetic elements should attract and repel, like parallel and anti-parallel solenoids and bar magnets.

12.3.8 Anchoring

Since the gas pressure inside a magnetic element is lower than outside, one would expect, at first sight, that the element would rise rapidly like a log immersed vertically in water. For example, if the average density inside an element is equal to one half that of the ambient gas then, neglecting viscosity, its upward vertical acceleration should be equal to one half the local g, or about 135 meters/second^2. This is incompatible with observed lifetimes of the order of 10^5 seconds. On second thought, the magnetic forces might possibly exert a net downward force on the element. That would be possible only if these forces exerted a net *upward* force on the ambient gas, which does not seem probable.

Before attempting to explain how magnetic elements are anchored in the photosphere, let us return to Eq. 12.6, in which $\boldsymbol{v}$ is the velocity of the plasma with respect to a fixed reference frame. If Q is the charge on a particle, the force $Q\boldsymbol{v} \times \boldsymbol{B}$ acts on the *particle*. Equation 12.6 therefore assumes implicitly that the charged particles have the velocity $\boldsymbol{v}$ of the plasma, in other words that their mean free path is small, say with respect to the radius of the magnetic element.

Also, the magnetic force on the convecting plasma, of density $\boldsymbol{J} \times \boldsymbol{B}$, points somewhat downward.

These simple facts possibly explain the anchoring in the following way. The partly ionized gas flows downward, and the charged particles spiral down with it, positive particles turning in the direction of $\boldsymbol{v} \times \boldsymbol{B}$, and electrons in the opposite direction. Thus the downwelling possibly overcomes the force of buoyancy.

See also Sect. 13.9.

12.3.9 A More Realistic Approximation

This second approximation is more realistic, but more elaborate. Since we have not actually performed the calculation, the procedure that follows can be no more that a rough guide for further work. This approximation is far from ideal. For example, it provides no information about the shape of the magnetic element, or about the boundary conditions, and it again sets the time derivative equal to zero, save for a rough calculation of a time constant.

We pursue our discussion of Sect. 11.2.

As previously, we set

$$\frac{\partial}{\partial t} = 0\,, \quad \frac{\partial}{\partial \phi} = 0\,, \quad v_\phi = 0\,, \quad B_\phi = 0\,, \tag{12.38}$$

but we now set

$$\frac{\partial}{\partial z} \neq 0\,, \quad B_\rho \neq 0\,. \tag{12.39}$$

So $\boldsymbol{v}$ and $\boldsymbol{B}$ both have ρ and z-components, which are both functions of ρ and z.

There is again zero space charge density because $\boldsymbol{\nabla}\cdot(\boldsymbol{v}\times\boldsymbol{B}) = 0$ (Sect. 7.2). Also $\boldsymbol{\nabla} V = \boldsymbol{0}$, and

$$\boldsymbol{J} = \sigma(\boldsymbol{v} \times \boldsymbol{B}) = \sigma(v_z B_\rho - v_\rho B_z)\hat{\boldsymbol{\phi}}\,. \tag{12.40}$$

There are no electric forces.

The independent variables are the coordinates ρ and z, and the unknowns are v_ρ, v_z, B_ρ, B_z. Since we have four unknowns, we need four equations. As already noted in Sect. 12.3.4, these unknowns are all constant *at a given point*, by hypothesis. This does not mean that a given ion has a constant velocity.

With the above poloidal $\boldsymbol{B}$, both $\boldsymbol{J}$ and $\boldsymbol{A}$ are azimuthal, so that $\boldsymbol{A}$ has only an azimuthal component. Then there are really only three unknowns, A_ϕ, v_ρ, v_z. This is not useful, however, because we have only one differential equation for $\boldsymbol{A}$, namely $\boldsymbol{\nabla}\cdot\boldsymbol{A} = 0$ which is an identity, while we need $\boldsymbol{B}$, as we shall see presently.

The equation $\boldsymbol{\nabla}\cdot\boldsymbol{J} = 0$ is also an identity, $\boldsymbol{J}$ being azimuthal and axisymmetric, like $\boldsymbol{A}$.

A) By conservation of mass,

$$\boldsymbol{\nabla}\cdot(N_{\mathrm{H}}\boldsymbol{v}) = 0\,, \tag{12.41}$$

where N_{H} is the number of hydrogen atoms per unit volume. Then

$$\frac{v_\rho}{\rho} + \frac{\partial v_\rho}{\partial \rho} + \frac{v_\rho}{N_{\mathrm{H}}}\frac{\partial N_{\mathrm{H}}}{\partial \rho} + \frac{\partial v_z}{\partial z} + \frac{v_z}{N_{\mathrm{H}}}\frac{\partial N_{\mathrm{H}}}{\partial z} = 0\,. \tag{12.42}$$

The quantities N_{H} and $\partial N_{\mathrm{H}}/\partial z$ are fairly well known functions of z (Allen, 1973), but both variables are also functions of ρ, like the gas pressure p. The gas pressure is a function of N_{H} and of the temperature, which is also fairly well known (Allen, 1973).

The flow pattern inside a magnetic element, represented here by the two functions v_ρ and v_z, depends on the magnetic force, of density $\boldsymbol{J}\times\boldsymbol{B}$, and hence on $\boldsymbol{J}$.

B) From the Maxwell equation for the divergence of $\boldsymbol{B}$,

$$\frac{B_\rho}{\rho} + \frac{\partial B_\rho}{\partial \rho} + \frac{\partial B_z}{\partial z} = 0 \, . \tag{12.43}$$

C) From another one of Maxwell's equations, with $\partial/\partial t = 0$,

$$\boldsymbol{\nabla} \times \boldsymbol{B} = \mu_0 \boldsymbol{J} \, , \tag{12.44}$$

$$\frac{\partial B_\rho}{\partial z} - \frac{\partial B_z}{\partial \rho} = \mu_0 \sigma (v_z B_\rho - v_\rho B_z) \, . \tag{12.45}$$

Here σ is a function of $B^2 = B_\rho^2 + B_z^2$ (Cambel, 1963).

D) Finally, we have the Euler equation of motion for an element of volume:

$$\widetilde{\boldsymbol{F}} - \boldsymbol{\nabla} p = \rho_{\text{gas}} \frac{D\boldsymbol{v}}{Dt} \, , \tag{12.46}$$

where $\widetilde{\boldsymbol{F}}$ is the sum of the gravitational and magnetic force densities, ρ_{gas} is the gas density, and the derivative on the right travels with the element of volume. See, for example, Rutherford (1959). This equation disregards viscosity, which is presumably permissible here.

The magnetic force density can be written

$$\widetilde{\boldsymbol{F}}_{\text{mag}} = \boldsymbol{J} \times \boldsymbol{B} = \frac{1}{\mu_0} (\boldsymbol{\nabla} \times \boldsymbol{B}) \times \boldsymbol{B} \, , \tag{12.47}$$

so as to eliminate the variable $\boldsymbol{J}$. Substituting the value of $D\boldsymbol{v}/Dt$,

$$\widetilde{\boldsymbol{F}} - \boldsymbol{\nabla} p = \rho_{\text{gas}} \left[(\boldsymbol{v} \cdot \boldsymbol{\nabla}) \boldsymbol{v} + \frac{\partial \boldsymbol{v}}{\partial t} \right] \, , \tag{12.48}$$

where the last term within the bracket is equal to zero under steady-state conditions.

This set of four simultaneous partial differential equations must be solved numerically.

Choosing an appropriate volume of integration is not simple. At first, one might seek solutions inside a cylindrical volume with $b = 7.5 \times 10^4$ meters and, say, $l = 1.5 \times 10^5$ meters to $l = 7.5 \times 10^5$ meters. However the radius b is presumably a function of the gas pressure, and therefore of the depth, for the following reason.

Return to Eq. 12.24. If the gas pressure is much smaller along the axis than outside, then

$$\mu_0 \frac{J^2 b^2}{2} = p_{\text{out}} \, . \tag{12.49}$$

The ambient gas pressure presumably decreases exponentially with height within the photosphere and thus, if the above assumption is correct, the product Jb, and B on the axis, from Eq. 12.10, also decrease exponentially with height. The above equation applies only if the current density J is uniform, and if the radius b increases slowly with height. Of course the current distribution inside a magnetic element depends on the flow pattern, and the radius b of a magnetic element increases with height. Then J would decrease even faster.

There is only one boundary condition for $\boldsymbol{B}$: the surface integral of its normal component over the whole surface of the magnetic element is equal to zero:

$$\oint \boldsymbol{B} \cdot \hat{\boldsymbol{n}}\, d\mathcal{A} = 0\,, \tag{12.50}$$

where $\hat{\boldsymbol{n}}$ is the unit vector normal to the element of area $d\mathcal{A}$.

Finally, we require boundary conditions for $\boldsymbol{v}$. If the volume of integration is a cylinder, one might set $v_z = 0$ and $v_\rho = -300$ meters/second at $\rho = b/2$ on the top surface. With b a function of altitude, one might set $v_z = 0$ over the top surface. One might also set the tangential component of velocity v_z equal to zero over the sides, and $v_\rho = 0$ over the bottom surface.

To calculate the dissipated power, the stored magnetic energy, and the time constant, one would presumably proceed as follows. Once $\boldsymbol{v}$ and $\boldsymbol{B}$ are known as functions of ρ and z inside the current distribution, one could calculate the conductivity σ from Cambel (1963), the current density $\boldsymbol{J}$ from Eq. 12.40, and the dissipated power. One would then calculate the magnetic field outside the current distribution from the value of $\boldsymbol{J}$ inside, say within a volume having a few times the volume of integration, and then sum the magnetic energies, inside and outside. The time constant would then follow from Eq. 12.36.

12.4 Summary

We consider magnetic elements in the solar photosphere as vertical magnetic flux tubes carrying azimuthal electric currents throughout their volume. These electric currents are generated by the inward radial motion of the photospheric plasma, and their polarity is such that they *amplify* any seed axial magnetic field of either polarity.

Under equilibrium conditions B at a given point is proportional to the square root of the power that drives the plasma inward against the magnetic force.

As a first approximation, we assume a uniform current density and calculate the corresponding magnetic flux density, the magnetic flux, the stored magnetic energy, the power dissipated by ohmic heating, and the time constant of the current distribution. We find that the ohmic power dissipation accounts for only a small part of the observed brightness. We also find a time constant of 80 seconds as a rough lower bound, on the assumption that the convection is "turned off" suddenly. Observed time constants are much longer.

Our model accounts for the existence of regions of opposite polarity immediately next to some magnetic elements and for the mutual attraction and repulsion between elements.

13 Case Study: Sunspots

Contents

Sunspots seem to be roughly vertical magnetic flux tubes, much like magnetic elements (Chapter 12). Like magnetic elements, they are paired, in that a spot of a given polarity is accompanied by one or more spots of the opposite polarity. But there are major differences: a) sunspots are larger than magnetic elements by an order of magnitude or more; b) sunspots are dark, while magnetic elements are bright; and c) the dark region of sunspots, called the *umbra*, is surrounded by a lighter ring, which is called the *penumbra*. We cannot, of course, address all these aspects, but we can attempt to understand, to a certain extent, how sunspots operate.[1]

[1] P. Lorrain and O. Koutchmy, *The sunspot as a self-excited dynamo*, Astronomy and Astrophysics **339**, 610–614 (1998).

13.1 Introduction

Sunspots (Fig. 13.1) are complex magnetic structures in the solar photosphere. They have been known for centuries. First, they were believed to be due to planets orbiting around the Sun, but Galileo (1564–1642) showed that they are part of the solar surface. There exists a wide variety of sizes and shapes. For a general introduction to sunspots, see Zirin (1988), as for magnetic elements (Chapter 12), or see Foukal (1990).

The diameter of the central dark region, the umbra, varies widely, but it is typically 5 to 10 megameters and B, at the center, is a few tenths of a tesla. The umbra is the top of the magnetic flux tube. It is far from uniform; a long exposure reveals a complex structure, with bright umbral dots, light bridges, etc.

The umbra is surrounded by a penumbra, which usually has bright and dark radial striations. The nature of the penumbra is still a matter of debate.

A sunspot of a given polarity is accompanied by one or more spots of the opposite polarity, showing that sunspots are somehow linked together below the photosphere (Cargill et al., 1994). Also, when coronal loops (Chapters 15 and 16) are rooted in pairs of sunspots of opposite polarities, the loops appear to follow magnetic field lines that link the sunspots. Often, the plane of the loop forms a large angle with the vertical. The flux tubes of the associated sunspots are presumably also tilted at a large angle.

Over one half century ago Cowling (1945) wrote "An adequate theory of a sunspot must explain the coolness of the spot, its magnetic field, and its dynamical properties." He would still not be satisfied today, despite the fact that the INSPEC lists on average, well over 100 new papers on sunspots every year.

To our knowledge, none account for the generation of the azimuthal currents that necessarily accompany the axial magnetic field.

13.2 Our Model

Our model is essentially the same as the one proposed by P. Lorrain and S. Koutchmy (1993) for magnetic elements (Chapter 12), and that serves for solar spicules (Chapter 14). Our model accounts for the existence of the azimuthal current and its associated axial magnetic field, and also for the Wilson depression. However, it does not account for the penumbra, nor for dots, light bridges, etc., in the umbra, nor for the outward Evershed flow around sunspots. One would hope that a more elaborate model could account for some of these other phenomena.

For simplicity, we assume a single, vertical, axisymmetric magnetic flux tube. This is a gross approximation. Sunspots are at best only roughly axisymmetric, and the magnitude of the magnetic field in the umbra, as a function of the radial position, depends on the complexity of the plasma flow. For

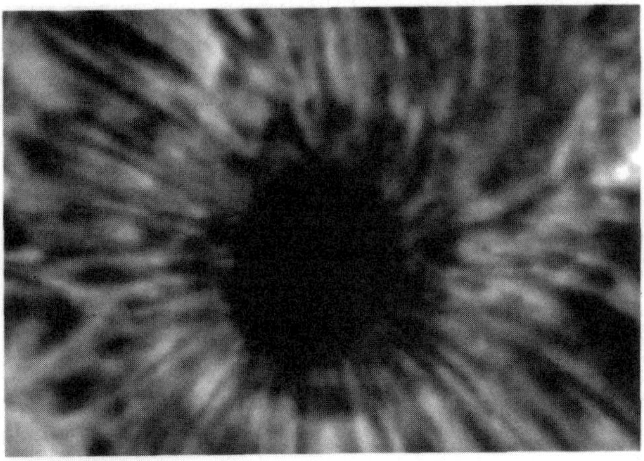

Fig. 13.1. A sunspot photographed near the disk center, at the center of the H_α line ($\lambda = 656.3$ nanometers). The dark central region is the *umbra*, while the surrounding region is the *penumbra* closer in, and the *superpenumbra* further out. The figure covers an area of 44 by 70 arcseconds, or about 32 by 51 megameters. For comparison, the diameter of the Earth is about 13 megameters. This image was obtained with the R.B. Dunn solar telescope at Sacramento Peak, New Mexico, USA, by Eugenia Christopoulou of the University of Patras, Greece, and collaborators

example, Lites et al. (1993) studied an umbral magnetic field that comprised five maxima, all of the same polarity.

13.3 Plasma Flows Above and Below Sunspots

Sunspots exhibit inward and downward flows in the umbra and in the penumbra. There are also supersonic downflows that extend far up into the chromosphere, and far into the surrounding region. Downwelling in and around sunspots has been reported by many authors. Zirin and Wang (1989, 1991) and Wang and Zirin (1992) observed inward and downward flows in the umbra and penumbra, and Balthasar et al. (1996) measured velocities in the moat, in the penumbra, and in the Evershed flow. Brekke et al. (1990) reported downflows above sunspots. Also, Wang and Zirin (1992) observed an inward drift of granules near growing pores, as one would expect with downwelling. Various groups have devised methods for observing flows in and around sunspots *below* the photosphere, for example by seismology and by tomography (Duvall et al., 1993; Duvall, 1995; Duvall et al., 1996; Kosovichev, 1996; Lindsey et al., 1996; Braun, 1995; Bogdan and Braun, 1995; Braun et al., 1996; Sun et al., 1997; Bogdan et al., 1998). Steiner et al. (1998) simulated the interaction between magnetic elements and granular and non-stationary convection.

Unfortunately, it seems that none of those authors correlated the plasma drift with the rate of change of the magnetic field.

Kosovichev (1996), Lindsey et al. (1996), Braun (1995), Bogdan and Braun (1995), and Braun et al. (1996) have observed *upwelling* and *outflows* around *decaying* active regions, as expected.

There are at least two types of force acting in the neighborhood of a sunspot: the convecting flow associated with downwelling or upwelling, and the magnetic braking force, of density $\boldsymbol{J} \times \boldsymbol{B}$.

As in Chapter 12, one expects an inward and downward flow, in and above the umbra and inner penumbra, because it is just the type of flow that can generate a vertical, self-excited magnetic flux tube (Chapter 11).

It is the inflow and downflow above and below sunspots, and the resulting magnetic flux tube, that we are concerned with here.

It is said that the plasma in the umbra and in the inner penumbra flows *inward*, while the outer part of the penumbra flows *outward*, both at about 0.5 kilometer/second. This is surprising because, visually, there is no demarcation between the inner and outer parts of a penumbra. In some cases the whole penumbra flows inward.

13.4 The Magnetic-flux-tube Dynamo

Sunspot umbras exhibit roughly vertical magnetic fields, and require a *local* azimuthal current distribution, like a solenoid, or like a solar magnetic element (Chapter 12). Many authors have discussed current distributions in sunspots, but without considering the generation of the currents. This is surprising because, early this century, Larmor (1919, 1929, 1934) and Cowling (1934) had proposed that sunspots are self-excited dynamos run by *up*welling plasma that would enter the spot at the base, giving the outward Evershed flow at the surface. That model was a move in the right direction, except that we now know that the flow of plasma is downward, not upward. Also, an upflowing plasma would have a positive radial velocity in cylindrical coordinates, because the ambient pressure decreases with height, and would give negative feedback, as we shall see below. See Chapter 11.

The component $-\boldsymbol{\nabla} V$ of the electric field $\boldsymbol{E}$ comes from volume and surface electric charges, if any, in the convecting plasma. See below. The component $-\partial \boldsymbol{A}/\partial t$ term is the electric field induced by a time-dependent magnetic field (Sect. 5.2). In convecting, conducting fluids, in the absence of an externally applied electric field, $E_{\text{typical}} \lesssim (vB)_{\text{typical}}$ (Eq. 7.15).

We use cylindrical coordinates ρ, ϕ, z, and assume a steady state and axisymmetry:

$$\frac{\partial}{\partial t} = 0 \text{ and } \frac{\partial}{\partial \phi} = 0 \,. \tag{13.1}$$

For the moment, we also set

$$B_\rho = 0 \,. \tag{13.2}$$

We remove this restriction in Sect. 13.8.2. Finally, we set

$$v_\phi = 0 \,, \quad B_\phi = 0 \,. \tag{13.3}$$

We disregard the Coriolis force, which certainly makes $v_\phi \neq 0$.

From Sect. 7.2, the electrostatic space charge density inside a conductor that moves at a velocity $\boldsymbol{v}$ in a magnetic field $\boldsymbol{B}$ is given by

$$\widetilde{Q}_\mathrm{f} = -\epsilon_0 \boldsymbol{\nabla} \cdot (\boldsymbol{v} \times \boldsymbol{B}) = -\epsilon_0 \boldsymbol{\nabla} \cdot [v_\rho \hat{\boldsymbol{\rho}} \times (B\hat{\boldsymbol{z}})] \equiv 0 \,. \tag{13.4}$$

Since there is zero electric space charge and presumably no surface charges, V is zero and, for a steady state,

$$\boldsymbol{J} = \sigma(\boldsymbol{v} \times \boldsymbol{B}) = -\sigma v_\rho B \hat{\boldsymbol{\phi}} \,, \tag{13.5}$$

where σ is the conductivity in a direction perpendicular to $\boldsymbol{B}$.

In Fig. 11.2, the radial component v_ρ of the plasma velocity $\boldsymbol{v}$ is negative, while B is positive. The field $\boldsymbol{v} \times \boldsymbol{B}$ and the induced current density $\boldsymbol{J}$ are azimuthal, in the direction shown.

The magnetic field of the induced azimuthal electric current points up *in the same direction as the assumed seed magnetic field*, as in Sect. 11.3.8. There is therefore positive feedback: the induced azimuthal current *amplifies* any existing axial magnetic field. We therefore have a self-excited dynamo, because positive feedback is *the* essential property of self-excited dynamos (Chapter 10).

If the seed magnetic field points down instead of up, then both $\boldsymbol{v} \times \boldsymbol{B}$ and $\boldsymbol{J}$ change direction, and the magnetic field of the induced azimuthal current points down, *again in the same direction as the seed field.*

We thus have the following situation: a downwelling plasma can be the seat of a self-excited dynamo whose axial magnetic field has the same sign as the seed field. That is the mode of operation of magnetic flux tubes (Chapter 11). The increase of the density with depth maintains the negative radial velocity of the plasma all along the flux tube.

So the inward and downward motion of the photospheric plasma in the vertical magnetic field of the sunspot gives an azimuthal $\boldsymbol{v} \times \boldsymbol{B}$ field that generates an azimuthal current whose magnetic field *adds* to any existing axial field, whatever its polarity. Thus, given a seed vertical magnetic field of either polarity, there is *positive feedback* and the axial magnetic field builds up exponentially. So B should soon become infinite, which is absurd. In fact, B tends to an asymptotic value that is a function of the power that drives the convection (Chapter 10).

Now suppose that the magnetic flux tube is established with $\boldsymbol{B}$ pointing up, and that the flow reverses sign: the downwelling becomes upwelling. As the upflowing plasma rises, it expands because of the decreasing ambient pressure, its radial velocity v_ρ is *positive*, and it generates a $\boldsymbol{v} \times \boldsymbol{B}$ field that points in the $-\hat{\boldsymbol{\phi}}$-direction and a corresponding current whose magnetic field *opposes* the existing field. Since the flux tube has an inherent time constant (Sect. 11.3.8), its magnetic field decreases more or less slowly. This is just what Kosovichev (1996) observed: there is an *upflow* in a *decaying* active region.

Anchoring is possibly the same as with magnetic elements (Sect. 12.3.8). But see Sect. 13.9.

Let us check the induction equation for this dynamo:

$$\frac{\partial \boldsymbol{B}}{\partial t} = \boldsymbol{\nabla} \times (\boldsymbol{v} \times \boldsymbol{B}) - \frac{\boldsymbol{\nabla} \times (\boldsymbol{\nabla} \times \boldsymbol{B})}{\mu_0 \sigma} \tag{13.6}$$

$$= \boldsymbol{\nabla} \times (\boldsymbol{v} \times \boldsymbol{B}) - \frac{\boldsymbol{\nabla} \times \boldsymbol{J}}{\sigma} . \tag{13.7}$$

For a steady state, the LHS is zero. Also, Eq. 13.5 applies. Then, if the conductivity σ is uniform, the induction equation reduces to the identity $\mathbf{0} \equiv \mathbf{0}$.

13.5 Why Are Sunspots Dark?

At first sight, the Ohmic power dissipation in a sunspot should make the umbra bright, and not dark. However, as we saw in the case of solar magnetic elements (Sect. 12.3.5), the extra Ohmic power density J^2/σ is completely negligible compared to the power radiated by the photosphere. The same applies to sunspots.

Magnetic elements are bright, we hypothesize, because the flux tube acts as a light guide, so that one can see the solar plasma some distance below the photosphere, where the temperature is higher (Sect. 11.5).

Then why are sunspots dark? One possible explanation is the following. Recall that a given sunspot is always associated with one or more sunspots of the opposite polarity. Say a sunspot S has a single companion S'. These spots S and S' presumably share a common flux tube, with plasma flowing down each one and diverging in the region where the two flux tubes meet. If the two spots do share a common flux tube, and if flux tubes act as light guides, then, when looking *down* spot S, one is actually looking *up* spot S'! Since the solar atmosphere is dark compared to the photosphere, spot S appears dark.

13.6 The Radial Distribution of J_ϕ

Weiss (1990), Jahn (1983), and others proposed that there exists a sharp transition between the magnetic flux tube and the external plasma. However, this does not mean that the equivalent solenoid is thin. Hirayama (1992) proposed a thin current sheet at the periphery of magnetic elements, and Solanki and Schmidt (1993), Jahn (1983), and Pizzo (1990) also proposed thin current sheets for sunspots. If J, and hence $v_\rho B$ in Eq. 13.5, are significant only near the periphery of the tube then, a short distance inside the flux tube, the plasma flows down along the lines of $\boldsymbol{B}$.

In favor of the thin current sheet model, one may argue that a) the inward radial velocity v_ρ of an element of volume of the plasma in Eq. 13.5 decreases as the element of volume moves in the direction of the axis, and b) the azimuthal conductivity σ in the same equation is lower near the axis, where the pressure p is lower and B is larger.

The case of pores is interesting. *Pores* are small, short-lived sunspots that often exhibit a bright "lining", as in Zirin (1988, see his Figures 10.7 b and c). The magnetic flux tube of a pore thus seems to be maintained by a current sheet, where there are Joule losses. Now, as we shall see in Sect. 13.8.2, the magnetic flux tube of a sunspot flares near the surface, while the bright lining of a pore seems to indicate that its flux tube is not flared, despite the steep vertical pressure gradient. This is surprising. Also, pores do not have penumbræ. Do penumbræ require a flared magnetic flux tube? See Sect. 12.3.

13.6.1 The Magnetic Field Configuration

Several authors have reported on the value of $\boldsymbol{B}$ as a function of the radius in the umbra. Solanki and Schmidt (1993) compiled observations by Beckers and Schröter (1989), Kawakami (1983), Lites and Skumanich (1990), Solanki et al. (1992), and Wittmann (1974). See also S. Koutchmy and Adjabshirizadeh (1981), Adjabshirizadeh and S. Koutchmy (1983), Arena et al. (1990), Title et al. (1993), Howard (1996), Sutterlin et al. (1996), Abramov Maksimov et al. (1996), Keppens and Martínez Pillet (1996), and Stanchfield et al. (1997).

One would therefore hope to gain information on a sunspot's magnetic flux tube by comparing the field of a sunspot with that of coils of various geometries. This is what we now do.

Although this procedure is promising, it is far from ideal.

1. The magnetic flux tubes of sunspots are often far from vertical, as noted in Sect. 13.1.

2. It is impossible to say at what altitude the magnetic field is measured, with respect to the current distribution.

3. Measurements of the magnitude and inclination of $\boldsymbol{B}$ as functions of the radial position are difficult and approximate. Moreover, the spots are usually not situated at the center of the disk, which complicates the geometry (Title et al., 1993).

4. Sunspots are highly complex structures. The umbra, which is the region that we are mostly concerned with, is far from uniform. See, for example, Adjabshirizadeh and S. Koutchmy (1983), and Livingston (1991). Considering

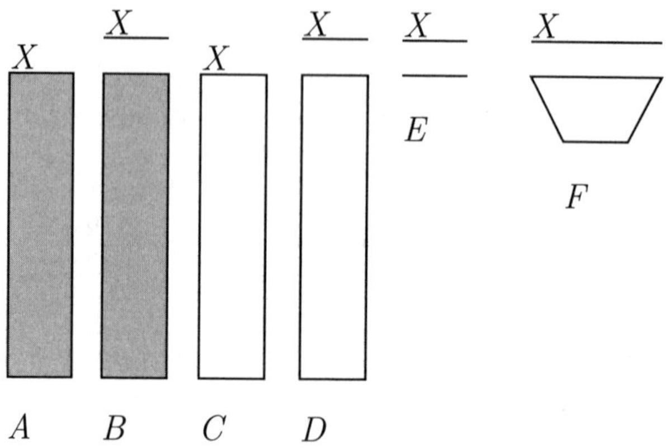

Fig. 13.2. Cross-sections of coils A to F, and the planes X where $B/B_{\rm m}$ and θ are calculated

a spot as the end of a vertical magnetic flux tube of circular cross-section, as we do here, is a rough approximation.

We nonetheless compare observed data on the magnitude of $\boldsymbol{B}$ and on the inclination θ of magnetic field lines with respect to the axis of symmetry as functions of the radial position, with the corresponding variables for six coils. We assume that the outer penumbral boundary corresponds to the outer radius of the coil.

We averaged the Solanki and Schmidt (1993) curves for $B/B_{\rm m}$ and for θ visually at seven values of $r/r_{\rm p}$, where $B_{\rm m}$ is the magnetic field at the center, θ is expressed in degrees, and $r_{\rm p}$ is the outer radius of the penumbra. See also Solanki (1990). With $x = r/r_{\rm p}$, a regression analysis with third-order polynomials yields the equations

$$\frac{B}{B_{\rm m}} = a + bx + cx^2 + dx^3 \,, \tag{13.8}$$

$$\theta = e + fx + gx^2 + hx^3 \,, \tag{13.9}$$

with

$$a = 1.00 \,, \quad b = -0.158 \,, \quad c = -1.13 \,, \quad d = 0.616 \,, \tag{13.10}$$

$$e = 0 \,, \quad f = 134 \,, \quad g = -53.3 \,, \quad h = -0.0807 \,. \tag{13.11}$$

We have shown three significant figures, but the data are much less accurate than that.

We calculated the same six coefficients for six coils of circular cross-section, defined as in Fig. 13.2 and Table 13.1. Coil E is a simple loop. Coil F is a thin cone, with the upper radius equal to twice the lower radius, and height equal to the upper radius. We calculated the field at one lower radius

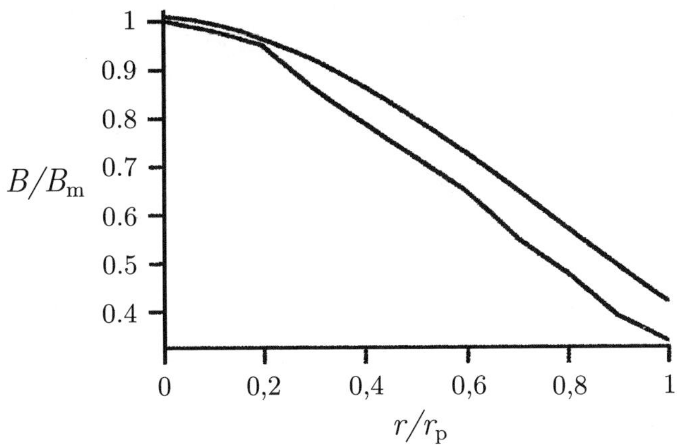

Fig. 13.3. The ratios B/B_m, where B_m is the value of B on the axis, as functions of r/r_p for the sunspot data compiled by Solanki and Schmidt (1993) (broken line) and for coil A (smooth line). The radius r_p is either the outer radius of the penumbra or the outer radius of the coil

above the top. We calculated this field because, as we shall see below, the flux tube is presumably conical near the surface.

Call R_0 and R the inside and outside radii respectively, L the length, and D the distance between the end of the coil and the plane where B and θ are calculated.

Table 13.1. Five coils

Coil	R_0	L	D
A	0	$20R$	0
B	0	$20R$	R
C	R	$20R$	0
D	R	$20R$	R
E	R	0	R

We calculated the angle θ between a magnetic field line and the axis of symmetry, for the same data. According to both sets of curves, J is larger than average near the axis.

Tables 13.2 and 13.3 show the coefficients for the six coils.

At the center, where $x = r/r_p = 0$, for all the fields, $B/B_m = 1$ and $\theta = 0$. A field is thus characterized by the values of B/B_m and of θ at $x = r/r_p = 1$. These two quantities are the sums $a+b+c+d$ and $e+f+g+h$. Table 13.4

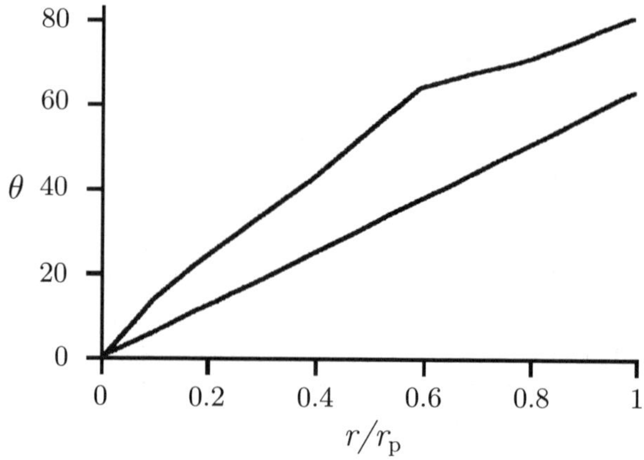

Fig. 13.4. The angles θ between a magnetic field line and the axis of symmetry, for the same data. According to both sets of curves, J is larger than average near the axis

shows the sums for the Solanki and Schmidt (1993) (S&S) data and for the six coils.

We did not calculate the field of the loop in its plane for the following reason. At one radius from the axis, θ is equal to 90 degrees, which is similar to the S&S value, but $B/B_{\rm m}$ is very large, contrary to the S&S value.

Table 13.2. The ratio $B/B_{\rm m}$

Coil	a	b	c	d
A	1.0	-0.0546	-0.931	0.396
C	1.0	-0.0964	0.445	-0.540
D	1.0	0.0102	-0.318	0.0530
E	1.0	-0.0600	0.360	-0.557
F	1.0	-0.00608	-0.0612	-0.0181

Table 13.3. The angle θ (degrees)

Coil	e	f	g	h
A	0.377	60.7	4.69	-1.94
B	-0.00423	40.8	0.113	-2.79
C	-0.0337	36.05	-9.87	27.93
D	0.0156	34.2	2.11	-1.36
E	-0.104	43.4	-23.6	42.2
F	-0.00312	23.7	0.119	1.51

The field that agrees best with the S&S data is A, the field in a plane at the end of a coil of zero inside radius, whose length is 20 times its radius. Figures 13.3 and 13.4 show plots of the fields S&S and A. So, according to these calculations, the azimuthal current density in a sunspot flux tube is approximately uniform from the axis to the outer radius of the penumbra, with J somewhat larger than average near the axis.

13.7 Magnetic Pressure and Gas Pressure

It is interesting to relate the gas pressure to the magnetic pressure at an arbitrary depth. Assume that the flux tube is a circular cylinder, and call p the internal gas pressure and B_0 the axial magnetic flux density. Each one of the four variables, p, v_ρ, J, B_0, is a function of the three others and of the coordinates ρ and z. The conductivity σ is a function of p and of B. Assume a uniform current density J, a steady state, and a zero net radial force on an element of volume. Then the radial pressure gradient is equal to the magnetic force density (Sect. 12.3.4). Both vectors point outward.

At the radius ρ,

$$B = \mu_0 \int_\rho^b J\, d\rho = \mu_0 J(b - \rho)\,, \quad J = -\frac{1}{\mu_0}\frac{\partial B}{\partial \rho}\,. \tag{13.12}$$

Then

$$\boldsymbol{\nabla} p = \frac{\partial p}{\partial \rho}\hat{\boldsymbol{\rho}} = JB\hat{\boldsymbol{\rho}} = -\frac{1}{\mu_0}\frac{\partial B}{\partial \rho}B\hat{\boldsymbol{\rho}} = -\frac{1}{2\mu_0}\frac{\partial (B^2)}{\partial \rho}\hat{\boldsymbol{\rho}} \tag{13.13}$$

and

$$\frac{\partial}{\partial \rho}\left(p + \frac{B^2}{2\mu_0}\right) = 0\,, \quad p + \frac{B^2}{2\mu_0} - p_{\text{ext}}\,, \tag{13.14}$$

where p_{ext} is the ambient gas pressure outside the flux tube.

Thus the sum of the gas pressure p and of the magnetic pressure $B^2/(2\mu_0)$ is independent of the radial position ρ, as one might expect. See Sect. 4.9.

Table 13.4. Sums of coefficients

Field	$a+b+c+d$ (B/B_m)	$e+f+g+h$ (θ)
S&S	**0.328**	**81.04**
A	0.420	63.83
B	0.668	38.12
C	0.809	54.08
D	0.745	34.97
E	0.743	61.90
F	0.915	25.33

With $B_{\rm m} = 0.25$ tesla at the center of the spot, the magnetic pressure, there, at the surface, is 2.5×10^4 pascals. This is comparable to the gas pressure of 1.8×10^4 pascals at -75 kilometers given by Vernazza et al. (1981).

Whatever the radial dependence of the azimuthal current density J, the magnetic flux density B is maximal on the axis, decreases with ρ inside the flux tube, and drops to zero at the edge of the tube: $B = 0$ outside the flux tube. So $p = p_{\rm ext}$ outside the tube. Inside the flux tube, the gas pressure p increases gradually with ρ and, at the surface of the tube, where B drops to zero, p rises to $p_{\rm ext}$.

13.8 The Flux Tube Radius As a Function of z

Our reference sunspot is the one observed by Stanchfield et al. (1997). Its radius was 3 megameters, and B at the center was 0.25 tesla. Using an average B of 0.125, $\Phi = 3.5 \times 10^{12}$ webers.

Since the gas pressure increases with depth below the photosphere, the radius of a vertical magnetic flux tube presumably decreases with increasing depth. We investigate this phenomenon in two stages. First, we consider levels below -36 megameters, where the Standard Solar Model (Bahcall and Pinsonneau, 1992) provides the pressure as a function of depth. Then we discuss the flux tube radius higher up.

13.8.1 The Radius b Below -36 Megameters

Call z the vertical coordinate normal to the photosphere, with the unit vector $\hat{z}$ pointing up, and with the origin at the photosphere. We calculate the flux tube radius b as a function of the external *pressure* $p_{\rm ext}$, and then we use the Standard Solar Model to deduce b as a function of the *depth* z, where z is negative.

Set

$$J = J_b \left(\frac{\rho}{b} \right)^n , \quad J_{\rho > b} = 0 , \tag{13.15}$$

where J_b is the azimuthal current density at the radius b. Assume that the exponent n is independent of z. If $n = 0$, then J is uniform. If $n = 5$, then $J_{b/2} = J_b/32$.

The Joule power density is J^2/σ watts/meter^3, with σ largest where the magnetic field is smallest.

Below -36 megameters, b varies slowly with z and we can use the analysis of P. Lorrain and Salingaros (1993). At the radius ρ,

$$B = \mu_0 \int_\rho^b J \, d\rho = \frac{\mu_0 J_b b}{n+1} \left[1 - \left(\frac{\rho}{b} \right)^{n+1} \right] . \tag{13.16}$$

As with a long solenoid, B outside is negligible because the return flux extends over a large region.

The magnetic flux in the tube is

$$\Phi = \int_0^b B 2\pi\rho \, d\rho = \frac{\pi\mu_0 J_b b^3}{n+3} \, . \tag{13.17}$$

If Φ and n are independent of z, then $J_b b^3$ is independent of z.

Call the magnetic flux density on the axis of the flux tube B_0, and set the magnetic pressure on the axis equal to a fraction K of the external pressure p_{ext}, where $K \leq 1$:

$$\frac{B_0^2}{2\mu_0} = \frac{1}{2\mu_0}\left(\frac{\mu_0 J_b b}{n+1}\right)^2 = K p_{\text{ext}} \, . \tag{13.18}$$

Eliminating J_b from Eqs. 13.17 and 13.18,

$$b = \left[\frac{(n+3)\Phi}{\pi(n+1)}\right]^{1/2} \frac{1}{(2\mu_0 K p_{\text{ext}})^{1/4}} = \frac{U}{p_{\text{ext}}^{1/4}} : \tag{13.19}$$

the magnetic flux tube radius b is proportional to the inverse fourth root of the external pressure. It is not very sensitive to the value of n.

Also,

$$\frac{db}{dz} = \frac{db}{dp_{\text{ext}}}\frac{dp_{\text{ext}}}{dz} = -\frac{U}{4p_{\text{ext}}^{5/4}}\frac{dp_{\text{ext}}}{dz} \, . \tag{13.20}$$

With the z-axis pointing up as above, the last derivative is negative, and the flux tube radius b increases with increasing z, since U, defined in Eq. 13.19, is positive.

For the Stanchfield et al. (1997) sunspot, with $K = 1$,

$$U = 4.6 \times 10^7 \ \text{ for } \ n = 0 \, , \ \text{ and } U = 3.1 \times 10^7 \ \text{ for } \ n = 5 \, . \tag{13.21}$$

From Eq. 13.18,

$$J_b = \frac{B_0(n+1)}{\mu_0 b} \tag{13.22}$$

and, from Eq. 13.15, the azimuthal current per meter of length is

$$\tilde{I} = \int_0^b J \, d\rho = \frac{J_b b}{n+1} = \frac{B_0}{\mu_0} \, . \tag{13.23}$$

So, with increasing pressure p_{ext}, or with increasing depth,

a) B_0 increases: $B_0 \propto p_{\text{ext}}^{1/2}$, Eq. 13.18;

b) b decreases: $b \propto 1/p_{\text{ext}}^{1/4}$, Eq. 13.19;

c) J_b increases: $J_b \propto 1/b^3 \propto p_{\text{ext}}^{3/4}$, Eqs. 13.17 and 13.19;

d) $\tilde{I}$ increases: $\tilde{I} \propto B_0 \propto p_{\text{ext}}^{1/2}$, Eqs. 13.18 and 13.23; and

e) $|v_\rho|$ decreases, from Fig. 11.2.

We recapitulate our assumptions.

1) We first assumed the current distribution of Eq. 13.15.

2) We assumed that n is independent of z. This is satisfactory, first because n can lie between 0 and, say, 10, and second because b is fairly insensitive to n.

3) Equation 13.16 assumes that the radius b of the flux tube does not change rapidly with z. As we shall see below, this assumption is justified, at least below about -10 megameters.

4) In writing Eq. 13.19, we assume that Φ is independent of z. That is plausible, again because b does not change rapidly with z.

To calculate b as a function of z, we now require the external pressure p_{ext} as a function of depth below the photosphere. Bahcall is the authority in this field (Bahcall and Ulrich, 1988; Bahcall and Pinsonneau, 1992), but see also Hendry (1993).

The Bahcall and Pinsonneau (1992) table extends from the center of the Sun up to -36 megameters, but the range -100 megameters to -36 megameters is sufficient for our purposes. A third-order regression analysis for this range yields

$$p_{\text{ext}} = -2.72 \times 10^{10} - 1470z - 2.22 \times 10^{-5} z^2 - 7.13 \times 10^{-13} z^3 \,. \quad (13.24)$$

The fit is near-perfect, but the enormous *negative* pressure at $z = 0$ is absurd. Recall that the atmospheric pressure at the surface of the Earth is about 10^5 pascals. Vernazza et al. (1981) quote a pressure of 1.2×10^4 pascals at $z = 0$, or six orders of magnitude smaller, and of the correct sign! Adding the point $z = 0$, $p_{\text{ext}} = 1.2 \times 10^4$ pascals before fitting the cubic, again gives a large negative pressure at $z = 0$. Thus the Bahcall and Pinsonneau (1992) table cannot be extrapolated to the photosphere. The table given in the earlier Bahcall and Ulrich (1988) paper does not extend above -56 megameters.

Figure 13.5 shows the flux tube radius b for the Stanchfield et al. (1997) sunspot as a function of z, using Eq. 13.19 with $\Phi = 3.5 \times 10^{12}$ webers, $K = 1$, and $n = 0$, for the range $z = -100$ megameters to $z = -40$ megameters.

Note that the b-scale covers only about 50 kilometers, while the z-scale covers 60 megameters. So b varies slowly with z at these depths. Indeed, $db/dz = 0.0012$ at -50 megameters.

For the Stanchfield et al. (1997) sunspot, $\Phi \approx 3.5 \times 10^{12}$ webers. According to the Bahcall and Pinsonneau (1992) table, the pressure at a depth of 50 megameters is 8×10^{10} pascals. Then, at that depth, from Fig. 13.5 or from Eq. 13.19, for $n = 0$, $K = 1$,

$$b = \frac{4.6 \times 10^7}{(8 \times 10^{10})^{1/4}} = 86 \text{ kilometers} \,. \quad (13.25)$$

From Eq. 13.18, with $n = 0$, $K = 1$,

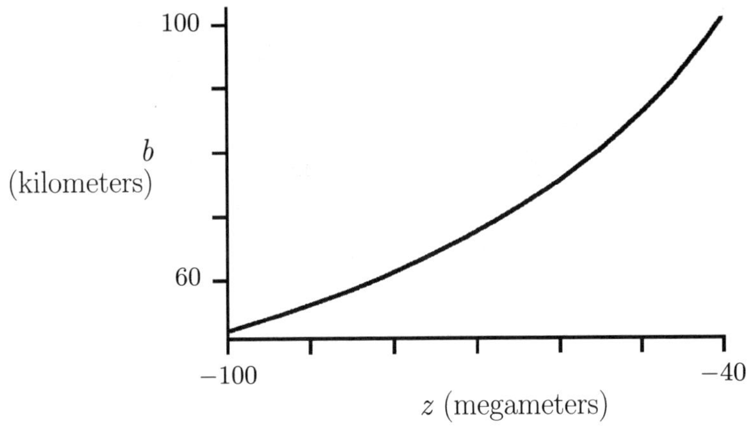

Fig. 13.5. Magnetic flux tube radius b as a function of depth for the Stanchfield et al. (1997) sunspot. Over this range, the flux tube radius b increases slowly with decreasing depth. The Stanchfield et al. (1997) sunspot had a radius of 3 megameters. Indeed, $db/dz = 0.0012$ at -50 megameters. Note that the b-scale covers only about 50 kilometers, while the z-scale covers 60 megameters

$$B_0 = (2\mu_0 K p_{\text{ext}})^{1/2} \approx 450 \text{ teslas} . \tag{13.26}$$

From Eqs. 13.22 and 13.23, again with $n = 0$, $K = 1$,

$$J_b \approx 4 \text{ kiloamperes/meter}^2 , \quad \tilde{I} \approx 4 \times 10^8 \text{ amperes/meter} . \tag{13.27}$$

The large value of the current density J_b is not unreasonable because

$$J = \sigma |v_\rho| B . \tag{13.28}$$

With $|v_\rho|$ equal to only 10 meters/second, and with $B = 400$ teslas, $\sigma \approx 1$, which is 8 orders of magnitude smaller than the conductivity of copper at room temperature. See P. Lorrain and S. Koutchmy (1993).

Recall that $\tilde{I}$ is the azimuthal current per meter of flux tube length. With $n = 0$, $\tilde{I} = J_b b$.

13.8.2 The Flux Tube Radius b Above −36 Megameters

At this time, it seems impossible to find an expression for the magnetic flux tube radius b that applies above −36 megameters. However, the following discussion might be useful for readers interested in pursuing this work.

According to the Bahcall and Pinsonneau (1992) table, the pressure gradient just below the photosphere is *very* large: the pressure increases from the Vernazza et al. (1981) value of 1.2×10^4 pascals at the surface to the

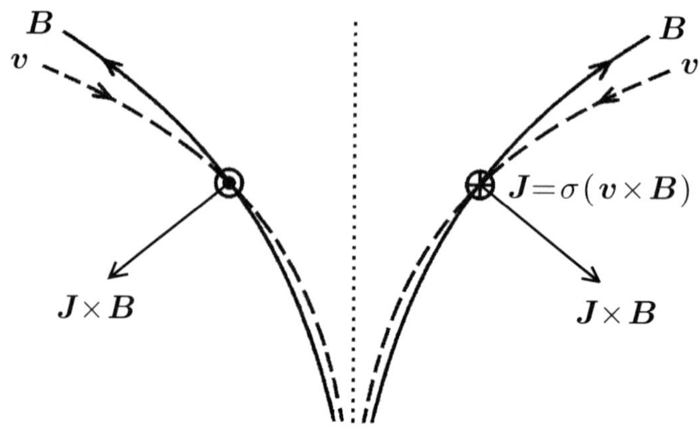

Fig. 13.6. The magnetic force density at the top of a magnetic flux tube, for example the flux tube of a sunspot. With a magnetic field $\boldsymbol{B}$ that points up, as above, the azimuthal current density $\boldsymbol{J}$, and the magnetic force density $\boldsymbol{J}\times\boldsymbol{B}$, point in the directions shown. If, instead, the magnetic field points down, then both $\boldsymbol{B}$ and $\boldsymbol{J}$ change sign, and the magnetic force density again points outward and downward, adding to the gravitational force and cancelling the force of buoyancy. The force of buoyancy comes from the fact that the flux tube is largely evacuated, as we saw in Sect. 11.3.2

Bahcall and Pinsonneau value of 3.0×10^{10} pascals at -36 megameters, or by a factor of 3×10^7! Over the next 36 megameters down, their table gives a pressure ratio of 8, and over the following 36 megameters their pressure ratio is 3.5. Therefore, above -36 megameters, from Eq. 13.19, the magnetic flux tube flares, and the calculations of Sect. 13.8.1 do not apply.

13.9 The Wilson Depression

It has long been known that sunspots lie at the bottom of a so-called Wilson depression 300 to 500 kilometers deep. The depression presumably results from the downward component of the magnetic force as in Fig. 13.6. The depression is only a few percent of the sunspot diameter, which is typically 5 or 10 megameters.

In Sect. 11.3.1 we investigated the magnetic force exerted at the top of a magnetic flux tube, and we showed that the magnetic force density $\boldsymbol{J}\times\boldsymbol{B}$ points downward and radially outward, as in Fig. 13.6. So the $\boldsymbol{v}\times\boldsymbol{B}$ field generates an azimuthal current of density $\boldsymbol{J} = \sigma(\boldsymbol{v}\times\boldsymbol{B})$ that increases any existing $\boldsymbol{B}$. The azimuthal current distribution is that of a conical coil, and one can expect a magnetic field line to be closer to the axis than the streamline associated with the azimuthal current $\boldsymbol{J}$. We assume, here, as usual, that $\boldsymbol{B}$ points up. The induced current, of density $\boldsymbol{J} = \sigma(\boldsymbol{v} \times \boldsymbol{B})$, flows in the direction shown and *increases* any existing $\boldsymbol{B}$. If, instead, $\boldsymbol{B}$ points down, then $\boldsymbol{J}$ changes sign and there is again positive feedback. In both cases, the magnetic force, of density $\boldsymbol{J} \times \boldsymbol{B}$ points outward and downward. The downward component adds to the gravitational force density and lowers the surface.

One expects that the relative orientations of the $\boldsymbol{v}$ and $\boldsymbol{B}$ vectors in the Wilson depression would be similar to those lower down in the flux tube, with $\boldsymbol{v}$ forming a larger angle with the vertical than $\boldsymbol{B}$: the converging flow generates a more or less vertical $\boldsymbol{B}$. See Sect. 11.3.

This magnetic force, of density $\boldsymbol{J} \times \boldsymbol{B}$, acts on the charge carriers of the azimuthal current in the flux tube. The charge carriers are electrons and positive ions, electrons drifting in one azimuthal direction and positive ions in the opposite direction. The electrons carry most of the current because of their smaller mass, higher speed, and longer mean-free-path. Since the mean-free-paths are finite, the charge carriers drag the ambient gas along with them until the gravitational force density, added to the magnetic force density, is equal to the gas pressure gradient:

$$-\widetilde{m}g\hat{\boldsymbol{z}} + \boldsymbol{J} \times \boldsymbol{B} = \boldsymbol{\nabla} p\,, \tag{13.29}$$

where $\widetilde{m}$ is the plasma mass density and p is the gas pressure.

The downward component of the magnetic force adds to the gravitational force, increasing the effective value of g and depressing the surface. As for magnetic elements (Chapter 12), the gas pressure in the flux tube is lower than ambient and the downward component of $\boldsymbol{J} \times \boldsymbol{B}$ serves to anchor the flux tube, cancelling the force of buoyancy. See Sect. 12.3.8.

While this is a plausible explanation for the Wilson depression, it is not too advisable to substitute numbers into Eq. 13.29, because, of the five variables, only g is known with accuracy. The flux tube flares rapidly near the surface.

13.10 Summary

Observations show that there is downflow in and around sunspots. We show that sunspots are self-excited magnetic-flux-tube dynamos run by this downflow. The magnetic field is similar to that of a solenoid, in that both require local azimuthal currents, according to the Maxwell equation for the curl of $\boldsymbol{B}$. The azimuthal current results from an azimuthal $\boldsymbol{v} \times \boldsymbol{B}$ field generated by the radial inflow of the downflowing plasma. The magnetic force points outward, and slightly downward. Comparison between various observations on sunspots with the fields of six different coils seem to indicate that the azimuthal current is distributed nearly uniformly over the umbra and penumbra, with a slight excess near the axis.

The tube radius, at depth, is inversely proportional to the fourth root of the ambient pressure. The Standard Solar Model of Bahcall and Pinsonneau (1992) provides the pressure as a function of depth below −36 megameters, while the Vernazza et al. (1981) model for the solar atmosphere provides the pressure at the level of the photosphere. It seems that the fanning out of the flux tube near the top causes the Wilson depression.

14 Case Study: Solar Spicules

Contents

Spicules grow like hair on the solar surface. Since astronomers can see them from the Earth, they are huge: up to 10 000 kilometers long. What are they? They seem to follow magnetic field lines, but they do not diverge on emerging from, or plunging into, the surface, where the magnetic field lines diverge rapidly. Also, they have a uniform diameter of about 150 to 200 kilometers, despite large changes in the ambient *B*. That does not make sense. According to our model, spicules are akin to the coronal loops of Chapters 15 and 16. [1]

14.1 Introduction

Solar spicules are luminous jets that erupt up to 10 000 kilometers from the surface of the Sun and that have diameters, all along their length, of

[1] P. Lorrain and S. Koutchmy, *Two dynamical models for solar spicules*, Solar Physics **165**, 115–137 (1996).

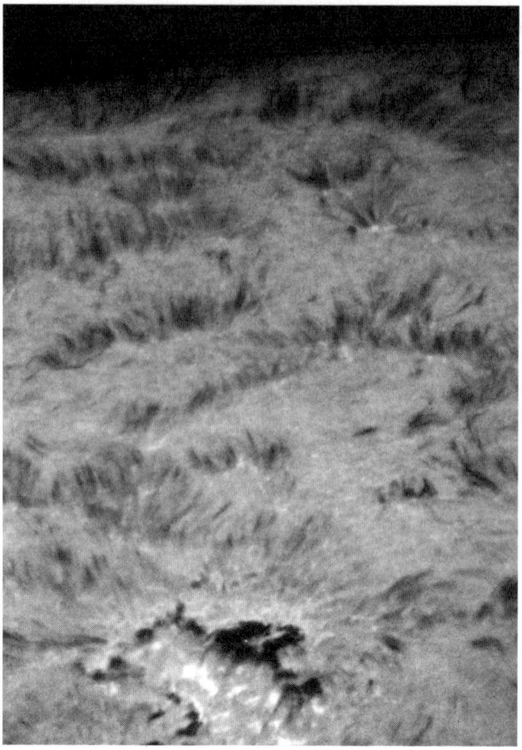

Fig. 14.1. Spicules on the solar surface, near the limb (edge)

only about 150 to 200 kilometers (Veselovsky et al., 1994; Dara et al., 1998; Zirin and Cameron, 1998). We assume here that a solar spicule is a self-channeled proton beam emitted by a magnetic element, and surrounded by a cold sheath. We assume that the beam originates in a self-excited dynamo that exploits a $\boldsymbol{v} \times \boldsymbol{B}$ field and that is situated below the element, where $\boldsymbol{v}$ is the local plasma velocity and $\boldsymbol{B}$ the local magnetic flux density. In falling back, the sheath provides a return current of protons that cancels the outgoing current. We discuss the channeling of charged-particle beams of very large cross-section, both in the present chapter and in Chapters 15 and 16. We propose velocity channeling, which is apparently a new concept. We assume a steady state and a hydrogen plasma.

As observed at the limb (edge), solar spicules are mostly straight luminous jets that erupt from the surface of the Sun. Spicules emit low-excitation emission lines like H_α. So-called *macrospicules*, or *spikes*, extend up to 30 000 kilometers (S. Koutchmy and Loucif, 1991). Spicules appear to originate in magnetic elements. Spicules rise rapidly to their maximum height, the root disconnects from the surface, the proper motion of the source of the main cool radiation becomes downward, and they disappear. They last about 5 or 15 minutes.

Magnetic elements (Chapter 12) are small regions in the photosphere that exhibit roughly vertical magnetic fields of typically 0.15 tesla.

Spicules are not perfectly vertical nor straight; at the poles and at the equator, they follow magnetic field lines (Secchi, 1877; Dunn, 1960; Foukal, 1990; Zirin, 1988; Zirin and Cameron, 1998).

Zirin and Cameron (1998) reported that many more spicules go up than come down, there are numerous double and multiple spicules, macrospicules start with an "Eiffel tower" shape, and there is some evidence of rotation in spicules.

The apparent velocity of the *head* of a spicule, as observed at the limb, is typically 25 kilometers/second, but it can be as large as 50 to 100 kilometers/second. This head velocity is surprisingly low: a particle having an initial vertical velocity of 25 kilometers/second could reach a maximum height of only about 1000 kilometers, disregarding collisions, with $g = 272$ meters/second^2. A video sequence of spicule H_α spectra shows Doppler-shift speeds of 50 to 100 kilometers/second. Brueckner's observation of the *hot* part of spicules yielded speeds of 100 to 400 kilometers/second.

The density of the cool sheath of a spicule is larger than ambient by one or two orders of magnitude, and the accelerations along its length appear to be synchronous.

The gravitational potential energy required to create a single spicule is impressive. Typically, a mature spicule has a height of 5000 kilometers, a diameter of about 150 kilometers, $N_H \approx 10^{18}$ per meter^3, $g = 272$ meters/second^2, and $mgh \approx 10^{17}$ joules. This energy appears over a period of about 5 minutes, which corresponds to a power of about 300 terawatts, or 3×10^{14} watts, or the equivalent of 3×10^5 large nuclear power stations for a single spicule. This rough calculation disregards the energy required to evacuate the ambient plasma from the path of the spicule (Sect. 14.9). At any given time, there are about 60 000 to 70 000 spicules on the solar surface (Zirin, 1988).

According to our model, spicules are beams of protons whose speed is about 400 kilometers/second and whose energy is about 1 kilolectronvolt. Then the proton current is

$$I = e\frac{dN_\rho}{dt} = eAv_\rho N_H \approx 10^{15} \text{ amperes} , \tag{14.1}$$

$$J = \frac{I}{A} \approx 64\,000 \text{ amperes/meter}^2 . \tag{14.2}$$

If m' is the mass ejected per second by a single spicule, then

$$m' = \frac{I}{e} m_H \approx 10^7 \text{ kilograms/second} . \tag{14.3}$$

14.2 Devising a Model

Authors generally agree that spicules are jets of some sort. See, for example, Haerendel (1992). Various authors have postulated explosions occurring within the photosphere, for example Dere et al. (1989) and Andreev and Kosovichev (1994). But this is incompatible with the spectacular channeling of spicules.

A model for spicules should ideally provide answers to most of the following questions.

1. What type of particle is involved, and are the particles charged or uncharged (Sect. 14.3)?
2. How can the channeling be so efficient (Sect. 14.4)?
3. What is the origin of the low-energy hydrogen that one sees when observing spicules through an H_α filter (Sect. 14.5)?
4. How can a convecting plasma like that of the photosphere generate a beam of particles? (Sect. 14.7)?
5. Why are spicules associated with magnetic elements (Sect. 14.7)?
6. Why do spicules disappear (Sect. 14.7)?
7. What is the energy of the particles (Sect. 14.8)?
8. How can the rate of advance of the head of a spicule be only about 25 kilometers/second, while the particle velocity, as measured by the Doppler effect, ranges from 50 to 400 kilometers/second (Sect. 14.9)?

Our model attempts to answer all these questions. We shall take them up one after the other, even though they overlap to some extent.

We consider a solar spicule to be a self-channeled particle beam emitted by a magnetic element and composed largely of protons, but also of electrons, neutrals, and heavy ions, surrounded by a cold sheath of hydrogen ions. The orientation of the beam depends on that of the magnetic element and on the ambient magnetic field. The beam originates in a self-excited charged-particle accelerator that exploits a $\boldsymbol{v} \times \boldsymbol{B}$ field below the element (Sect. 11.6). Here $\boldsymbol{v}$ is the local plasma velocity, and $\boldsymbol{B}$ is the local magnetic flux density.

We assume a steady state and a hydrogen plasma.

Within a spicule the Coriolis acceleration is $2\boldsymbol{v} \times \boldsymbol{\Omega}$, where $\boldsymbol{v}$ is the *proton* velocity, 50 to 400 kilometers/second, and $\boldsymbol{\Omega}$ the angular velocity of the Sun. The resulting displacement is unobservable.

14.3 What Type of Particle?

Since spicules more or less follow magnetic field lines, they must be charged-particle beams. But the net vertical electric current must be zero, right up to the top of the spicules, for otherwise the electric potential of the photosphere and of the chromosphere would either increase or decrease indefinitely with time, and spicule emission would soon stop.

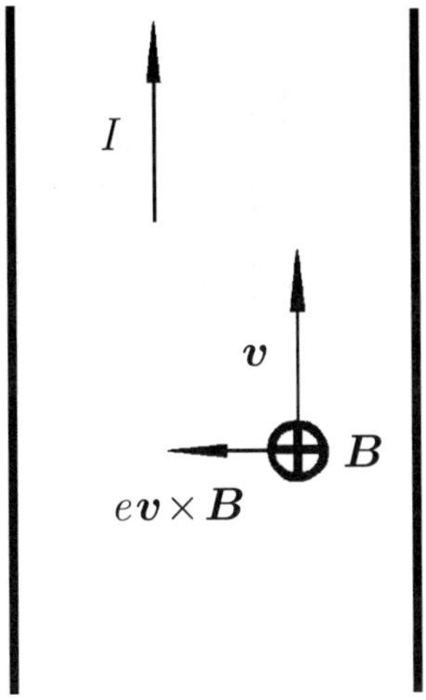

Fig. 14.2. The magnetic pinching force $e\boldsymbol{v} \times \boldsymbol{B}$ on a beam proton

Assuming a hydrogen plasma and disregarding molecular ions, there are three types of particle present: H_0, H_+, and electrons. We disregard H_- ions because their extra electron has a binding energy of only about one electron-volt and is soon stripped (Sweetman, 1971; Prelec, 1977).

As we shall see in Sect. 14.5, spicules appear to be essentially proton beams. At the head, the protons ram heavy ions, neutrals, and electrons of the ambient plasma forward. This might explain the excess concentration of heavy ions such as Na^+ in the solar atmosphere (Meyer, 1991).

The return current is presumably provided by the downward flow of the cold sheath under the action of gravity. Since the flow follows the magnetic field lines, the cold sheath comprises low-energy charged particles, presumably protons. See also Sects. 14.5 and 14.9.2.

14.4 Repulsion and Pinching

In a vacuum, a charged-particle beam tends to blow up because of the repulsive electrostatic force between the particles. But the beam is also pinched by the attractive magnetic force exerted by the beam's azimuthal magnetic field, as in Fig. 14.2. This latter effect is in agreement with the rule that parallel currents attract.

In a vacuum, the net force on a particle of charge e at the periphery of a charged-particle beam is

$$\boldsymbol{F} = \boldsymbol{F}_\mathrm{e} + \boldsymbol{F}_\mathrm{m} = \frac{e\widetilde{Q}}{2\pi\epsilon_0 b}\left(1 - \frac{v^2}{c^2}\right)\hat{\boldsymbol{\rho}}\,, \tag{14.4}$$

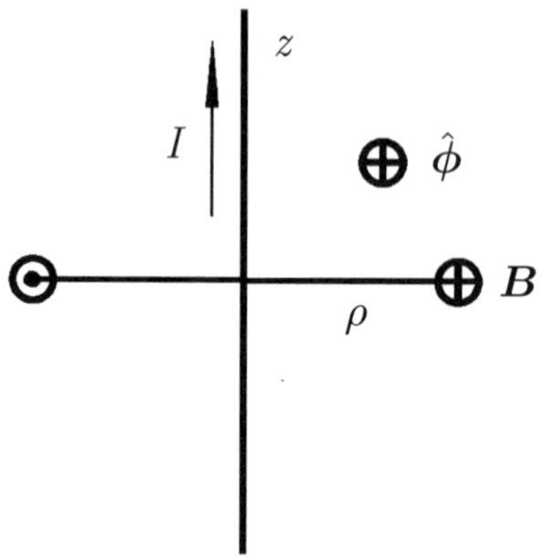

Fig. 14.3. Magnetic field B of a current I, and cylindrical coordinates ρ, ϕ, z. The quantities I and B have the same sign, and $B = \mu_0 I/(2\pi\rho)$

where $\widetilde{Q} = I/v$ is the electric charge per meter, I the beam current, v the particle speed, b the beam radius, and c the speed of light. This relation is valid even at relativistic speeds. The repulsive electrostatic force $\boldsymbol{F}_e$, given by the first term, prevails if $v^2 \ll c^2$. The second term is the *magnetic pinching force* of Fig. 14.2. The net force $\boldsymbol{F}$ is weak at relativistic speeds, when v^2 approaches c^2, but it is always repulsive.

In a low-pressure plasma the propagation of a charged-particle beam is more favorable. When a proton beam propagates in a hydrogen plasma, plasma protons drift away from the beam by electrostatic repulsion, and plasma electrons fall in until the cylindrical region occupied by the beam is nearly neutral. Then the net force $\boldsymbol{F}$ on a proton is equal to zero and we have *self-focusing.* Impact ionization at the beam head produces protons that also drift away, and electrons that remain trapped.

14.5 Channeling

Figures 14.2 and 14.3 show the variables involved in the discussion that follows. Several phenomena concur to channel a proton beam of very large cross-section in a hydrogen plasma, despite scattering.

The protons propagate in the $+z$-direction. We assume that the electrostatic charge density $\widetilde{Q}$ is zero, which makes the electrostatic repulsive force also zero: the first term in Eq. 14.4 is negligible.

a) Consider a proton that is situated at the periphery of the beam and that has a velocity $\boldsymbol{v}$ in the general direction of the z-axis, but pointing somewhat outward. The magnetic field $\boldsymbol{B}$ is azimuthal. The magnetic pinching force

$e(\boldsymbol{v} \times \boldsymbol{B})$ has a negative radial component that reorients the proton in the direction of the axis without slowing it down, since the magnetic force remains orthogonal to the velocity.

Call the proton radial and axial velocities v_ρ and v_z. Assuming a uniform current density J_z, with $\widetilde{Q} = 0$ as above, the magnetic pinching force on a proton at the periphery of a beam of radius b is proportional to b:

$$F_\rho = -ev_z B_\phi = -ev_z \mu_0 \frac{\pi b^2 J_z}{2\pi b} = -\frac{e\mu_0 v_z J_z b}{2} . \tag{14.5}$$

So, other things being equal, the broader the beam, the larger the magnetic pinching force on a peripheral beam proton.

b) Since the pinching force is proportional to the axial proton speed v_z, slow protons leak out at the periphery of the beam. They provide the return current referred to in Sects. 14.3, 14.5, and 14.9.2, losing their gravitational potential energy in falling back.

When observing solar spicules at the $\mathrm{H}\alpha$ wavelength, one sees the shroud of cool hydrogen atoms.

c) The efflux of protons at the periphery, per meter of length, is equal to $2\pi b n v_\rho$, where n is the proton number density inside the spicule, and v_ρ is the radial velocity of a proton. Thus

$$\frac{\partial}{\partial z}(\pi b^2 n v_z) = 2\pi b n v_\rho . \tag{14.6}$$

Since the beam radius b changes slowly with z, we can write that

$$\frac{\partial}{\partial z}(n v_z) = 2n \frac{v_\rho}{b} . \tag{14.7}$$

So the rate of proton loss at the periphery is inversely proportional to the beam radius b; the broader the beam, the fewer protons are lost at the periphery.

d) Radial momentum gradually disappears with increasing z according to a) and c) above. Recall that momentum is conserved during collisions. One would therefore expect the spicule radius to decrease somewhat with altitude, an effect that has not been observed to date.

e) The *transverse temperature* of a particle beam is the temperature measured by an observer moving with the beam at the average particle velocity. The transverse temperature of the beam decreases with increasing z, like the radial momentum.

f) At the periphery, the magnetic force $-e(\boldsymbol{v} \times \boldsymbol{B})$ on the *electrons* that are rammed forward points outward, and they spill out. The beam remains neutral by attracting slow electrons from the ambient plasma.

g) *Velocity channeling* occurs in the following way. This is apparently a new concept. See also Chapter 16.

A proton inside the beam is subjected to a force

$$\boldsymbol{F} = e[\boldsymbol{E} + (\boldsymbol{v} \times \boldsymbol{B})] . \tag{14.8}$$

Immediately after a collision the proton starts off with a random velocity, and the above force applies in the intervals between collisions. Thus, at first sight, the average energy acquired by a proton should be of the order of Fl, where l is the proton mean-free-path. For a steady state, and in the present context, that is incorrect for the following reason.

Imagine that we apply an $\boldsymbol{E}$ field to a hydrogen plasma. Protons move in the direction of $\boldsymbol{E}$, together with the neutrals and electrons that they ram forward. The accelerating region thus becomes a channel inside which matter flows forward, and protons lose little energy in collisions, as if their mean free path were very long. It is this phenomenon that we call *velocity channeling.*

Velocity channeling can be expected if the beam is long and has a large diameter. It is probably unimportant in laboratory-sized particle beams.

Electrons flow backward in the $\boldsymbol{E}$ field, but they have little effect on the forward moving protons because of their relatively small mass.

Velocity channeling in a hydrogen plasma does not apply to an electron beam for the following reason. For a given velocity, the kinetic energy of a particle is proportional to its rest mass. Thus, if a proton and an electron have the same velocity, the proton kinetic energy is about 2000 times larger than that of the electron. Imagine an electron beam in a hydrogen plasma. A velocity channel for the electrons cannot be established because the electron-to-proton energy transfer in a collision is highly inefficient, and the protons never acquire the electron velocity.

h) *Hole boring* (Briggs et al., 1977; Fernsler et al., 1991) probably contributes to channeling at the higher ambient pressures, either below the photosphere or at low altitudes: the beam heats the plasma and reduces the density of scatterers. This effect gains importance with increasing beam radius b, the heating being proportional to the cross-section, or to b^2, while the loss of heat at the periphery is proportional to the circumference, or to b.

i) Observations on solar spicules appear to show that they are fairly immune from the many types of instabilities that affect high-current electron and ion beams generated in laboratories, although some mature spicules have irregular diameters. This immunity of spicules to instabilities is presumably due to their large diameters.

Readers will think of other phenomena that favor the channeling of very broad proton beams in a hydrogen plasma. We return to this subject in Chapter 16.

14.6 Self-excited Dynamos

As we saw in Chapter 10, a *dynamo* is a device that generates an electric current by moving a conductor in a magnetic field. It therefore uses a $\boldsymbol{v} \times \boldsymbol{B}$

field. A *self-excited dynamo* uses its own current to produce the required magnetic field: given a seed magnetic field of outside origin, a system can, under proper conditions, generate an electric current and a further magnetic field, and eventually become self-sustaining after removal of the seed field. A self-excited disk dynamo starts spontaneously as soon as the seed magnetic field is applied. We learned about self-excited dynamos in Chapters 10 and 11.

It has been known since Larmor (Sect. 1.7) that the Sun's magnetic fields originate in self-excited dynamos located in the convecting plasma. The Earth's main magnetic field also originates in self-excited dynamos. Those are situated in the liquid part of the core.

If there were an infinite amount of power available, then the electric current and the magnetic field would both grow exponentially with time. However, with a finite amount of power available, the current and the magnetic field build up, after removal of the seed field, to steady values that correspond to the power input (Chapter 10). There is, of course, conservation of energy; the energy required to set up the magnetic field and to overcome ohmic and other losses comes from the thermal source that drives the convection.

14.7 A Self-excited Accelerator for Spicules

We now investigate a hypothetical $\boldsymbol{v}\times\boldsymbol{B}$ accelerator at the base of a magnetic element that could generate a proton beam. Refer to Sect. 11.6. The accelerator starts spontaneously, as soon as a seed *electric current* appears. The seed current must originate in some outside source. The power source for the accelerator comes from the thermal forces that drive the convection in the photosphere. It is doubtful that such accelerators could exist in laboratory-sized plasmas.

We follow the notation of Fig. 14.4, and use a subscript "a" for the J's and B's in the accelerator.

This accelerator utilizes the vertical component of the local $\boldsymbol{v}\times\boldsymbol{B}_a$ field, where $\boldsymbol{v}$ is the plasma velocity and $\boldsymbol{B}_a$ the magnetic flux density in the accelerator region. Since the accelerator depends on the vagaries of the convection, it operates by bursts, and the beam energy varies with time. The termination of a burst signals the disappearance of the spicule.

The accelerator that we propose here is axisymmetric. This contradicts the Cowling "theorem", according to which a self-excited dynamo cannot be axisymmetric. However, that is not important, because, after over 60 years, there still do not exist valid proofs of the "theorem" (Kolm and Mawardi, 1961; Shercliff, 1965; Fearn et al., 1988; Alexeff, 1989; P. Lorrain, 1991; Ingraham, 1995).

Within the accelerator, protons accelerated upward give spicules, while protons accelerated downward disappear by scattering.

Electrons and protons accelerate in opposite directions. Now, for a given energy, a proton has about 43 times more momentum than an electron. Some

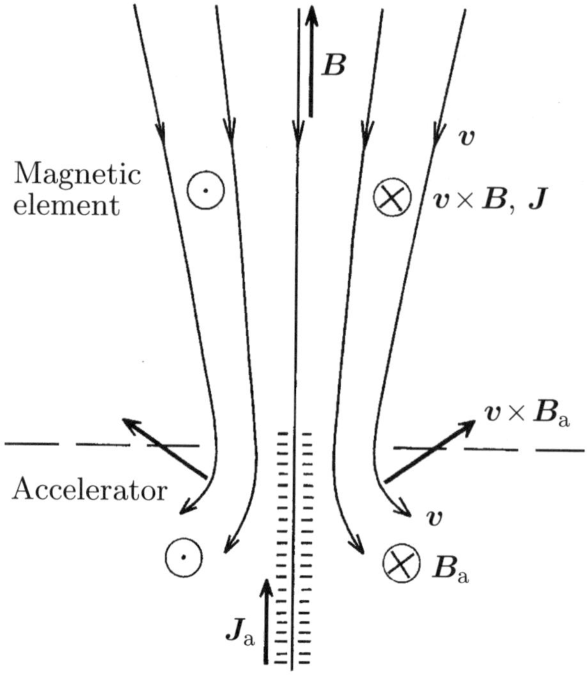

Fig. 14.4. In the accelerator, the downwelling plasma of the magnetic element is on its way back to the surface

of the counter-streaming electrons thus rebound and acquire a velocity component in the direction of the proton velocity, thereby partly canceling the proton current. They are expelled by the magnetic force.

As we saw in Chapter 12, a magnetic element is a more or less vertical magnetic flux tube located in the photosphere. It is a self-excited dynamo that generates an azimuthal electric current and an axial magnetic field, thanks to the *inward* plasma velocity.

Figure 14.4 shows plausible streamlines for the photospheric plasma within and below a magnetic element. The magnetic element extends over the region where the radial component of the plasma velocity is negative. Below the element, the plasma flows outward, and then upward back to the surface.

The region below the magnetic element is the seat of a second self-excited dynamo that is run by the *outward* plasma velocity and that accelerates protons in the upward direction (Sect. 14.7.1). Presumably, thanks to velocity channeling (Sect. 14.5), a proton beam generated in or below the photosphere can emerge.

The axial magnetic field of the magnetic element above the accelerator, as in Fig. 14.4, acts as a gun barrel for the proton beam in the following manner. From Sect. 14.8, the proton energy is of the order of 1 kiloelectronvolt or less. Consider a proton whose velocity has a radial component. The radius of curvature of its trajectory, projected in a plane perpendicular to the axial magnetic field of about 0.15 tesla, is of the order of three centimeters or less. The proton trajectories are thus helices that wrap tightly around the axial field lines of the magnetic element. Above the magnetic element the element's magnetic field vanishes rapidly as at the end of a solenoid, and the protons thereafter follow the Sun's outer magnetic field.

This rotation of the protons about the element's magnetic field lines decreases somewhat the ambient axial magnetic field. Also, the axial component of the proton current contributes an azimuthal magnetic field, which transforms the flux tube of the magnetic element into a flux rope (Chapter 11).

In traversing the element, the proton beam loses part on its energy by scattering. This explains in part the excess thermal radiation from magnetic elements. But see also Sect. 11.5.

14.7.1 J_{a} in the Accelerator, Self-excitation

We return to our general analysis of axisymmetric fields in convecting conducting media (Sect. 11.2), where we found that the current density is given by

$$\boldsymbol{J} = \sigma\left(-\frac{\partial V}{\partial \rho} - v_z B_\phi\right)\hat{\boldsymbol{\rho}} + \sigma(v_z B_\rho - v_\rho B_z)\hat{\boldsymbol{\phi}} + \sigma\left(-\frac{\partial V}{\partial z} + v_\rho B_\phi\right)\hat{\boldsymbol{z}} \,. \quad (14.9)$$

The equations given there are rigorous, but applying them to the accelerator requires various approximations.

We consider successively the above three components of the electric current density $\boldsymbol{J}_{\mathrm{a}}$, starting with $J_{\mathrm{a},\phi}$.

The Azimuthal Current Density $J_{\mathrm{a},\phi}$

First,

$$J_{\mathrm{a},\phi} = \sigma(v_z B_{\mathrm{a},\rho} - v_\rho B_{\mathrm{a},z}) \,. \quad (14.10)$$

Now $J_{\mathrm{a},\phi}$ and $B_{\mathrm{a},z}$ have the same sign. For example, if $J_{\mathrm{a},\phi}$ points in the positive ϕ-direction, then $B_{\mathrm{a},z}$ points in the positive z-direction. With the radial component of the plasma velocity v_ρ positive, the $-v_\rho B_{\mathrm{a},z}$ term above provides *negative* feedback: assuming an initial $B_{\mathrm{a},z}$ and its associated $J_{\mathrm{a},\phi}$, that term tends to reduce both $B_{\mathrm{a},z}$ and $J_{\mathrm{a},\phi}$ to zero. Similarly, the first term in Eq. 14.10, $v_z B_{\mathrm{a},\rho}$, also provides negative feedback; for example, in Fig. 14.4, $v_z < 0$ and $B_{\mathrm{a},\rho} > 0$. We therefore set

$$J_{\mathrm{a},\phi} \approx 0 \,, \quad B_{\mathrm{a},z} \approx 0 \,. \quad (14.11)$$

Since $\boldsymbol{J} = \boldsymbol{\nabla} \times \boldsymbol{B}/\mu_0$, this means that $\partial B_{\mathrm{a},\rho}/\partial z \approx 0$. Also, from Eq. 14.10, $B_{\mathrm{a},\rho} \approx 0$.

The Axial Current Density $J_{\mathrm{a},z}$

Second,

$$J_{\mathrm{a},z} = \sigma \left(-\frac{\partial V}{\partial z} + v_\rho B_{\mathrm{a},\phi} - v_\phi B_{\mathrm{a},\rho} \right) . \tag{14.12}$$

We have just seen that, within the accelerator, $B_{\mathrm{a},z} \approx 0$. Thus $\boldsymbol{B}_\mathrm{a}$ has only ρ and ϕ-components. Then the axial current density $J_{\mathrm{a},z}$ is perpendicular to $\boldsymbol{B}_\mathrm{a}$.

We set $\partial V/\partial z \approx 0$ for the following reason. We assume that the axial component of $\boldsymbol{v} \times \boldsymbol{B}_\mathrm{a}$ accelerates charged particles in the axial direction without accumulating charges on the way, because the return path through the ambient plasma has a very large cross-section, and hence a very low resistance. Then

$$J_{\mathrm{a},z} \approx \sigma \left(v_\rho B_{\mathrm{a},\phi} - v_\phi B_{\mathrm{a},\rho} \right) . \tag{14.13}$$

Assume a seed axial current density $J_{\mathrm{a},z}$ that is positive, or that points up, as in Figures 14.3 and 14.4. Its magnetic field $\boldsymbol{B}_\mathrm{a}$ is azimuthal, in the positive direction, as in the figures. Then, if the radial component v_ρ of the plasma velocity is *positive*, the currents induced by the $v_\rho B_{\mathrm{a},\phi}$ term also point up, and the dynamo action provides an *aiding* axial *current*, as required for self-excitation.

If the seed $J_{\mathrm{a},z}$ is negative, then the axial current points down and its magnetic field $B_{\mathrm{a},\phi}$ points in the negative ϕ-direction. With a *positive* v_ρ, the $v_\rho B_{\mathrm{a},\phi}$ term is then negative, and it again provides an *aiding* axial *current* of density $\sigma v_\rho B_{\mathrm{a},\phi}$.

Therefore, an *outward* motion of the plasma, in cylindrical coordinates, *amplifies* any axial electric *current*, whatever its polarity. Since we have positive feedback, we have a self-excited dynamo.

As for the self-excited disk dynamo (Chapter 10), there is no threshold below which self-excitation does not operate. Once the dynamo has started, σ decreases because of the increasing $B_{\mathrm{a},\phi}$, and the outward v_ρ decreases because of the braking action of the magnetic force, of density $\boldsymbol{J}_\mathrm{a} \times \boldsymbol{B}_\mathrm{a}$ (Sect. 14.7.2), which points inward.

If there were an infinite amount of power available to drive the convection, then the $v_\rho B_{\mathrm{a},\phi}$ term would soon make $J_{\mathrm{a},z}$ infinite. We assume that this positive feedback is overwhelming. So, finally, since $B_{\mathrm{a},\rho} \approx 0$,

$$J_{\mathrm{a},z} \approx \sigma v_\rho B_{\mathrm{a},\phi} . \tag{14.14}$$

Within our assumptions, it is the positive radial component v_ρ of the plasma velocity that generates the axial current density $J_{\mathrm{a},z}$. The field $B_{\mathrm{a},\phi}$ is a function of the coordinates ρ and z, of v_ρ, and of σ, which is itself a function of $B_{\mathrm{a},\phi}^2$ (Cambel, 1963; Braginskii, 1965).

The lines of $\boldsymbol{B}_{\mathrm{a}}$ are conical helices with small $B_{\mathrm{a},\rho}$ and $B_{\mathrm{a},z}$-components.

If the current density $J_{\mathrm{a},z}$ were uniform out to a radius b then we would have

$$2\pi\rho B_{\mathrm{a},\phi} = \mu_0 \pi \rho^2 J_{\mathrm{a},z}\ , \quad B_{\mathrm{a},\phi} = \mu_0 J_{\mathrm{a},z}\frac{\rho}{2}\ . \tag{14.15}$$

The Radial Current Density $J_{\mathrm{a},\rho}$

Finally,

$$J_{\mathrm{a},\rho} = \sigma\left(-\frac{\partial V}{\partial \rho} - v_z B_{\mathrm{a},\phi}\right)\ . \tag{14.16}$$

The first term in the parenthesis is an electric field; the second term is the radial component of the $\boldsymbol{v} \times \boldsymbol{B}_{\mathrm{a}}$ field.

From Figure 14.4, the axial component of the plasma velocity v_z within the accelerator is negative. Say the axial current density $J_{\mathrm{a},z}$ points in the positive direction of the z-axis as in Figure 14.2. Then the azimuthal component of the magnetic flux density $B_{\mathrm{a},\phi}$ points in the positive direction of ϕ, and $-v_z B_{\mathrm{a},\phi}$ is positive. Then that term gives a positive radial current density $J_{\mathrm{a},\rho}$, which means that positive charges drift away from the axis, and negative charges move toward the axis. This makes the region near the axis negative.

The resulting radial electrostatic field $-\partial V/\partial \rho$ opposes the $-v_z B_{\mathrm{a},\phi}$ term. Under steady-state conditions, the electrostatic space charge density near the axis of symmetry is constant, and

$$J_{\mathrm{a},\rho} = 0\ : \tag{14.17}$$

there is zero net radial force on the protons of the beam. The above equation applies, and the proton beam is neither pinched nor expanded, radially.

14.7.2 The Magnetic Force Density in the Accelerator

From Eq. 14.11, with $J_{\mathrm{a},\phi} \approx 0$, $B_{\mathrm{a},z} \approx 0$ as in Eq. 14.11, and $J_{\mathrm{a},\rho} = 0$, the magnetic force density reduces to

$$\boldsymbol{J} \times \boldsymbol{B} = J_{\mathrm{a},z}\hat{\boldsymbol{z}} \times (B_{\mathrm{a},\rho}\hat{\boldsymbol{\rho}} + B_{\mathrm{a},\phi}\hat{\boldsymbol{\phi}}) = -J_{\mathrm{a},z}B_{\mathrm{a},\phi}\hat{\boldsymbol{\rho}} + J_{\mathrm{a},z}B_{\mathrm{a},\rho}\hat{\boldsymbol{\phi}}\ . \tag{14.18}$$

Within our approximations, the magnetic force density has no axial component.

Since $J_{\mathrm{a},z}$ and $B_{\mathrm{a},\phi}$ have the same sign, the radial term $-J_{\mathrm{a},z}B_{\mathrm{a},\phi}$ points inward. It brakes the outward plasma velocity, and makes the pressure higher inside the accelerator than outside. If p is the internal pressure, then, for a steady state,

$$\boldsymbol{\nabla} p = -J_{\mathrm{a},z}B_{\mathrm{a},\phi}\hat{\boldsymbol{\rho}}\ . \tag{14.19}$$

The pressure gradient points inward. The beam current heats the gas, which tends to reduce the density.

The azimuthal component of the magnetic force density in Eq. 14.18 can be either positive or negative, and it can either brake or amplify any azimuthal plasma velocity.

If $J_{a,z}$ were uniform, then from Eq. 14.15, at the radius ρ we would have

$$\boldsymbol{\nabla} p = -\frac{\mu_0}{2} J_{a,z}^2 \rho \hat{\boldsymbol{\rho}} \,, \quad p = p_{ext} + \frac{\mu_0}{4} J_{a,z}^2 (b^2 - \rho^2) \,. \tag{14.20}$$

14.8 The Proton Energy

Speeds of up to 400 kilometers/second have been observed in spicules, as we saw in the Introduction to this chapter. At 400 kilometers/second, a proton has an energy of 835 electronvolts and, were it not for collisions, it could rise to an altitude of 510 megameters above the photosphere, assuming a uniform g. No spicule rises to that height, presumably because the proton beam loses most of its energy within the velocity channel.

Near the poles, where $B \approx 1 \times 10^{-4}$ tesla, proton trajectories wrap around the magnetic field lines. This is in agreement with an energy of the order of 1 kiloelectronvolt: at that energy the radius of curvature of a proton trajectory, in a plane perpendicular to the axis of a spicule, is very much smaller than the observed spicule radii.

Within the accelerator, a proton gains kinetic energy $\mathcal{E}$ at the rate

$$\frac{d\mathcal{E}}{dz} = e(\boldsymbol{v} \times \boldsymbol{B}_a)_z = e v_\rho B_{a,\phi} \tag{14.21}$$

and, at the radius ρ within the accelerator,

$$B_{a,\phi} = \frac{\mu_0}{\rho} \int_0^\rho J_{a,z} \, \rho \, d\rho \,. \tag{14.22}$$

The proton energy and the current density are both functions of the coordinates ρ and z.

14.9 The Beam Head

Figure 14.5 shows the current I, the vector potential $\boldsymbol{A}$ and its time derivative at the beam head. At the beam head, the incident beam protons lose their forward energy in blasting their way through the ambient plasma and in establishing the velocity channel; they are sacrificed. The protons that follow extend the channel and are again sacrificed. Plasma electrons surge into the beam head to cancel the excess positive charge accumulated by the incident protons. See also Sect. 16.4.

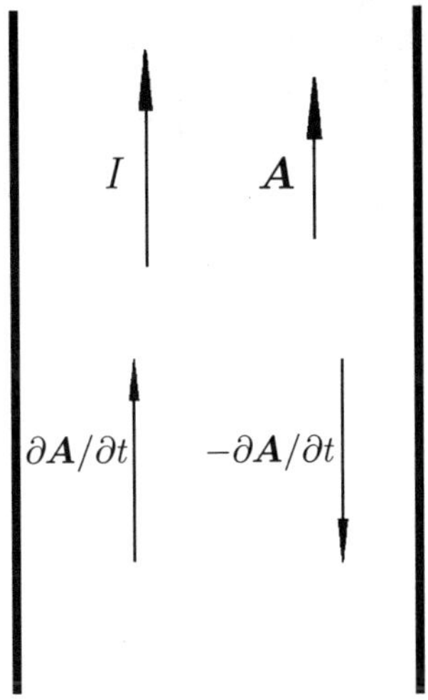

Fig. 14.5. Near the beam head. The vector $\boldsymbol{A}$ points in the same direction as I, and the electric field $-\partial \boldsymbol{A}/\partial t$ points backward

At the maximum height the incident proton energy is exhausted.

It has been suggested that spicules are invisible beyond a certain height because they become too hot for Balmer α emission. This is contradicted by radiometry measurements at 1.3 millimeters performed during the 1991 eclipse and that show no signal beyond an altitude of 7500 kilometers (Lindsey et al., 1992; Foukal, 1992).

14.9.1 The Beam Head Speed

We can find an approximate theoretical expression for the beam head speed in the following way. Assume that the incident protons all have the same axial speed v_z. The beam radius is b. During the time dt the beam head progresses by a distance dz, and the energy required to evacuate the volume $\pi b^2\, dz$ is $\pi b^2 p_{\mathrm{amb}}\, dz$, where p_{amb} is the ambient gas pressure. This requires the kinetic energy stored in a length $v_z\, dt$ of the beam. Thus, if n is the average number density of the protons in the beam and m the proton mass,

$$\pi b^2 p_{\mathrm{amb}}\, dz = \left(\tfrac{1}{2} n m v_z^2\right) \pi b^2 v_z\, dt \;, \tag{14.23}$$

and the speed of the beam head is

$$\frac{dz}{dt} = \frac{n m v_z^3}{2 p_{\mathrm{amb}}} \;. \tag{14.24}$$

In this equation, the right-hand side is the kinetic energy flux density in the beam, divided by the ambient pressure. This beam-head speed is an upper limit because we have disregarded the kinetic energy of the expelled protons.

This expression is interesting, but it is not too useful for a number of reasons. a) The proton number density n inside a spicule is an unknown function of the altitude. b) The axial speed of the incident protons v_z decreases with altitude because of the gain in potential energy and because of scattering. Measurements of the proton speed vary between 50 and 400 kilometers/second, or by a factor of 8, and $8^3 \approx 500$. c) The ambient pressure decreases with altitude (Vernazza et al., 1976, 1981). d) Reported head speeds vary from 25 to 100 kilometers/second, or by a factor of 4.

14.9.2 The Return Current at the Beam Head

The vector potential $\boldsymbol{A}$ of the proton beam points forward as in Fig. 14.5, in the direction of the axial proton velocity, and is largest on the axis. Within the beam head, $\partial \boldsymbol{A}/\partial t$ points forward, and the resulting electric field $-\partial \boldsymbol{A}/\partial t$ points backward. The charged particles of the incident beam, as well as those of the ambient plasma, are all subjected to this backward-pointing axial electric field. The incident protons slow down and the plasma *protons* give a return current of protons, but those are expelled from the beam by the magnetic force because of their retrograde motion.

Plasma *electrons* accelerate forward in this field and transfer their forward momentum to the ambient plasma. They are also expelled by the magnetic force, which points outward because of their negative charge.

14.10 Summary

Solar spicules are proton beams that originate at the base of magnetic elements, where the solar plasma flows outward, and then upward on its way back to the surface. Spicules carry a sheath of excited hydrogen, electrons, and various ions, that one sees when observing at the H_α wavelength. Because the proton beam is very wide, about 150 to 200 kilometers, several phenomena channel the beam and render it more and more homogeneous.

15 Case Study: Solar Coronal Loops as Self-Channeled Proton Beams I

Contents

Solar coronal loops are spectacular luminous arcs that extend upward from the surface of the Sun up to altitudes of 400 megameters, which is 30 times the diameter of the Earth. The beam diameter is about one megameter, when observed at the H_α wavelength, but 0.3 megameter at X-ray wavelengths. How can that be? What are they? Are they charged particle beams? If so, what kind of particle are they? And what is the particle energy, what is the beam current? Just how much power is involved in one loop? In this chapter we attempt to answer all these questions, and the next chapter poses several other questions. We refer to solar coronal loops, but the main properties of loops possibly apply also to fibrils, coronal jets, streamers, threads, filamentary structures in prominences, to a host of thread-like structures in the solar atmosphere, and to solar-like stars.[1]

15.1 Introduction

Various authors consider a coronal loop as an *LRC* circuit, with the electric current generated by the flow of photospheric plasma, and with the circuit

[1] See P. Lorrain, D. Redžić, S. Koutchmy, J. McTavish, O. Koutchmy, *Solar coronal loops as self-channeled proton beams*, Solar Physics, submitted (2005).

Fig. 15.1. Typical solar coronal loops

closing in the subphotosphere. We go further: according to our model, both spicules (Chapter 14) and coronal loops are proton beams. The next chapter accounts for the spectacular channeling.

Our discussion applies equally well to hot and cold, and to flare and non-flare loops.

Loops, like spicules, seem to be self-channeled beams composed mostly of protons and hydrogen atoms, all moving forward at the same velocity, accompanied by thermal electrons and heavy ions, and surrounded by a sheath of excited hydrogen and of various ions. Slow electrons from the ambient plasma that penetrate into the beam are propelled forward by impact, and expelled by the magnetic field of the beam protons.

Particle acceleration possibly occurs in the lower chromosphere, or within or below the photosphere, as proposed by several authors, for example by Zaitsev et al. (1998a,b). The flow pattern proposed by these authors is the same as the one that we proposed in Sect. 14.7 and P. Lorrain and S. Koutchmy (1996). That flow pattern generates the proton beam. Acceleration can be due either to $-\partial \boldsymbol{A}/\partial t$ or to $\boldsymbol{v} \times \boldsymbol{B}$ fields. Acceleration possibly occurs at each turn, somewhat as in a synchrotron, and the energy of the beam protons increases with time. The accelerator is presumably self-excited.

15.2 Observations

There exists an abundant literature on loops. The first half of the book by Bray et al. (1991) provides an excellent review of the observations up to that time. See also the long list of papers by Bray and Loughhead (1983, 1985) entitled "High Resolution Photography of the Solar Chromosphere". For more recent papers see Melrose (1991), Packer (1991), Ulmschneider et al. (1991), Golub (1991), Sams et al. (1992), Klimchuk et al. (1992), Somov (1992), D'Silva (1993), Cargill et al. (1994), Priest (1994), Rušin (1994), Porter and Klimchuk (1995), Kano and Tsuneta (1995), Doschek and Feldman (1996), November and S. Koutchmy (1996). We refer to many papers on specific aspects of loops that were published between 1995 and 2004 throughout the rest of this paper.

There exists a broad range of loop sizes and shapes, and the time scales range from seconds to months (Fleck et al., 1994). Loops are observed both at the limb and on the disk, at H_α, EUV and X ray wavelengths, from Earth-based observatories, from rockets, and from satellites. Temperatures are of the order of millions of degrees. It seems that observed loops are often, in fact, sheafs of thinner loops.

Loops often emerge from the penumbra of sunspots. Sams et al. (1992) observed dozens of loops emerging from the penumbra of a single spot.

Some loops extend to heights of more than 400 megameters, which is 30 times the diameter of the Earth. Some are just as large, but nearly horizontal. Hiei (1994a,b) saw the formation of a loop whose head propagated at a speed of 20 to 40 kilometers/second, eventually giving an arc about the length of one solar radius. Tsuneta (1992) reported the same phenomenon. Aschwanden et al. (1999b,c) have investigated the 3D geometry of loops.

Satellites provide X-ray images of loops on the disk. Those images show highly excited heavy ions such as Fe XXV. The beam width is about 0.3 megameter. Their resolution is inferior to that of H_α images, but there is so little X-ray emission from the photosphere that the contrast at those wavelengths is excellent, and loops appear on the disk as thin bright filaments.

Klimchuk et al. (1992), and then Watko and Klimchuk (2000) measured the widths of loops on the solar disk, and concluded that there is no significant variation of cross-section along the length of a loop. This confirmed earlier observations by Bray et al. (1991). McClymont and Mikic (1994) and Klimchuk et al. (2000) arrived at the same conclusion. Schrijver (2001) found that loop cross-sections are independent of height. Kjeldseth et al. (1998) found that loops appear to be isothermal along their lengths.

As a rule, new loops are small and active. They last for hours, fade, and others grow at a higher and higher altitude (Bray et al., 1991; Gary et al., 1996).

Loops observed at the H_α wavelength have widths of 1 or 2 megameters, except for two thin loops observed by Bray et al. (1991) (see Fig. 16.2) that had diameters of about 130 kilometers, just above a sunspot. See also Sect. 16.5.2.

Low-energy hydrogen, plus presumably protons, electrons, and heavy ions form a sheath around the proton beam. As a rule, sheaths descend along loop legs, driven by gravity and shepherded by the charged particles along the local magnetic field. The resulting electric currents are negligible because the speed of the sheath is lower than that of the beam protons by orders of magnitude.

15.3 What Are They?

Clearly, loops are related to the magnetic field between sunspots or between active regions, but just what is it that one sees? One sees the radiation from excited atoms, but why should that radiation be localized along field lines? A steady magnetic field of course does not ionize a gas.

We must first dispel some common misconceptions. Coronal loops are referred to under a variety of names, for example as "magnetic tubes", "magnetic loops", "magnetic flux tubes", "flux bundles", etc. Are those terms appropriate?

Magnetic field lines are of course invisible because they are not material objects (Sect. 4.1). They are an invaluable "thinking crutch", but no more: they simply indicate the orientation of the vector $\boldsymbol{B}$ at a given point. The above names are therefore inappropriate. In particular, a loop is definitely not a sheaf of individual magnetic field lines.

The term "magnetic flux tube" is equally inappropriate because a magnetic flux tube is a self-excited dynamo that requires an inward plasma flux all along its length (Chapter 11). The concept of magnetic flux tube applies equally well to solar magnetic elements (Chapter 12) and to sunspots (Chapter 13), but it does not make sense in the case of coronal loops.

Figure 15.2 shows field lines for a pair of current-carrying coils. These field lines are somewhat similar to the loops of Fig. 15.1, but many questions come to mind.

1. Near an active region, or near a sunspot, there is a magnetic field $\boldsymbol{B}$ at every point, and space is filled with magnetic field lines. So why should the magnetic field stand out only along those specific field lines occupied by loops?

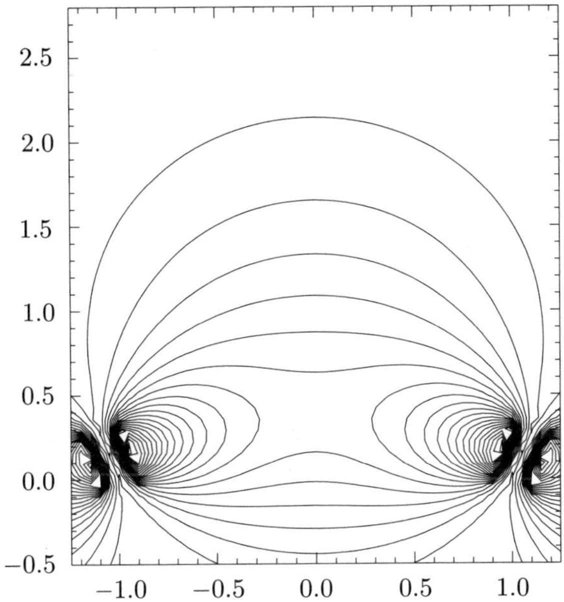

Fig. 15.2. Magnetic field lines for a pair of slanted current-carrying coils. Near the coils, the lines fan out and pinch

2. Just what is it that one sees? One sees radiation from excited atoms, but why should that radiation be localized along field lines? A steady magnetic field of course does not ionize a gas.

3. Could the radiation be thermal bremsstrahlung? If so, then one would have to explain the temperature excess in a loop.

4. Could coronal loops be beams of charged particles that follow magnetic field lines? Maybe so: particle beams are invisible, but they become visible if they ionize the ambient medium, and loops radiate at the H_α wavelength and all the way up to EUV. A particle beam can also radiate if there is bremsstrahlung, as in a synchrotron, but that seems to be out of the question here, the particle energies are much too low.

5. The fact that loops are often horseshoe-shaped shows that the two feet are somehow linked together below the photosphere, but how? Cargill et al. (1994) hypothesize that loops close below the photosphere. So do Zaitsev et al. (1998a,b), and others, who consider a loop to be an *LRC* circuit. The sub-photosphere link, which certainly exists, is reminiscent of sunspots, a sunspot of a given polarity being accompanied by one or more spots of the opposite polarity.

6. If loops are particle beams that more or less follow magnetic field lines, the particles must be charged, but are the particles electrons, or protons, or heavy ions?

7. There is a huge amount of power involved: judging from the fact that loops are visible in the far ultraviolet, we calculate in Sect. 15.7 that the particles must have energies of the order of 100 kiloelectronvolts, or possibly much more. Then the power associated with our reference loop of Sect. 15.7 is of the order of a million gigawatts or more.

8. It seems impossible to distinguish between the two legs of a loop. This possibly means that particle acceleration occurs at both ends.

9. A sheaf of field lines, such as those of Fig. 15.2 increases in diameter by a large factor as one moves away from the source, while loops have the same diameter all along their lengths (Klimchuk and Porter, 1995; Klimchuk et al., 1992, 2000). There are two problems, here. First, how can the beam diameter be independent of the altitude, despite the fact that the magnetic field, the temperature, and the density of the solar atmosphere all vary by many orders of magnitude? Second, how can the beam maintain its diameter close to the photosphere, where the magnetic field lines diverge or converge rapidly? So the ambient magnetic field guides the beam, but only to a certain extent.

10. Another striking characteristic of coronal loops of a given family is that they do not interact at all, as is obvious in Figs. 15.1 and 15.3. Now a particle beam has a magnetic field of its own, with the result that parallel particle beams attract each other. So a family of loops should quickly coalesce into a single one, contrary to the observations.

11. Also, there are cases where two families of loops repel each other, as in Fig. 15.3. How can parallel beams repel each other, if they do not attract each other?

Our model attempts to answer all these questions.

As we shall see, coronal loops seem to be self-channeled beams consisting mostly of protons accompanied by electrons, heavy ions, and neutrals. The acceleration possibly occurs below the photosphere at both feet. The acceleration seems to occur at each turn, somewhat as in a synchrotron, and the energy of the beam protons increases with time. The accelerators are presumably self-excited.

Assuming that we have answered all these question, many problems remain. To name just three: why do loops have a specific diameter, of the order of one megameter? And how can the penumbra of a single sunspot harbor dozens of loops? That seems to require a hugely complex convection in the penumbra and below. And why do the two "loops" of Fig. 16.2 diverge?

Fig. 15.3. Two families of coronal loops repelling each other (TRACE mission from NASA)

15.4 Loops As Charged-particle Beams

Considering loops as charged-particle beams is not a new concept. Many years ago Jefferies and Orrall (1965) suggested that loops are beams of fast particles originating in the photosphere. Such fast particles would not be visible, but an enveloping sheath of excited hydrogen and ions would be. Foukal (1978), and later Priest (1981) and Leha et al. (1996), concluded that coronal loops must carry a considerable current. Packer (1991) found evidence suggesting the existence of a net current flowing along coronal loops, from one foot to the other. According to Melrose (1991), measurements of the orientation of the magnetic field suggest that a current flows along loops, from one foot to the other. J. Chen (1989, 1990), Cargill et al. (1994), Demoulin and Klein (2000), and Klimchuk et al. (2000) observed that the magnetic field lines of loops are twisted, and concluded that there exists a field-aligned current along loops. McClymont and Mikic (1994), using Yohkoh vector magnetogram data, also concluded that loops carry electric currents. Various authors have proposed that coronal loops can be considered as LRC circuits, assuming therefore that a loop carries a current and that the circuit closes below the photosphere.

Clearly, flows above and below footpoints should be investigated (P. Lorrain and O. Koutchmy, 1998).

The direction of twist observed on loop sheaths provides the direction and magnitude of the loop current I. The longitudinal magnetic field in a loop is the guiding field minus the field of the orbiting protons, as explained in Sect. 16.2.2. The azimuthal component is $\mu_0 I/(2\pi\rho)$, where ρ is the radial position.

The sum of the net vertical currents in the two legs of a loop is zero, for otherwise the solar electrostatic charge would drift, and loop formation would soon stop.

Much work has been done on the propagation of particle beams. See, for example, the paper by Bekefi et al. (1980) on particle-beam weapons, and a later paper by Hughes and Godfrey (1984) on small-angle scattering in charged-particle beams. Peratt (1986) has written an extensive paper in two parts on filamentary electric currents in cosmic plasmas.

If loops are indeed particle beams, then the beam emerges from the photosphere at one foot, and plunges back into the photosphere at the other foot. Since it seems impossible to distinguish between the two legs of a loop, the energy expended by the proton beam in forming the sheath must be negligible. Then the loop resistance R is zero, or nearly so. The circuit presumably closes below the photosphere, but the beam is not appreciably affected by traveling a few hundred megameters: relatively few protons are lost on the way, and the protons lose a relatively small amount of their kinetic energy.

There exists an abundant literature on laboratory-sized charged and uncharged particle beams, but laboratory experiments seem to be irrelevant here because loops have such huge dimensions, coronal pressures are so low, and the temperatures are so high.

15.5 The Beam Current and the Beam Power

Tables 2.2, 2.9, and 4.17 of Bray et al. (1991) provide observed values for magnetic fields in loop sheaths, for mature loops. Let us calculate the corresponding value of the beam current I. The calculation is highly approximate, first because the observed values vary widely and, second, because the sheath current, if any, adds to the beam current. But see Sect. 16.8. Also, as noted above, because of the finite resolution, what seems to be a single loop might be a sheaf of several thinner loops.

The ion beam carries a current I and its magnetic field is $\boldsymbol{B}$, as in Fig. 14.2. At X-ray wavelengths, beams have half-widths of about 0.3 megameter. However, their magnetic fields are measured within the sheath. For $B \approx 3$ milliteslas at a radius of one megameter,

$$B = \mu_0 \frac{I}{2\pi b}\,, \quad I = \frac{2\pi b B}{\mu_0} \approx \frac{2\pi \times 10^6 \times 0.003}{4\pi \times 10^{-7}} \approx 10^{10} \text{ amperes}\,. \qquad (15.1)$$

In view of the wide variety of coronal loops, this current is in agreement with Zaitsev et al. (1998a,b), who find a current of about 6×10^{10} to 1.4×10^{12} amperes. With our figures, the proton flux is about 10^{29} protons/second, or about 170 kilograms/second.

What is the beam power? Assuming a proton energy of 100 kiloelectronvolts (Sect. 15.7), the beam power is about 10^{15} watts, or 10^6 gigawatts.

As mentioned above, new loops are usually small and active; they last for hours, fade, and others grow at a higher and higher elevation (Bray et al., 1991; Gary et al., 1996). It is during this stage that the magnetic and kinetic energies build up.

15.6 The Magnetic Energy

What is the inductance L of a coronal loop? Assume that the loop is plane, and that it forms a complete circle (Cargill et al., 1994). Then we can use formulas given in Engineering handbooks for the inductance of a closed loop of wire. For a circular loop of mean radius a and wire radius b,

$$L = 39.37a \left[7.353 \log_{10} \left(\frac{8a}{b} \right) - 6.386 \right] \times 10^{-8} \text{ henrys} . \tag{15.2}$$

Setting $a = 100$ megameters and $b = 1$ megameter,

$$L \approx 600 \text{ henrys} . \tag{15.3}$$

For a current of 10^{10} amperes, the energy stored in the magnetic field that links such a loop is

$$\mathcal{E}_{\text{mag}} \approx \frac{1}{2} L I^2 \approx 3 \times 10^{22} \text{ joules} , \tag{15.4}$$

or the energy produced by a one-gigawatt reactor in about one million years.

At the center of the above loop,

$$B = \frac{\mu_0 I}{2a} \approx 6 \times 10^{-5} \text{ tesla} \approx 1 \text{ gauss} . \tag{15.5}$$

15.7 The Proton Energy

There are a number of ways of estimating the order of magnitude of the proton energy in loops and in spicules. We discuss a few here, but readers will probably think of others.

1. First, what are the thermal speeds of a proton and of an electron at a temperature of one million degrees? For the z-component of the velocity, $mv^2/2 = kT/2 = eV$ joules, where e is the magnitude of the charge of the electron, and where the voltage V is loosely called the energy, in electronvolts:

$$v_{\text{proton, thermal}} = \left(\frac{kT}{m} \right)^{1/2} \approx \left(\frac{1.4 \times 10^{-23} \times 10^6}{1.7 \times 10^{-27}} \right)^{1/2} \tag{15.6}$$

$$\approx 10^5 \text{ meters/second} , \tag{15.7}$$

$$v_{\text{electron, thermal}} \approx 4 \times 10^6 \text{ meters/second} . \tag{15.8}$$

In both cases, $V \approx 40$ electronvolts.

2. What about the gravitational potential energy? Consider a singly-charged particle of mass m. It emerges from the photosphere at a speed v_0 and with an energy eV joules. It rises to an altitude H. Then, with $g = 272$ meters/second^2 at the surface of the Sun,

$$\tfrac{1}{2}mv_0^2 \geq mgH \ , \quad v_0 \geq (2gH)^{1/2} \ , \quad v_0 \geq 23H^{1/2} \ , \tag{15.9}$$

$$eV \geq mgH \ . \tag{15.10}$$

With $H = 100$ megameters and for a proton,

$$v_0 \geq 233 \text{ kilometers/second} \approx 200 \text{ kilometers/second} \ , \tag{15.11}$$

$$V \geq \frac{mgH}{e} \approx 300 \text{ electronvolts} \ . \tag{15.12}$$

So an energy of only 300 electronvolts is enough to drive a proton up to an altitude of 100 megameters above the photosphere. (If one takes into account the variation of g with altitude, one finds an energy of 248 electronvolts.)

Since, as we shall see, the beam proton kinetic energies are much larger than 300 electronvolts, we can ignore the gravitational force. We can also disregard the Coriolis force $m(2\boldsymbol{v} \times \boldsymbol{\Omega})$, where $\boldsymbol{\Omega}$ is the angular velocity of the Sun.

3. Now the Sun carries a net charge of roughly +80 coulombs, and the electrostatic potential at the surface is roughly +1000 volts (Sect. 3.9). The reason for the existence of this charge is that electrons and protons are at about the same temperature and that there are more electrons than protons that have the *gravitational* escape speed. With a positive surface potential, the escape speed for electrons increases, the escape speed for protons decreases, and the net escaping current is zero. The net current must be zero, for otherwise the electrostatic potential of the Sun would drift one way or the other indefinitely. See Sect. 3.9.

So what is the kinetic energy gained by a proton that rises above the surface of the Sun? A rough calculation will suffice. Half way up a 100-megameter loop, at an altitude of 50 megameters, the solar radius being 700 megameters,

$$E \approx \frac{80}{4\pi\epsilon_0(7.5 \times 10^8)^2} \approx 1.3 \times 10^{-6} \text{ volt/meter} \ . \tag{15.13}$$

Then, in moving upward a distance of 100 megameters from the surface of the Sun, a proton gains a kinetic energy of about 130 electronvolts because of the repulsive electrostatic force, while it loses a kinetic energy of 300 electronvolts because of the attractive gravitational force, for a net loss of about 170 electronvolts.

4. One can measure the rate of advance of the head. In the case of spicules, that speed is typically 25 kilometers/second (P. Lorrain and S. Koutchmy, 1996). Now, at 25 kilometers/second, a proton could only reach an altitude of about 1 megameter, and loops reach a few hundred megameters. As discussed in P. Lorrain and S. Koutchmy (1996), and as we shall see below in Sect. 16.4, the head speed is much lower than the speed of the beam protons.

The speed of propagation of the loop beam head reported by Hiei (1994a,b) was 20 to 40 kilometers/second.

5. It is possible to measure speeds along loops, either from the Doppler shift of the H_α line, or by observing the motion of "knots", or condensations, along the loop (Bray et al., 1991). Those speeds range from 20 to 150 kilometers/second. But both methods yield the speeds of the *sheath*, and not the speed of the protons in the beam.

The observed velocities along coronal loops are up one leg (Wood et al., 1998) and down the other, but usually down both legs (Schrijver, 2001), never up both legs. It is easy to understand why the sheath velocity can be downward in both legs: the sheath falls in the gravitational field. One can also expect an upward velocity along one leg if, during the upward part of the trajectory of the loop, the expelled protons and hydrogen atoms start with a sufficiently large upward velocity. With this interpretation, the sheath cannot flow up both legs. In the case of spicules, the sheath velocity is either upward or downward.

6. The fact that coronal loops are commonly observed at X-ray wavelengths provides a reliable lower limit for the proton energy. Doschek et al. (1995) observed loops in the wavelength band 0.3 to 1 nanometer. Now, if λ is the wavelength in meters,

$$V = \frac{h\nu}{e} = \frac{hc}{e\lambda} \approx \frac{10^{-6}}{\lambda} \,. \tag{15.14}$$

That group therefore observed transitions in excited heavy ions in the energy range 1 to 3 kiloelectronvolts. Then what is the energy of the beam protons? In view of the many unknowns, the following will suffice. Say the proton energy is E_p, and assume that, in an inelastic head-on collision, the heavy ion, of mass K times larger than the proton mass, is at rest before the collision. The excitation energy of the heavy ion is E_i. Conservation of energy and of momentum implies that excitation of the heavy ion cannot occur if E_p lies below the threshold value

$$E_\mathrm{p} = E_\mathrm{i}\frac{K+1}{K} \approx E_\mathrm{i} \,. \tag{15.15}$$

So the proton energy is also about 3 kiloelectronvolts. This value applies all along a loop because the gravitational potential energy is negligible.

This 3 kiloelectronvolts is a *minimum* value for the kinetic energy of the beam protons because coronal loops are presumably also visible at shorter wavelengths. So the beam protons probably have kinetic energies that are much larger than either the thermal, or gravitational, or electrostatic energy. This explains, in part, why loop cross sections are independent of the altitude.

Say the proton energy is 100 kiloelectronvolts. Then

$$v = \left(\frac{2eV}{m}\right)^{1/2} \approx 1.37 \times 10^4 V^{1/2} \tag{15.16}$$

$$\approx 4.3 \times 10^6 \text{ meters/second} \tag{15.17}$$

for our reference loop. If the loop has a visible length of 200 megameters, a proton takes about 40 seconds to travel from one foot to the other. It then continues below the photosphere for, say, the same distance. Then it takes roughly one minute to go around once. Below the photosphere the beam presumably traverses self-excited particle accelerators [Sect. 14.7 above; P. Lorrain and S. Koutchmy (1996)] and emerges once more.

With $I = 10^{10}$ amperes, $b = 0.3$ megameter, and $v = 5 \times 10^6$ meters/second, the proton number density in a loop is

$$n = \frac{I}{\pi b^2 e v} \approx \frac{10^{10}}{\pi \times (0.3 \times 10^6)^2 \times 1.6 \times 10^{-19} \times 5 \times 10^6} \tag{15.18}$$

$$\approx 4 \times 10^{10}/\text{meter}^3 \,. \tag{15.19}$$

15.8 Our Reference Loop

Our *reference loop* has the following characteristics:

loop mean radius $a \approx 100$ megameters,
beam radius $b \approx 0.3$ megameter,
current $I \approx 10^{10}$ amperes,
$B_{\text{sheath}} \approx 3$ milliteslas,
proton density $n \approx 4 \times 10^{10}/\text{meter}^3$,
proton energy ≈ 100 kiloelectronvolts,
proton speed $\approx 5 \times 10^6$ meters/second,
proton flux $\approx 10^{29}$/second,
magnetic energy $\mathcal{E}_{\text{mag}} \approx 3 \times 10^{22}$ joules,
beam mechanical power $IV \approx 10^{15}$ watts,
beam mass for the semi-circular part of the loop ≈ 7 tonnes.

15.9 Summary

Solar coronal loops appear to be self-channeled beams composed mostly of protons, but accompanied by electrons, heavy ions, and neutrals. Our model attempts to explain the main features of loops: it explains both the high-energy particles observed at X-ray wavelengths and the low-energy particles observed at H_α wavelengths. Judging by the fact that loops are visible in the far ultraviolet, the proton energy could be as large as 100 kiloelectronvolts or more. We calculate order-of-magnitude numerical values for a typical reference loop for the current, the proton density, the proton flux, the magnetic energy, the beam power, and the beam mass.

16 Case Study: Solar Coronal Loops as Self-Channeled Proton Beams II

Contents

Although this chapter is a continuation of the discussion of solar coronal loops of Chapter 15, it can be read independently. As we shall see, the channeling of loops is even more astonishing than what we found in the previous chapter. The huge size of the phenomenon makes loop Physics totally inaccessible to laboratory experiments.[1]

16.1 Introduction

Possibly the most striking property of loops is the fact that the beam radius is the same along their lengths, over a distance that can be as large

[1] See P. Lorrain, D. Redžić, S. Koutchmy, J. McTavish, O. Koutchmy, *Solar coronal loops as self-channeled proton beams*, Solar Physics, submitted (2005).

as 600 megameters. This is despite the fact that they go through highly inhomogeneous magnetic fields, and that the magnetic field, the pressure, the temperature, and the density of the ambient medium all vary by many orders of magnitude over their lengths. Now, save in exceptional conditions, laboratory-sized particle beams *always* diverge. Moreover, loops of a given family do not interact at all, while loops of two different families repel. As we shall see, a broad proton beam is efficiently channeled in a hydrogen plasma.

16.2 Channeling a Broad Proton Beam

Refer to Sect. 11.3 for a discussion of Magneto-Fluid Dynamics in axisymmetric convecting conducting fluids.

We use cylindrical coordinates ρ, ϕ, z, with the z-axis pointing along the beam, in the direction of the proton velocity. We set $\partial/\partial z = 0$ and $\partial/\partial \phi = 0$. We neglect relativistic effects, so that $v^2 \ll c^2$, and we assume that the protons have zero energy spread (S. Koutchmy and Livshits, 1992; Koutchmy et al., 1994; S. Koutchmy et al., 1997; Ferreira and Mendoza-Briceno, 1997; Drake et al., 2000).

We also assume a steady state, and no specific beam profile: the proton density is an arbitrary function of the radial position. We disregard oscillations (Aschwanden et al., 1999a; Nakariakov et al., 1999; Robbrecht et al., 2001).

A proton beam can propagate over long distances in the solar atmosphere only if the channeling is highly efficient.

The cross-section for ionization by a proton is a function of the proton energy. It is of the order of 10^{-20} meter^2 for a proton energy of a few tens of electronvolts, and rapidly decreases with increasing proton energy (Freeman and Jones, 1974). Since the beam proton kinetic energy is much larger than that of the ambient hydrogen atoms, inelastic scattering is highly improbable. So only proton-proton and proton-hydrogen atom electric scattering matter.

Proton-hydrogen atom elastic collisions are usually described by the hard-spheres model, leading to the above cross-section, which is independent of the proton energy. In the photosphere, the hydrogen atom number density n_H is about 10^{23}/meter^3 (Vernazza et al., 1981) and the mean-free-path is only about 1 millimeter. Above, in the chromosphere, the mean-free-path is very much longer. For example, at an altitude of 2.7 megameters, n_H is about 5×10^{14}/meter^3 (Vernazza et al., 1981), and the mean-free-path is approximately 200 kilometers. Recall that loops rise up to as much as 400 megameters above the photosphere.

As we shall see, many phenomena contribute to the channeling of a broad proton beam in the solar atmosphere (Chapter 14). Some prevent the beam from either fanning out or pinching (Sect. 16.5.2), while others make it more and more homogeneous.

16.2.1 Electric and Magnetic Forces on a Proton Beam, Without Rotation

Most of the discussion in this section also applies to Sect. 16.2.2, where we take rotation into account.

Consider a proton beam of arbitrary radius b in a hydrogen plasma, in a region remote from the beam head. The only magnetic field, aside from the guiding field, is that of the proton current. In this section we assume individual proton trajectories that are rectilinear:

$$v_\rho = 0\,, \quad v_\phi = 0\,, \quad v_z = v \neq 0\,. \tag{16.1}$$

We also assume, for the moment, that the beam contains only protons, of number density n and speed v, electrons of number density n_e of unspecified speed v_e, and hydrogen atoms. The hydrogen atoms inside the beam form a moving medium. The beam protons and the beam hydrogen atoms flow without resistance through the sea of electrons, because of the large kinetic energy of the protons and hydrogen atoms, and also because of the great mass difference: the electron sea is completely transparent for the heavy beam particles.

With a net charge density $\widetilde{Q}_f$, a net current density $\boldsymbol{J}$, a proton velocity $\boldsymbol{v}$, and a gas pressure P, the force per unit volume on protons inside the beam is the sum of three terms:

$$\widetilde{\boldsymbol{F}}_p - \widetilde{Q}_p(\boldsymbol{E} + \boldsymbol{v} \times \boldsymbol{B}) - \boldsymbol{\nabla} P_{p \mid H}\,, \tag{16.2}$$

where the subscript "p" stands for protons and the subscript "H" stands for hydrogen, where $\widetilde{Q}_p = en$, and where n is the number of protons per cubic meter.

Contrary to magnetic flux tubes (Chapter 11), which are partly evacuated, the total gas pressure inside a coronal loop is the sum of two non-negative independent terms, the proton plus hydrogen pressure P_{p+H}, and the electron pressure P_e. As we shall see in Sect. 16.8, the absence of interaction between neighboring loops is an indication that $\boldsymbol{\nabla} P_{p+H}$ is equal to zero. The vanishing of $\boldsymbol{\nabla} P_{p+H}$ inside the beam is also corroborated by the conclusions reached in Sect. 16.2.2 below. Therefore, to calculate the force on a beam proton, we need only take into account the electric and magnetic forces.

1. Let us calculate the radial force $\boldsymbol{F}$ on a proton at the radius ρ. Since we have assumed axisymmetry, the sheath has zero electric and magnetic fields inside the beam. As we shall see below, again in Sect. 16.8, the sheath fields seem to be negligible.

There is a radial electric force on a beam proton that can be either attractive or repulsive, depending on the sign of the net charge density $\widetilde{Q}_f$, and there is an attractive magnetic force. By hypothesis, the two forces cancel:

$$\boldsymbol{F} = e[\boldsymbol{E} + (\boldsymbol{v} \times \boldsymbol{B}))] = \boldsymbol{0}\,. \tag{16.3}$$

The electric radial force $e\boldsymbol{E}$ at the radius ρ depends only on the net charge per unit length inside that radius. Similarly, the radial magnetic force $e(\boldsymbol{v} \times \boldsymbol{B})$ at ρ depends only on the beam current inside that radius. We disregard the magnetic field of the electrons that are impelled forward; that field cancels part of the magnetic field of the proton beam, but those electrons are soon expelled by the magnetic field of the beam protons.

It seems reasonable to assume that the proton axial velocity is constant, but our calculations show that there is in fact a small oscillatory term.

Assuming a constant proton axial velocity v, we can find the net space charge density $\widetilde{Q}_{\mathrm{f}}$ as follows. Assuming that v is independent of ρ, then, from one of Maxwell's equations and from Eq. 16.3,

$$\widetilde{Q}_{\mathrm{f}} = \epsilon_0 \boldsymbol{\nabla} \cdot \boldsymbol{E} = -\epsilon_0 \boldsymbol{\nabla} \cdot (\boldsymbol{v} \times \boldsymbol{B}) \tag{16.4}$$

$$= -\epsilon_0 \boldsymbol{\nabla} \cdot (-v B_\phi \hat{\boldsymbol{\rho}}) = \epsilon_0 \frac{1}{\rho} \frac{\partial}{\partial \rho} (\rho v B_\phi) , \tag{16.5}$$

with

$$B_\phi = \frac{\mu_0}{2\pi\rho} \int_0^\rho nev\, 2\pi\rho\, d\rho = \frac{\mu_0 nev\rho}{2} . \tag{16.6}$$

Thus

$$\widetilde{Q}_{\mathrm{f}} = \epsilon_0 \frac{1}{\rho} \frac{\partial}{\partial \rho} \left(\frac{\mu_0 nev^2 \rho^2}{2} \right) = \epsilon_0 \frac{1}{\rho} \mu_0 e v^2 n \rho = en \frac{v^2}{c^2} . \tag{16.7}$$

Recall that $\epsilon_0 \mu_0 = 1/c^2$. By hypothesis, the net charge density $\widetilde{Q}_{\mathrm{f}}$ is independent of both ρ and the beam diameter b. The net charge density is also very much smaller than the proton charge density: in the present case, $v^2/c^2 \approx 3 \times 10^{-4}$. So the electron number density n_{e} is only slightly smaller than the proton number density n:

$$n_{\mathrm{e}} = \left(1 - \frac{v^2}{c^2} \right) n . \tag{16.8}$$

The beam must be slightly positive to provide a repulsive electric force on a proton that cancels the attractive magnetic force.

The radial electric force on a proton is

$$F = \frac{e^2}{2\pi\epsilon_0 \rho} \int_0^\rho (n - n_{\mathrm{e}}) 2\pi\rho\, d\rho - \frac{\mu_0}{2\pi\rho} v \int_0^\rho e^2 vn\, 2\pi\rho\, d\rho . \tag{16.9}$$

Since we have assumed that the proton speed v is independent of ρ, the v in front of the second integral can go under its integral sign and

$$F = \frac{e^2}{\epsilon_0 \rho} \int_0^\rho \left[\left(1 - \frac{v^2}{c^2} \right) n - n_{\mathrm{e}} \right] \rho\, d\rho = 0 . \tag{16.10}$$

We assume that v is independent of ρ. This is probably not a bad approximation in view of points 2 and 6 below.

At first sight, the above result seems unrealistic because one expects that the ambient electrons should swarm in to ensure neutrality. That proves to be incorrect. See Sect. 16.7.

We return to Eq. 16.10 later on in Sect. 16.6.

Note that the hydrogen atoms inside the beam form a uniformly moving, non-magnetic, medium. The polarization vector is

$$\boldsymbol{P} = \epsilon_0(\epsilon_\mathrm{r} - 1)[\boldsymbol{E} + (\boldsymbol{v} \times \boldsymbol{B})] \,. \tag{16.11}$$

Taking into account Eq. 16.3 and the definition

$$\boldsymbol{D} = \epsilon_0 \boldsymbol{E} + \boldsymbol{P} \,, \tag{16.12}$$

one finds that $\boldsymbol{D} = \epsilon_0 \boldsymbol{E}$ inside the beam [Sect. 7.2 above; Redžić (2001); P. Lorrain (2001)]. Thus the first equality in Eq. 16.5, as well as the first term in Eq. 16.9, are consequences of the Maxwell equation for $\boldsymbol{D}$.

Now consider our reference loop of Sect. 15.8, with $B \approx 3$ milliteslas and $v \approx 5 \times 10^6$ meters/second. The attractive magnetic force on a proton at the periphery is then about 2×10^{-15} newtons. The electric force is of the same magnitude, but repulsive. Now such a force, if applied to a proton (mass 1.7×10^{-27} kilogram), would give it an inward acceleration of about 10^{12} meters/second^2! So the electric and magnetic radial forces on a beam proton are both huge, but they balance *exactly*, and the beam is remarkably stable.

The ambient plasma contains heavy ions, and the beam protons propel them forward (P. Lorrain and S. Koutchmy, 1996; Meyer, 1991). Because heavy ions are slow and heavily charged, they contribute disproportionately to the space charge.

What happens to the rare ambient protons and electrons that manage to drift into the beam? Those protons are propelled forward, and most are expelled, as we show below. Ambient electrons that fall into the beam are also propelled forward. They feel an attractive electric force, and a repulsive magnetic force. The radial electric force on an electron points inward and has the same magnitude as the outward electric force on a proton. The radial magnetic force on a forward-moving electron points outward. Do those two forces cancel as in the case of the protons? Definitely not, because a forward-moving electron has been impelled forward by a proton and has a much higher speed than the proton. So forward-moving electrons are expelled by the beam's magnetic field.

Electrons, protons, and hydrogen atoms that are rammed roughly forward form a sheath composed of excited hydrogen atoms, heavy ions, and electrons. The sheath cools by radiation. Radiation energy losses were investigated by Jefferies and Orrall (1965), among others.

The contribution of the heavy ions to the beam current is appreciable, but we neglect the heavy ion current in what follows.

2. *Velocity channeling* applies to broad, long, proton beams in a medium consisting mostly of hydrogen: the proton beam creates a channel inside

which protons and hydrogen atoms all flow forward at the same axial velocity [Sect. 14.5 g) above; Hughes and Godfrey (1984); P. Lorrain and S. Koutchmy (1996)]. This makes the proton mean-free-path in the forward direction very long, even in a dense plasma. This also increases the thermalization time (Jefferies and Orrall, 1965) by a large factor.

Velocity channeling does not occur with electron beams in hydrogen for the following reason. Since the mass of an electron is much smaller than that of a proton or of a hydrogen atom, fast forward-moving electrons transfer only a small fraction of their forward velocity to protons and hydrogen atoms, and electrons cannot establish a velocity channel in a hydrogen plasma.

3. *Hole boring* possibly contributes to the channeling, especially below the photosphere: the beam heats the ambient plasma, and hence reduces its density, increasing the mean-free-path. If hole-boring does occur, then it increases with beam radius b because the heat input is roughly proportional to b^2, while the heat loss at the periphery is proportional to b. Thus, other things being equal, the plasma should be less dense inside a broad beam than inside a narrow one.

4. Assuming a uniform proton number density n, and a uniform proton forward velocity v, the radial *inward* magnetic force on a proton at the periphery of a beam of radius b is proportional to v^2:

$$F_{\mathrm{mag}} = -evB_\phi = -ev\frac{\mu_0 \pi b^2 nev}{2\pi b} = -e^2 \mu_0 n v^2 \frac{b}{2} \, . \tag{16.13}$$

Thus slow protons leak out at the periphery of the beam.

5. The rate of proton loss at the periphery is inversely proportional to the beam radius b: for a given proton radial velocity v_ρ,

$$\frac{\partial}{\partial z}(\pi b^2 nv) = -2\pi b n v_\rho \, , \quad \frac{\partial}{\partial z}(nv) = -\frac{2nv_\rho}{b} \, . \tag{16.14}$$

This favors broad beams. We have assumed that the beam radius b does not change with z. The fact that loop diameters are uniform all along their lengths shows that proton losses are negligible.

6. The above equation also shows that the rate of proton loss at the periphery is proportional to v_ρ. This filters out wayward protons, gradually ridding the beam of its radial momentum and its transverse temperature. (The transverse temperature is the temperature that would be measured by an observer moving at the average forward velocity of the protons.)

So the beam transverse temperature is probably much lower than ambient. The electron temperature is presumably the same as the ambient temperature.

If the proton radial velocity $\boldsymbol{v}_\rho$ is positive, then the magnetic force $e(\boldsymbol{v}_\rho \times \boldsymbol{B})$ points forward; if $\boldsymbol{v}_\rho$ is negative, then the magnetic force points backward.

Readers will probably think of other phenomena that foster the propagation of broad proton beams in a hydrogen plasma.

16.2.2 Magnetic Forces on a Beam Proton, with Rotation

Here is a more realistic situation. Section 16.2.1 still applies, but now the protons describe helices in the ambient axial magnetic field $\boldsymbol{B}_{a,z}$.

The proton helices have random radii, and most are off-axis. Each proton is tied to its helix; it can change helix only by exchanging energy with another proton or with a heavy ion. We assume that $v_\perp \ll v_\parallel$.

How large is the helix radius R? For our reference beam (Sect. 15.8), with $B \approx 3$ milliteslas in the sheath, set $B_a \approx 1$ millitesla and

$$v_\perp = |v_\rho^2 + v_\phi^2|^{1/2} \ll v_\parallel \approx 5 \times 10^6 \text{ meters/second} , \tag{16.15}$$

$$v_\perp \leq 10^3 \text{ meters/second} , \tag{16.16}$$

$$R = \frac{m v_\perp}{B_a e} \approx \frac{1.7 \times 10^{-27} \times 10^3}{10^{-3} \times 1.6 \times 10^{-19}} \approx 1 \text{ centimeter} . \tag{16.17}$$

So the helix radius R is roughly 8 orders of magnitude smaller than the beam radius b, which is of the order of one megameter.

Assume overall axisymmetry, and assume that the sum of the magnetic fields of all the helices is the same as that of an axial coil in which the current density has an appropriate radial dependence. Then the rotation of the protons, each on its own helix, generates an axial axisymmetric magnetic field B_{rot}.

Now $v_\perp$ and B_z have opposite signs for the following reason. Say B_z is positive. Then, if $v_\perp$ is also positive, the magnetic force $\boldsymbol{v} \times \boldsymbol{B}$ on a proton points away from the helix axis, and the proton escapes. But if $v_\perp$ is negative, then $\boldsymbol{v} \times \boldsymbol{B}$ points inward and the proton stays on its helix. Similarly, if B_z is negative, $v_\perp$ must be positive. So the product $v_\perp B_z$ is always negative for protons in the beam. In other words, for an observer looking in the direction of the beam proton velocity, or in the positive direction of $\boldsymbol{v}$, the protons rotate clockwise if B_z is positive, and counterclockwise if B_z is negative. Multiplying $\boldsymbol{v} \times \boldsymbol{B}$ by e gives the centripetal force. Since $v_\perp$ and B_z have opposite signs, the orbiting protons generate a magnetic field B_{rot} that cancels part of the ambient axial field B_a:

$$B_z = B_a + B_{rot} , \quad |B_z| = |B_a| - |B_{rot}| . \tag{16.18}$$

The net B_z is minimum at $\rho = 0$. The existence of B_{rot} does not affect Eq. 16.10.

The proton angular velocity is the so-called "cyclotron frequency"

$$\boldsymbol{\omega} = -\frac{B_z e}{m} \hat{\boldsymbol{z}} . \tag{16.19}$$

Setting $B_z \approx 1$ millitesla,

$$\omega \approx 10^5 \text{ radians/second} . \tag{16.20}$$

The helix radius R, the period of rotation $T = 2\pi/\omega$, and the pitch P are all inversely proportional to $|B_z|$. For our reference beam (Sect. 15.8),

$$R = \frac{mv_\perp}{|B_z|e} \approx 1 \text{ centimeter} , \tag{16.21}$$

$$T = \frac{2\pi}{\omega} = \frac{2\pi m}{|B_z|e} \approx 60 \text{ microseconds} , \tag{16.22}$$

$$P = \frac{2\pi}{\omega} v_\| = \frac{2\pi m}{|B_z|e} v_\| \approx 300 \text{ meters} . \tag{16.23}$$

Since B_z is not uniform, ω is not uniform and the proton helices are not circular: the radius of curvature of a helix, projected on a plane perpendicular to the axis of symmetry, is smallest near the periphery of the beam. The protons do not rotate as a solid, and the helices precess about the axis of symmetry.

Since the beam width is unaffected by altitude (Klimchuk et al., 1992; Watko and Klimchuk, 2000; Klimchuk et al., 2000; Schrijver, 2001), the net value of B_z is seemingly unaffected by the ambient magnetic field B_a. The ambient field nonetheless guides the beam as in Sect. 16.5 below.

If the ambient magnetic field outside the Sun were the same as that of an ideal magnetic dipole at the center, with its $1/r^3$ radial dependence, then the magnitude of the field at an altitude of 400 megameters would be about one third of its value at the surface. One would rather expect that the ambient field would decrease very much faster than that because the magnetic field seems to originate in currents flowing close to the surface, in the photosphere.

The equation of motion of a beam proton is

$$m\frac{d\boldsymbol{v}}{dt} = e[\boldsymbol{E} + (\boldsymbol{v} \times \boldsymbol{B})] . \tag{16.24}$$

Here, $\boldsymbol{E}_\| = \boldsymbol{0}$: the electric field cannot have a longitudinal component, for otherwise the voltage difference between the two ends of a loop would be near infinite. So

$$\boldsymbol{E} = \boldsymbol{E}_\perp \tag{16.25}$$

and

$$\boldsymbol{v} \times \boldsymbol{B} = (\boldsymbol{v}_\| + \boldsymbol{v}_\perp) \times (\boldsymbol{B}_\| + \boldsymbol{B}_\perp) \tag{16.26}$$

$$= \boldsymbol{v}_\| \times \boldsymbol{B}_\perp + \boldsymbol{v}_\perp \times \boldsymbol{B}_\| + \boldsymbol{v}_\perp \times \boldsymbol{B}_\perp , \tag{16.27}$$

since $\boldsymbol{v}_\|$ and $\boldsymbol{B}_\|$ are parallel, while the angle between $\boldsymbol{v}_\perp$ and $\boldsymbol{B}_\perp$ can have any value. Thus

$$m\frac{d\boldsymbol{v}}{dt} = e(\boldsymbol{E}_\perp + \boldsymbol{v}_\| \times \boldsymbol{B}_\perp + \boldsymbol{v}_\perp \times \boldsymbol{B}_\| + \boldsymbol{v}_\perp \times \boldsymbol{B}_\perp) . \tag{16.28}$$

Since the net *radial* force is zero,

$$\boldsymbol{E}_\perp + \boldsymbol{v}_\parallel \times \boldsymbol{B}_\perp = \boldsymbol{0} \,, \tag{16.29}$$

and the first two terms in Eq. 16.28 cancel.

The third term on the right of Eq. 16.28 points in a direction perpendicular to the helix axis, and its contribution is zero, on average. As we saw above, the period of the helical motion is only about 60 microseconds.

The fourth term, $e(\boldsymbol{v}_\perp \times \boldsymbol{B}_\perp)$, points in either the $+\hat{\boldsymbol{z}}$ or the $-\hat{\boldsymbol{z}}$-direction. So it either accelerates or decelerates the proton in the $\hat{\boldsymbol{z}}$-direction and it fluctuates in the same manner as the third term. Its net effect is zero.

The $e(\boldsymbol{v} \times \boldsymbol{B})$ force does not affect the kinetic energy of the protons because it is orthogonal to $\boldsymbol{v}$. Neither does the $e\boldsymbol{E}$ force because

$$e\boldsymbol{E} \cdot \boldsymbol{v} = e\boldsymbol{E}_\perp \cdot \boldsymbol{v}_\perp \,,$$

which averages to zero. So the proton kinetic energy is conserved. The angular momentum of the beam is similarly conserved because there is no externally applied torque.

16.3 Diamagnetism of the Proton Beam

It is interesting to consider the proton beam as a magnetized medium with a "magnetization" $\boldsymbol{M}$. Each proton, in rotating along its helix, generates a magnetic field much like atoms in a magnetized medium. Think of a loop of wire of radius R that carries a current I. Its magnetic moment is $\pi R^2 I$. In the case of an orbiting proton, the current is its charge e divided by the period of rotation, $2\pi/\omega$. So a proton that orbits on a circular trajectory of radius R has a magnetic moment

$$\boldsymbol{m} = \frac{e\omega}{2\pi} \pi R^2 \hat{\boldsymbol{z}} = \frac{ev_\perp^2}{2\omega} \hat{\boldsymbol{z}} \,. \tag{16.30}$$

Recall that $\omega = -eB_z/m$. The magnetic moment per unit volume, or the "magnetization" is n times larger:

$$\boldsymbol{M} = \frac{ne\omega\pi R^2}{2\pi} \hat{\boldsymbol{z}} = \frac{nev_\perp^2}{2\omega} \hat{\boldsymbol{z}} \,, \tag{16.31}$$

whatever the sign of B_z. The magnetization can be a function of the radial coordinate ρ, since n, B_z, and R can all be functions of ρ. One can also write that

$$\boldsymbol{M} = -\frac{n\mathcal{E}_\perp}{B_\parallel^2} B_\parallel \hat{\boldsymbol{z}} \,, \tag{16.32}$$

where $\mathcal{E}_\perp$ is the proton kinetic energy associated with the transverse motion. Note that $\boldsymbol{M}$ and $\boldsymbol{B}_\parallel$ have opposite signs: the medium is diamagnetic.

The equivalent current density associated with the helical motion of the protons is

$$\boldsymbol{\nabla} \times \boldsymbol{M} = -\frac{d}{d\rho}\left(\frac{nev_\perp^2}{2\omega}\right)\hat{\boldsymbol{\phi}} . \tag{16.33}$$

We have assumed that the beam is both axisymmetric and infinite in the z-direction, so that $\partial/\partial z = 0$, $\partial/\partial\phi = 0$, and $\partial/\partial\rho = d/d\rho$. The sum of the magnetic fields of the orbiting protons is the same as that of an axisymmetric, ρ-dependent volume distribution of azimuthal currents. Thus, from the Maxwell equation for the curl of $\boldsymbol{B}$,

$$\boldsymbol{\nabla} \times (B_{\mathrm{rot}}\hat{\boldsymbol{z}}) = \mu_0 \boldsymbol{\nabla} \times \boldsymbol{M} \tag{16.34}$$

and

$$\frac{dB_{\mathrm{rot}}}{d\rho} = \mu_0 \frac{d}{d\rho}\left(\frac{nev_\perp^2}{2\omega}\right) . \tag{16.35}$$

Integrating, and using the fact that $B_{\mathrm{rot}} = 0$ for $n = 0$,

$$B_{\mathrm{rot}} = \frac{\mu_0 nev_\perp^2}{2\omega} . \tag{16.36}$$

Substituting $\omega = -e(B_{\mathrm{a}} + B_{\mathrm{rot}})/m$,

$$B_{\mathrm{rot}} = -\frac{\mu_0 nv_\perp^2 m}{2(B_{\mathrm{a}} + B_{\mathrm{rot}})} , \tag{16.37}$$

and

$$B_{\mathrm{rot}}^2 + B_{\mathrm{a}}B_{\mathrm{rot}} + \tfrac{1}{2}\mu_0 nv_\perp^2 m = 0 , \tag{16.38}$$

so that

$$B_{\mathrm{rot}} = -\frac{B_{\mathrm{a}}}{2} \pm \frac{B_{\mathrm{a}}}{2}\left(1 - \frac{2\mu_0 nv_\perp^2 m}{B_{\mathrm{a}}^2}\right)^{1/2} . \tag{16.39}$$

What about the $\pm$ sign? If $v_\perp$ is zero, then $B_{\mathrm{rot}} = 0$, so one must choose the plus sign:

$$B_{\mathrm{rot}} = -\frac{B_{\mathrm{a}}}{2} + \frac{B_{\mathrm{a}}}{2}\left(1 - \frac{2\mu_0 nv_\perp^2 m}{B_{\mathrm{a}}^2}\right)^{1/2} . \tag{16.40}$$

Outside the beam, $n = 0$, $B_{\mathrm{rot}} = 0$, and the net axial field is

$$B_z = \frac{B_{\mathrm{a}}}{2} + \frac{B_{\mathrm{a}}}{2}\left(1 - \frac{2\mu_0 nv_\perp^2 m}{B_{\mathrm{a}}^2}\right)^{1/2} . \tag{16.41}$$

It follows from Eq. 16.39 that B_{rot} and B_{a} have opposite signs, whereas Eq. 16.41 implies that B_{a} and B_z have the same sign. So B_{rot} and B_z have opposite signs, as expected.

The above derivation of Eqs. 16.39 and 16.41 is valid in the general axisymmetric case, with n, $v_\perp$, and B_z arbitrary functions of ρ. Assuming that n is a decreasing function of ρ and that $v_\perp$ is uniform, then $|B_{\mathrm{rot}}|$ decreases and $|B_z|$ increases with ρ. If $\boldsymbol{M}$ is uniform, then the corresponding current density vanishes, and there is a uniform surface current at the periphery of the beam, generating a uniform axial field B_{rot}. However, since one cannot expect a uniform net axial magnetic field B_z over the cross-section of the beam, ω is not uniform and the protons do not rotate in synchronism.

16.4 At the Beam Head

Occasionally, one observes a loop that starts at one foot, describes an arc, and plunges at what becomes the other foot of the loop (Hiei, 1994a,b). Let us see what happens at the beam head. See also Sect. 14.9.

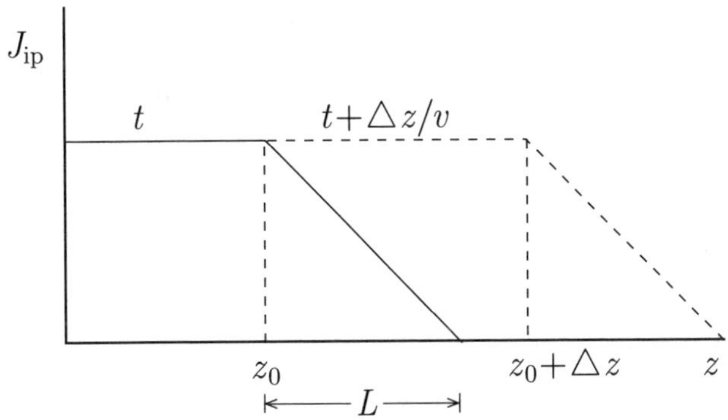

Fig. 16.1. The incident proton current density J_{ip} as a function of the time t at the beam head. We have shown a linear drop in J_{ip} with z, as an approximation for an exponential decrease

Choose the z-axis parallel to the beam, pointing in the direction of the proton velocity, which is either parallel or antiparallel to the local magnetic field. See Fig. 16.1. The impact zone at the beam head starts at $z = z_0$ and extends over a length L. For $z < z_0$, the beam is well established. Inside the impact zone, the current density J_{ip} is a function of both z and t. Set

$$J_{ip} = J_{ip0} \exp\left(-\frac{z - z_0}{L}\right) . \tag{16.42}$$

One might call the length L the attenuation distance. The beam head progresses at the speed dz_0/dt, which is much lower than the incident proton speed v.

Inside the impact zone, the incident protons clear the channel by colliding with ambient protons, electrons, heavy ions, and hydrogen atoms. The net positive charge in the beam, as in Eq. 16.7, supplies a forward-pointing macroscopic electric field. The incident protons in the impact zone lose most of their forward velocity, and the protons that follow extend the channel.

The net current density $\boldsymbol{J}_{net}$ and the vector potential $\boldsymbol{A}$ in the impact zone both point forward. At a given point, $\boldsymbol{A}$ increases with time, and $\partial \boldsymbol{A}/\partial t$ points forward, in the direction of $\boldsymbol{J}_{net}$. So the corresponding electric field $-\partial \boldsymbol{A}/\partial t$ points backward. This backward electric field accelerates electrons forward, slows down the incident protons, and accelerates both ambient and rebounding protons backward. Since $\boldsymbol{A}$ is maximum on the axis of symmetry, the retrograde current is maximum on the axis, which reduces B_ϕ near the axis, in the beam head.

16.5 Guiding the Proton Beam

16.5.1 Near-axial Ambient Magnetic Field

We consider first a single proton that is shot roughly along the z-axis into a uniform, axial magnetic field. The proton describes a helix. In Cartesian coordinates,

$$\boldsymbol{v}_0 = v_{0x}\hat{\boldsymbol{x}} + v_{0y}\hat{\boldsymbol{y}} + v_{0z}\hat{\boldsymbol{z}}\ , \quad \boldsymbol{B} = B_z\hat{\boldsymbol{z}}\ , \tag{16.43}$$

with

$$v_{0x} \ll v_{0z}\ , \quad v_{0y} \ll v_{0z}\ . \tag{16.44}$$

Assume that v_{0z} and B_z are both positive.

The proton is situated at (x, y, z). It describes a helix of radius R whose axis is parallel to the z-axis and situated at $x = x_0$, $y = 0$.

We now add a small x-component to the ambient magnetic field:

$$\boldsymbol{B} = B_z\hat{\boldsymbol{z}} + B_x\hat{\boldsymbol{x}}\ , \ \text{with}\ \ B_x \ll B_z\ . \tag{16.45}$$

We can show that the axis of the helix lies on the local magnetic field line as follows. The magnetic force on a proton is

$$\boldsymbol{F} = e(v_x\hat{\boldsymbol{x}} + v_y\hat{\boldsymbol{y}} + v_z\hat{\boldsymbol{z}}) \times (B_z\hat{\boldsymbol{z}} + B_x\hat{\boldsymbol{x}}) \tag{16.46}$$

$$= e\left[v_yB_z\hat{\boldsymbol{x}} - (v_xB_z - v_zB_x)\hat{\boldsymbol{y}} - v_yB_x\hat{\boldsymbol{z}}\right] \tag{16.47}$$

$$\approx e\left[v_yB_z\hat{\boldsymbol{x}} - (v_xB_z - v_zB_x)\hat{\boldsymbol{y}}\right]\ . \tag{16.48}$$

Then

$$F_x = m\frac{dv_x}{dt} = ev_yB_z\ , \tag{16.49}$$

$$F_y = m\frac{dv_y}{dt} = e(v_zB_x - v_xB_z)\ . \tag{16.50}$$

Assume that B_z, B_x, and v_z are constant. Then

$$v_x = v_0\sin\left(\frac{eB_z}{m}t + \phi\right) + \frac{v_zB_x}{B_z}\ , \tag{16.51}$$

$$v_y = v_0\cos\left(\frac{eB_z}{m}t + \phi\right)\ , \tag{16.52}$$

where ϕ is an arbitrary constant.

The average v_x is proportional to B_x. It shifts the helix in the direction of the local magnetic field line. If the helix overshoots, then F_x changes sign.

We now investigate the behavior of the beam in a non-uniform magnetic field.

16.5.2 Diverging or Converging Ambient Magnetic Field

Coronal loops originate and end in regions where electric currents flow, in and below the photosphere. Now the magnetic field of a current distribution disappears rapidly, away from the distribution. One good example is the magnetic field at the end of a solenoid. Think of a solenoid of radius R and length L. Inside, the magnetic flux density is B_0 and, outside, on the axis, at a distance $D \gg L$,

$$\frac{B_D}{B_0} = \frac{R^2 L}{2D^3} \,. \tag{16.53}$$

Say $L = R$ and $D = 10R$, then $B_D/B_0 = 5 \times 10^{-4}$: the magnetic field lines diverge rapidly. See also Fig. 15.2.

The fact that loops neither expand nor contract close to the photosphere is therefore surprising. A closer examination shows, however, that *axial* ambient magnetic field variations involve no change in the beam radius along a loop, even if the field varies by orders of magnitude. See Fig. 16.2. The explanation is as follows.

The helix radius R for our reference loop is 1 centimeter for $B_a = 1$ millitesla, and 10 meters for $B_a = 1$ microtesla. Then the helix radius is 5 orders of magnitude smaller that the beam radius b. So the beam protons generate electric and magnetic fields as in Sect. 16.2.1. The increase of the helix radius with decreasing B_a does not affect the beam radius: the protons still travel inside a cylinder 1 megameter in radius. A proton trajectory, projected in a plane perpendicular to the beam axis, is not a circle. Presumably, a change in the helix radius R only affects the frequency of proton-proton and proton-hydrogen collisions, and those collisions only affect the transverse momenta of the particles, without affecting their axial speed.

Let us look at the problem in another way. Let the ambient magnetic field have a ρ-component. Is the beam radius affected? Set

$$\boldsymbol{B}_a = B_{a,\rho}\hat{\boldsymbol{\rho}} + B_{a,z}\hat{\boldsymbol{z}} \quad \text{with} \quad B_{a,\rho} \ll B_{a,z} \,, \tag{16.54}$$

and assume axisymmetry. Then the equation of motion is (with $B_\phi = 0$)

$$\boldsymbol{F}_\rho = e(E_\rho + v_\phi B_z)\hat{\boldsymbol{\rho}} \,, \tag{16.55}$$

$$\boldsymbol{F}_\phi = e(v_z B_\rho - v_\rho B_z)\hat{\boldsymbol{\phi}} \,, \tag{16.56}$$

$$\boldsymbol{F}_z = e(-v_\phi B_\rho)\hat{\boldsymbol{z}} \,, \tag{16.57}$$

with $B_\rho = B_{a,\rho}$ and $B_z = B_{a,z}$.

Disregard the z-component of the magnetic force, which either accelerates or decelerates the protons, somewhat. Disregard also the ϕ-component of the force, which either increases or decreases the azimuthal velocity v_ϕ. Then we need only consider $\boldsymbol{F}_\rho$.

Both B_z and v_z can be either positive or negative. Recall that the product $v_\phi B_z$ is always negative, for protons to remain in the beam, as we saw above.

The electric force eE_ρ on a proton points outward. So the net outward force on a proton is

$$F_\rho = e(v_\phi B_z) + eE_\rho$$
$$= e(v_\phi B_z)\,. \tag{16.58}$$

Clearly, B_ρ has no effect on the radial force F_ρ.

In other words, B_ρ does not give a radial force: whether B_ρ is positive or negative, the proton beam does not fan out on emerging from the photosphere, and it does not pinch on plunging back into the photosphere.

Observe that, when a proton enters a region where the magnetic field diverges,

$$\omega = \frac{B_z e}{m} \text{ decreases and } R = \frac{m v_\phi}{eB_z} \text{ increases.}$$

16.6 What Determines the Beam Diameter?

H_α images show the sheaths, that have diameters of one or two megameters. But, at X-ray wavelengths, one sees the loop itself, and the diameters are only about 0.6 megameter.

Remarkably, none of our calculations yield the beam diameter, but the above phenomenon is possibly important here. Some readers might wish to investigate what happens if the beam radius b increases or decreases by db. See Item 5 in Sect. 16.2.1. We wish to know what is the feedback mechanism that keeps the beam diameter of the order of one megameter, despite the fact that the ambient B varies by many orders of magnitude. The following might be of some help.

At first sight, a plot of the total energy of the beam per meter as a function of ρ should have a minimum at the observed diameter. That proves to be wrong. Call the beam radius b, the net electric charge density $\widetilde{Q}$, and the beam current density J, with both $\widetilde{Q}$ and J independent of ρ. Then the electric energy per meter inside the beam is

$$\mathcal{E}_e = \int_0^b 2\pi\rho\, d\rho\, \frac{\epsilon_0 E^2}{2}\,, \tag{16.59}$$

with

$$E = \frac{1}{2\pi\epsilon_0\rho}\int_0^\rho 2\pi\rho'\, d\rho'\, \widetilde{Q} = \frac{\rho}{2\epsilon_0}\widetilde{Q}\,, \tag{16.60}$$

so that

$$\mathcal{E}_e = \frac{\pi}{16\epsilon_0}\widetilde{Q}^2 b^4\,. \tag{16.61}$$

There is no minimum.

Similarly, the magnetic energy per meter inside the beam is

$$\mathcal{E}_{\mathrm{m}} = \frac{\pi\mu_0}{16} J^2 b^4 \,. \tag{16.62}$$

There is again no minimum. We have disregarded the energy density of the ambient $\boldsymbol{B}$, which is independent of the beam radius b.

Outside the beam, the electric and magnetic energies are both proportional to $\ln(\infty/b)$, assuming that the fields extend to infinity. Again, there is no minimum.

What about the kinetic energy? Whether one assumes rectilinear or helical motion, the kinetic energy is independent of $\boldsymbol{B}$ because the magnetic force is orthogonal to $\boldsymbol{v}$.

What prevents the beam from becoming infinitely thin? The high temperature? No, the beam temperature is probably lower than ambient. Also, the inward $\boldsymbol{v} \times \boldsymbol{B}$ force is countered by the outward centrifugal force.

So what determines the beam diameter? At this stage, one can only surmise that the value of b is fixed below the photosphere, in particle accelerators similar to the one that presumably generates spicules (Chapter 14).

16.7 The Forces Outside the Beam

We now address the fact that coronal loops maintain their identity, despite being submerged in a plasma. We calculate the forces exerted on ambient protons and electrons situated *outside* the beam.

At the radius ρ outside, with a beam of protons of speed v, radius b, and a uniform net volume charge density $\widetilde{Q} = ne(v^2/c^2)$,

$$\boldsymbol{E} = \frac{\pi b^2 ne(v^2/c^2)}{2\pi\epsilon_0\rho}\hat{\boldsymbol{\rho}} \,, \tag{16.63}$$

$$\boldsymbol{B} = \mu_0 \frac{I}{2\pi\rho}\hat{\boldsymbol{\phi}} = \mu_0 \frac{\pi b^2 nev}{2\pi\rho}\hat{\boldsymbol{\phi}} = \frac{\pi b^2 nev}{(\epsilon_0 c^2)2\pi\rho}\hat{\boldsymbol{\phi}} \,. \tag{16.64}$$

So $E/B = v$. For our reference beam of Sect. 15.8,

$$\frac{\mu_0 I}{2\pi b} \approx 3 \times 10^{-3} \text{ tesla} = 30 \text{ gauss} \,. \tag{16.65}$$

We have disregarded the electric and magnetic fields of the sheath. Since $\widetilde{Q}$ is positive (Eq. 16.7), the electric force on a *proton* points outward. So ambient protons are repelled. The magnetic force, which depends on the proton's velocity vector, is oriented at random and can be disregarded.

The electric force on an *electron* is attractive because the net charge density $\widetilde{Q}$ in the beam is positive, and the magnetic force $e(\boldsymbol{v}_{\mathrm{e}} \times \boldsymbol{B})$ is orthogonal to the trajectory. Say the electron has a velocity $\boldsymbol{v}_{\mathrm{e}}$ that points in the direction of the beam, at a distance ρ, at a given moment. At that point, the electron curves either up or down, and the radius of curvature of its trajectory is $R = mv_e/(Be)$.

Say $B \approx 0.3$ millitesla, or about one tenth of the B of Sect. 15.8, ten radii outside the beam, and set $v_e \approx 10^6$ meters/second. Then, at that point, the radius of curvature of the electron's trajectory is of the order of a centimeter. So the magnetic field acts as a shield that prevents ambient electrons from reaching the beam, much as the Earth's magnetic field shields it from most cosmic rays. So the beam carries a net positive charge, despite the fact that it lies in a sea of electrons.

We have disregarded the electric and magnetic fields of the sheath, but the above radius of curvature is so small compared to the beam radius that, even if B were smaller by orders of magnitude, electrons would still be prevented from reaching the beam.

16.8 No Interaction Within a Given Family?

One of the many striking characteristics of coronal loops is that loops of given family do not interact at all. Similarly, spicules, fibrils, and filaments in prominences do not interact. It has long been known that laboratory-sized ion beams sometimes exhibit a similar behavior when a beam splits into two or more seemingly independent filaments (Weibel, 1959; Lee and Lampe, 1973; Miller, 1982). This is surprising because one expects an interaction between the electric and magnetic fields. So there must be both channeling and separation, which seems contradictory.

Bray et al. (1991), in their remarkable Fig. 2.5, which we reproduce here in Fig. 16.2, show two exceptionally thin loops, 130 kilometers wide, that either emerged from a sunspot, or that plunged into it, and that lasted only a few minutes. The loops show no sign of interaction, except for repulsion in the rapidly diverging magnetic field. Bray and Loughhead (1985) have discussed these loops in detail. But see Sect. 16.5.2.

Why is it that neighboring loops, or spicules, or ion filaments, do not coalesce? Consider a straight ion beam. Splitting it into two decreases the electrostatic energy, because ions of a given type repel, and increases the magnetic energy because parallel currents attract. Note that, if a beam splits, the charge and the current distributions inside each beam shift somewhat. Say the two beams are separated by a distance D. You increase the spacing by dD. A little thought will show that the new *electrostatic* force of repulsion is somewhat larger than if the charge distribution within each beam had not shifted. On the contrary, the new *magnetic* force of repulsion is somewhat less than if the current distributions had not shifted. So the repulsive force decreases somewhat more slowly than as $1/D$, while the attractive force decreases somewhat faster than as $1/D$. In other words, as the loops move away from each other, the repulsive force becomes slightly larger than the attractive force.

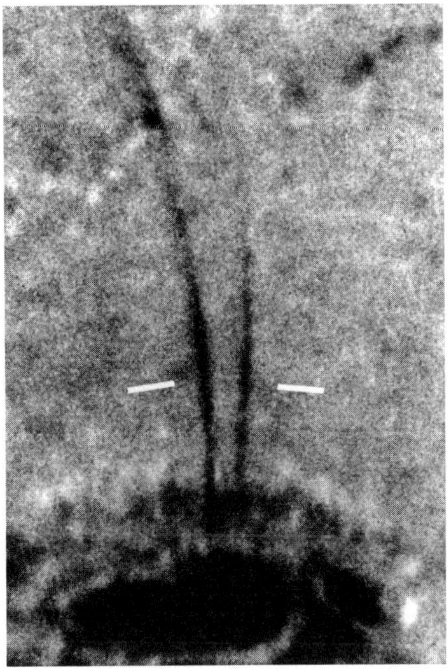

Fig. 16.2. Pair of loops, each 130 kilometers wide, observed by Bray et al. just above a sunspot (Fig. 2.5 of Bray et al., 1991, reproduced with permission)

Here is another way of looking at the problem of the lack of interaction between loops of a given family. Consider the straight portions of two neighboring loops, A and B. At a distance R outside loop A, $\boldsymbol{E}$ and $\boldsymbol{B}$ are given by Eqs. 16.63 and 16.64. We again disregard the fields of the sheaths. A proton moves at a speed v in loop B situated at the distance R. The proton velocities in the two loops point in the same direction. The electric force on a proton of loop B is repulsive, and the magnetic force is attractive, and they cancel. So, neglecting the fields of the sheaths, there is zero force between two parallel loops, with equal velocities pointing in the same direction.

What if one takes into account the fields of the sheaths? Since neighboring loops do not interact, one is led to believe that the sheaths carry a negligible charge, and a negligible current. The negligible current is easy to understand: the sheaths fall at relatively low speeds, under the action of gravity. It seems that the electric charge is also negligible.

Mok et al. (2001) simulated the magnetic interaction between two loops, without realizing that they would merge. They took no account of the above.

16.9 Two Families of Loops Repel?

Occasionally, one sees two families of loops that collide and repel, as in Fig. 15.3.

Let us turn again to the case of the straight portions of two neighboring loops A and B. If the currents in loops A and B flow in opposite directions, then the electric and magnetic forces are both repulsive, and loops A and B repel. So, if there is repulsion, the proton beams in the two families must flow in opposite directions. Say the proton beams in the left-hand family all point in the general direction away from the observer. Then the proton beams in the right-hand family all flow toward the observer.

16.10 Summary

Our model attempts to explain the main characteristics of solar coronal loops. They appear to be proton beams accompanied by electrons, heavy ions, and neutrals. Given their huge width of about one megameter, there is self-channeling: several phenomena increase the mean-free-path and render the beam more and more homogeneous. This maintains both the identity and the width of the beams over hundreds of megameters. The model also explains why loops follow magnetic field lines, despite the fact that the ambient pressure, magnetic field, and density all vary by many orders of magnitude over the distances traveled. It explains why the proton beams do not fan out on emerging from the photosphere, or pinch on returning to the photosphere. So the ambient magnetic field guides the beam, but only to a certain extent. Our model also proposes explanations for the fact that loops of a given family do not interact, while loops of two different families repel.

Part V

Appendices

A Characteristic Lengths and Times, a Justification[1]

Contents

It is common practice in subjects pertaining to Magneto-Fluid-Dynamics, for instance in Geophysics and in Astrophysics, to evaluate the order of magnitude of a quantity by referring to "characteristic" lengths and times.

A characteristic length $\mathcal{L}$ is defined as a distance over which the field studied "changes by an appreciable fraction."[2]

A characteristic time $\mathcal{T}$ is defined similarly.

---

[1] This appendix is a slightly longer version of F. Lorrain (2004). Both are abridged versions of F. Lorrain (2003). Author's address: Département de mathématiques et d'informatique, Collège Jean-de-Brébeuf, 3200 chemin de la Côte-Sainte-Catherine, Montréal (QC), Canada H3T 1C1. E-mail: florrain@brebeuf.qc.ca – The author wishes to thank Paul Lorrain, who first motivated him to do research on this subject and write about it, and who wrote the first paragraphs of the introduction to this appendix. Thanks are also owed to Louis-Philippe Giroux and André Joyal, who helped with some of the finer points of the mathematics behind it.

[2] Spitzer (1956), p. 40. Another definition: "a length characteristic of the scale of variation of the field quantities" (Ferraro and Plumpton, 1966, p. 15).

Characteristic lengths and times are used to estimate the order of magnitude of various derivatives that occur in physical equations. For example, one writes that the divergence or curl of a quantity, say $\boldsymbol{B}$, is of the order of $B/\mathcal{L}$. Or that $d\boldsymbol{B}/dt$ is of the order of $B/\mathcal{T}$. If one is concerned with the magnetic field of the Earth, one might set $\mathcal{L}$ equal to the Earth's radius and $\mathcal{T}$ equal to a certain large number of years.

Clearly, this is just common sense, but is it? See below.

A.1 Introduction

Characteristic lengths and times can obviously be used in many fields of Physics.[3] They are useful, and apparently reasonable. However, since the dependent variables, usually the fields, are relatively arbitrary functions of the independent variables, usually the space coordinates and the time, the procedure is not clearly reliable. Indeed, applied blindly, it can lead to absurd results. For instance, if $\boldsymbol{B}$ is the magnetic flux density, then, from one of Maxwell's equations, $\boldsymbol{\nabla} \cdot \boldsymbol{B} = 0$. Then, with the above reasoning, $B/\mathcal{L} \sim 0$, which is nonsense, because $\boldsymbol{B}$ can have any arbitrary value allowed by Maxwell's equations.

Strangely enough, there is almost no reference in the literature to a mathematical justification of the above procedure. Among 120 books and journal volumes on Fluid Mechanics, Magnetofluid Dynamics, or Plasma Physics, published from 1932 to 2002, the author found only one that contains brief hints that Fourier analysis might provide such a justification.[4]

As we shall see, Fourier analysis can indeed be used to justify the procedure, but with important caveats. We shall find that, under certain conditions, the order of magnitude of quantities such as df/dt, $d\boldsymbol{A}/dt$, $\boldsymbol{\nabla} f$, $\boldsymbol{\nabla} \cdot \boldsymbol{A}$, or $\boldsymbol{\nabla} \times \boldsymbol{A}$ can indeed be evaluated with respect to the orders of magnitude of the functions f or $\boldsymbol{A}$, through equalities or *inequalities* such as the following:

$$\text{norm}(df/dt) = \text{norm}(f)/\mathcal{T}\ , \qquad (A.1)$$

$$\text{norm}(\boldsymbol{\nabla} f) = \text{norm}(f)/\mathcal{L}\ , \qquad (A.2)$$

$$\text{norm}(\boldsymbol{\nabla} \cdot \boldsymbol{A}) \leq \text{norm}(\boldsymbol{A})/\mathcal{L}\ , \qquad (A.3)$$

$$\text{norm}(\boldsymbol{\nabla} \times \boldsymbol{A}) \leq \text{norm}(\boldsymbol{A})/\mathcal{L}\ , \qquad (A.4)$$

[3] See for example Backus et al. (1996), pp. 29–35, 248–249; Davidson (2001), pp. 30, 51–52, 82–83, 96; Ferraro and Plumpton (1966), pp. 15–18, 29, 31, 33, 143; Landau and Lifshitz (1987), pp. 56–58, 61, 69, 86, 131–132, 142; Parker (1979), pp. 31–33, 35, 42–43.

[4] Backus et al. (1996), pp. 30, 33, 249. This book contains an interesting discussion entitled "On judicious neglect of terms in equations" (ibid., pp. 29–35).

where $\mathcal{T}$ and $\mathcal{L}$ are appropriate constants, which vary from case to case. The *norm* of a function is defined in Sect. A.2: it is equal to, or related to, its root-mean-square value.

There are two important points to note.

(a) Characteristic lengths and times sometimes lead to *inequalities*, instead of equalities.[5] In particular, A.3 is true even if $\nabla \cdot \boldsymbol{A} = 0$, which avoids the absurd situation noted above. Similarly, A.4 is true even if $\nabla \times \boldsymbol{A} = \boldsymbol{0}$. We shall see in Sects. A.3 and A.6.3 that the inequalities in A.3 and A.4 arise from the elementary vector inequalities $|\boldsymbol{k} \cdot \boldsymbol{w}| \leq |\boldsymbol{k}||\boldsymbol{w}|$ and $|\boldsymbol{k} \times \boldsymbol{w}| \leq |\boldsymbol{k}||\boldsymbol{w}|$.

(b) The above equalities and inequalities do not apply to the values of df/dt and f at a given time, nor to the values of $\nabla \cdot \boldsymbol{A}$ and $\boldsymbol{A}$ at a given point: Eq. A.1 applies to the *norm* of df/dt and the norm of f, and so on.

Evaluating the order of magnitude of derivatives with the help of characteristic lengths and times is not as simple a matter as might be thought. Some authors are aware of some of the difficulties involved, particularly those authors who study boundary layers in fluid flows.[6] Here we attempt to clarify a few basic aspects of the subject. Certainly, other fields of Physics, besides Fluid Mechanics and Magneto-Fluid-Mechanics, would gain from using this method of evaluation, with due precaution.

Section A.2 defines and discusses two different measures of the order of magnitude of a function: its *amplitude* and its *norm*.

In Sect. A.3 we consider the case of simple sinusoidal functions.

Section A.4 presents a few basic equations of Fourier analysis, which we use later on. Our main results are derived in Sects. A.5 to A.7; though these results are on the whole quite simple in themselves, the reader must be warned that they can be proved only with some effort. This effort is worthwhile, however, for, through it, one gains a keen sense of the limits of the method of characteristic lengths and times.

Section A.8, finally, presents a simplified version of our results, with examples.

A.2 Amplitude and Norm of a Function

The simplest measure of the order of magnitude of a function is its amplitude, defined as follows.

Definition 1. *The* ***amplitude*** *of a scalar function $g(u)$ is the maximum, or lowest upper bound, of its absolute value if it is a real function, or of its modulus if it is a complex function:* amplitude$(g) \equiv \sup |g(u)|$.

Similarly, the ***amplitude*** *of a vector function $\boldsymbol{G}(u)$ is the maximum, or lowest upper bound, of its magnitude or modulus:* amplitude$(\boldsymbol{G}) \equiv \sup |\boldsymbol{G}(u)|$, *where of course*

[5] Backus et al. (1996, pp. 33, 249) also use inequalities.
[6] Choudhuri (1998), pp. 280–281; Tritton (1988), pp. 96–97, 101, 123–126.

$$|\boldsymbol{G}(u)| \equiv \sqrt{|G_x(u)|^2 + |G_y(u)|^2 + |G_z(u)|^2}\,. \tag{A.5}$$

The components of $\boldsymbol{G}(u)$ *may be either real or complex.*

Similar definitions apply to functions of several variables such as x, y, *and* z.

The concept of amplitude must be used with care. The amplitude of F is not necessarily a typical value of F. One must therefore be cautious when comparing the amplitudes of two functions.

Definition 2. *The* ***norm*** *of a periodic function* $F(u)$ *of period* T *is*

$$\operatorname{norm}(F) \equiv \sqrt{\frac{1}{T}\int_0^T |F(u)|^2\, du} \quad (F\ \textit{periodic}). \tag{A.6}$$

The ***norm*** *of a nonperiodic function is*

$$\operatorname{norm}(F) \equiv \sqrt{\int_{-\infty}^{\infty} |F(u)|^2\, du} \quad (F\ \textit{nonperiodic}). \tag{A.7}$$

Here F *may be real or complex.*

The norm of a vector-valued function is defined similarly. So is the norm of a function of several variables.

Note that, in the periodic case, the norm is the root-mean-square value of F. It can be shown in this case[7] that the norm of F is never smaller than the *average* value of $|F|$.

But in the nonperiodic case the norm is *not* a kind of average value of F, since one cannot divide by an infinite T.

In both cases, however, the norm of F measures to what extent the function F differs from the function with constant value 0. In fact $\operatorname{norm}(F)$ can be seen as a measure of the *distance* between F and the zero function; in particular, the norm, just like the absolute value or the modulus, satisfies the *triangle inequality*:[8]

$$\operatorname{norm}(F+G) \leq \operatorname{norm}(F) + \operatorname{norm}(G)\,. \tag{A.8}$$

If F is a periodic function of finite amplitude, then its norm A.6 is finite. On the other hand, if F is a *non*periodic function of finite amplitude, its norm A.7 could be infinite. *However, if the phenomenon studied is well localized, in such a way that the integral of* $|F(u)|^2$ *is practically zero outside a finite region, then, with a finite amplitude, the norm A.7 will also be finite.* If we compute the norms of several functions that are all negligible in this same fashion outside the same finite interval of length T, then their norms A.7 are

---

[7] See F. Lorrain (2003), Sect. 2.1.

[8] Courant and John (1989), Vol. I, pp. 14, 612–613.

equal to $\sqrt{T}$ times their root-mean-square values A.6 over that interval. In such a case, the norms A.7 of the functions are *proportional* to their root-mean-square values.

In fact, any function F with a finite norm of type A.7 can be treated in this fashion: there necessarily exists a finite region, outside of which the integral of F^2 is negligible!

The concepts of amplitude and norm each have advantages and disadvantages. We have already seen that amplitudes must be used with circumspection. For a nonsinusoidal function F, the norm may be more representative of the values of F than the amplitude, since the norm is calculated by integrating the square of F. Nevertheless, one must be just as careful when comparing the norms of two functions as when comparing their amplitudes. In addition, we shall see in Sect. A.7 that one must be careful when considering the norm of the product of two functions.

Of course, in many physical situations, the norm A.6 of F is a fairly good measure of the order of magnitude of F in general.

A.3 $\mathcal{L}$ and $\mathcal{T}$ of a Sinusoidal Function

Consider a simple sinusoidal function of the time t, of period T: $f(t) = f_0 \cos\omega t$ (where $f_0 > 0$, $\omega > 0$). Here amplitude$(f) = f_0$ and amplitude$(f') = \omega f_0$. With the norm A.6 we have norm$(f) = f_0/\sqrt{2}$ and norm$(f') = \omega f_0/\sqrt{2}$ (where $f' = df/dt$). Then, with $\mathcal{T} = 1/\omega$,

$$\text{amplitude}(f') = \text{amplitude}(f)/\mathcal{T}\ , \tag{A.9}$$
$$\text{norm}(f') = \text{norm}(f)/\mathcal{T}\ . \tag{A.10}$$

The **characteristic time** of f is $\mathcal{T} = 1/\omega = T/2\pi$. *Note that $\mathcal{T}$ is not the period of f, but its period divided by 2π.*

Similar results apply to a sinusoidal *vector* function $\boldsymbol{A}(t) = \boldsymbol{A}_0 \cos\omega t$.

Let us now study the case of a sinusoidal scalar function of $\boldsymbol{r} = (x, y, z)$, of wavelength $\lambda = 2\pi/k$: $f(\boldsymbol{r}) = f_0 \cos(\boldsymbol{k}\cdot\boldsymbol{r})$ (where $f_0 > 0$, $|\boldsymbol{k}| = k$). Then, again with the norm A.6, and with $\mathcal{L} = 1/k$, we have

$$\text{amplitude}(\boldsymbol{\nabla} f) = \text{amplitude}(f)/\mathcal{L}\ , \tag{A.11}$$
$$\text{norm}(\boldsymbol{\nabla} f) = \text{norm}(f)/\mathcal{L}\ . \tag{A.12}$$

The **characteristic length** of f is $\mathcal{L} = 1/k = \lambda/2\pi$. *Note that $\mathcal{L}$ is not the wavelength of f, but its* radian length.

The most interesting case is that of a sinusoidal *vector* function of $\boldsymbol{r}$, again of wavelength $\lambda = 2\pi/k$: $\boldsymbol{A}(\boldsymbol{r}) = \boldsymbol{A}_0 \cos(\boldsymbol{k}\cdot\boldsymbol{r})$ (where $|\boldsymbol{k}| = k$). Let amplitude$(\boldsymbol{A}) = A_0$. Then $\boldsymbol{\nabla}\cdot\boldsymbol{A} = -\boldsymbol{k}\cdot\boldsymbol{A}_0 \sin(\boldsymbol{k}\cdot\boldsymbol{r})$ and $\boldsymbol{\nabla}\times\boldsymbol{A} = -\boldsymbol{k}\times\boldsymbol{A}_0 \sin(\boldsymbol{k}\cdot\boldsymbol{r})$, so that

$$\text{amplitude}(\boldsymbol{\nabla} \cdot \boldsymbol{A}) = |\boldsymbol{k} \cdot \boldsymbol{A}_0| \le kA_0 \,, \tag{A.13}$$
$$\text{amplitude}(\boldsymbol{\nabla} \times \boldsymbol{A}) = |\boldsymbol{k} \times \boldsymbol{A}_0| \le kA_0 \,. \tag{A.14}$$

There are similar equations for the norm. Finally, again with $\mathcal{L} = 1/k$,

$$\text{amplitude}(\boldsymbol{\nabla} \cdot \boldsymbol{A}) \le \text{amplitude}(\boldsymbol{A})/\mathcal{L} \,, \tag{A.15}$$
$$\text{norm}(\boldsymbol{\nabla} \cdot \boldsymbol{A}) \le \text{norm}(\boldsymbol{A})/\mathcal{L} \,, \tag{A.16}$$
$$\text{amplitude}(\boldsymbol{\nabla} \times \boldsymbol{A}) \le \text{amplitude}(\boldsymbol{A})/\mathcal{L} \,, \tag{A.17}$$
$$\text{norm}(\boldsymbol{\nabla} \times \boldsymbol{A}) \le \text{norm}(\boldsymbol{A})/\mathcal{L} \,. \tag{A.18}$$

Note again that the characteristic length of $\boldsymbol{A}$ is its wavelength divided by 2π. But *note especially the inequalities that inevitably appear in A.15–A.18, due to the fundamental vector inequalities* $|\boldsymbol{k} \cdot \boldsymbol{A}_0| \le |\boldsymbol{k}||\boldsymbol{A}_0|$ *and* $|\boldsymbol{k} \times \boldsymbol{A}_0| \le |\boldsymbol{k}||\boldsymbol{A}_0|$.

However, a remarkable *equality* can be formed with the divergence and curl together: Eqs. A.13–A.14 imply that

$$\begin{aligned} &\text{amplitude}^2(\boldsymbol{\nabla} \cdot \boldsymbol{A}) + \text{amplitude}^2(\boldsymbol{\nabla} \times \boldsymbol{A}) \\ &= k^2 \text{amplitude}^2(\boldsymbol{A}) = \text{amplitude}^2(\boldsymbol{A})/\mathcal{L}^2 \,. \end{aligned} \tag{A.19}$$

There is a similar equality for the norm, which we shall encounter again later on (Eq. A.52).

The function $\boldsymbol{A}(\boldsymbol{r})$ can be seen as a waveform. When the field $\boldsymbol{A}$ is *transversal* ($\boldsymbol{A}_0 \perp \boldsymbol{k}$), then, on the one hand, $\boldsymbol{\nabla} \cdot \boldsymbol{A} = 0$ and $\text{norm}(\boldsymbol{\nabla} \cdot \boldsymbol{A}) = 0 < \text{norm}(\boldsymbol{A})/\mathcal{L}$; but, on the other hand, $\boldsymbol{\nabla} \times \boldsymbol{A} \ne \boldsymbol{0}$ and $\text{norm}(\boldsymbol{\nabla} \times \boldsymbol{A}) = \text{norm}(\boldsymbol{A})/\mathcal{L}$.

However, when the field $\boldsymbol{A}$ is *longitudinal* ($\boldsymbol{A}_0 \parallel \boldsymbol{k}$), the situation is reversed: on the one hand, $\boldsymbol{\nabla} \cdot \boldsymbol{A} \ne 0$ and $\text{norm}(\boldsymbol{\nabla} \cdot \boldsymbol{A}) = \text{norm}(\boldsymbol{A})/\mathcal{L}$; but, on the other hand, $\boldsymbol{\nabla} \times \boldsymbol{A} = \boldsymbol{0}$ and $\text{norm}(\boldsymbol{\nabla} \times \boldsymbol{A}) = 0 < \text{norm}(\boldsymbol{A})/\mathcal{L}$.

It is interesting to see what happens when $\boldsymbol{k}$ and $\boldsymbol{A}_0$ are neither parallel nor perpendicular. Unless the angle between them is close to 0, $\pi/2$, or π, the absolute values of their scalar and vector products will be of the same order of magnitude as kA_0; in other words, $|\boldsymbol{k} \cdot \boldsymbol{A}_0| \sim |\boldsymbol{k} \times \boldsymbol{A}_0| \sim kA_0$ (where the sign $\sim$ means "is of the same order of magnitude as").[9] Then, though A.15–A.19 of course remain true, we also have

$$\text{norm}(\boldsymbol{\nabla} \cdot \boldsymbol{A}) \sim \text{norm}(\boldsymbol{\nabla} \times \boldsymbol{A}) \sim \text{norm}(\boldsymbol{A})/\mathcal{L} \,, \tag{A.20}$$

and similar equations for the amplitudes.

Simple sinusoidal functions obey equations and inequalities such as A.9–A.12, A.15–A.19, and sometimes A.20. All these equations are valid as much for the norm as for the amplitude. With the help of Fourier analysis, we

[9] Boyd and Sanderson (1969), p. 48, footnote: they warn that, in their approximations, "the vector character of the equations is ignored; this *may* be a dangerous simplification and should be checked in particular instances."

shall be able to generalize these results to a wide class of functions – but only with the *norm*, not with the amplitude. Unfortunately, however, these results cannot be generalized to arbitrary functions.

A.4 Fourier Expansions

Let t be any real variable (such as time) and $f(t)$ a real-valued function. Under quite general conditions, $f(t)$ can be expressed as the sum of sinusoidal functions of the form $c\,e^{j\omega t}$. The sum in question can be a series such as

$$f(t) = \sum_{n=-\infty}^{\infty} c_n\, e^{jn\omega t} \quad \text{(where } c_{-n} = c_n^*\text{)} \tag{A.21}$$

if f is periodic of period $T = 2\pi/\omega$, or if f is defined only on the interval from t_0 to $t_0 + T$. The sum can also be an integral such as

$$f(t) = \frac{1}{\sqrt{2\pi}} \int_{-\infty}^{\infty} c(\omega)\, e^{j\omega t}\, d\omega \text{ [where } c(-\omega) = c(\omega)^*\text{]}. \tag{A.22}$$

Detailed sufficient conditions for the existence of such Fourier expansions of $f(t)$ can be found in standard textbooks.[10] Note that, although the terms of the series A.21 are complex, the sum is real and equal to $f(t)$. The same holds for the integral A.22. We use this complex notation because the derivative of $f(t)$ then takes a simple form, very close to that of $f(t)$.

The sum A.21 is a **Fourier series**. The integral A.22 is a **Fourier integral**. The complex functions $c_n\, e^{jn\omega t}$ and $c(\omega)\, e^{j\omega t}$ that appear in A.21 and A.22 are the **Fourier components** of f.

We shall need a few basic results from the theory of Fourier series and integrals.

Suppose that a real-valued periodic function $f(t)$ can be expanded as a Fourier series (Eq. A.21).[11] Then it also usually satisfies **Parseval's equality**

$$\text{norm}^2(f) \equiv \frac{1}{T} \int_0^T |f(t)|^2\, dt = \sum_{n=-\infty}^{\infty} |c_n|^2 \tag{A.23}$$

and $f(t)$ can usually be differentiated simply as follows:

$$f'(t) = \sum_{n=-\infty}^{\infty} jn\omega c_n\, e^{jn\omega t}\,. \tag{A.24}$$

[10] See the following footnotes.

[11] Sufficient conditions for A.21, A.23, and A.24 (Courant and John, 1989, Vol. I, pp. 35, 587–588, 594–595, 604, 614, 607–608): f is periodic, the three functions f, f', and f'' are everywhere defined and continuous, f''' is “sectionally continuous”.

Parseval's equality is essential to our purposes, because it provides a simple relationship between the norm of a function and the norms of its Fourier components.

Suppose now that a nonperiodic real-valued function $f(t)$ can be expanded as a Fourier integral (Eq. A.22). Among the set of sufficient conditions given by Courant and John for such an expansion,[12] note especially the following one: that $\int_{-\infty}^{\infty} |f(t)|\, dt < \infty$. *This means that the physical phenomenon we are studying is a relatively localized one.*

If this $f(t)$ has a Fourier expansion, it then usually satisfies equations similar to A.23–A.24:[13]

$$\text{norm}^2(f) \equiv \int_{-\infty}^{\infty} |f(t)|^2\, dt = \int_{-\infty}^{\infty} |c(\omega)|^2\, d\omega \text{ (Parseval's equality)}, \quad \text{(A.25)}$$

$$f'(t) = \frac{1}{\sqrt{2\pi}} \int_{-\infty}^{\infty} j\omega c(\omega)\, e^{j\omega t}\, d\omega\ . \quad \text{(A.26)}$$

If norm(f) is to be of any use to us, it must be finite. *Once again we see that the phenomenon studied must be relatively localized.*

There are similar equations for real-vector-valued functions $\boldsymbol{A}(t)$.

For a real-valued scalar function $f(\boldsymbol{r})$, where $\boldsymbol{r} = (x, y, z)$, we have[14]

$$f(\boldsymbol{r}) = \frac{1}{\sqrt{8\pi^3}} \iiint_{\mathbf{R}^3} c(\boldsymbol{k})\, e^{j\boldsymbol{k}\cdot\boldsymbol{r}}\, dk_x\, dk_y\, dk_z\ , \quad \text{(A.27)}$$

where $\boldsymbol{k} = (k_x, k_y, k_z)$, $c(-\boldsymbol{k}) = c(\boldsymbol{k})^*$, and $\mathbf{R}^3$ is the totality of three-dimensional space. Also

$$\text{norm}^2(f) \equiv \iiint_{\mathbf{R}^3} |f(\boldsymbol{r})|^2\, dx\, dy\, dz \quad \text{(A.28)}$$

$$= \iiint_{\mathbf{R}^3} |c(\boldsymbol{k})|^2\, dk_x\, dk_y\, dk_z\ , \quad \text{(A.29)}$$

$$\boldsymbol{\nabla} f(\boldsymbol{r}) = \frac{1}{\sqrt{8\pi^3}} \iiint_{\mathbf{R}^3} c(\boldsymbol{k}) \left(\boldsymbol{\nabla} e^{j\boldsymbol{k}\cdot\boldsymbol{r}}\right) dk_x\, dk_y\, dk_z \quad \text{(A.30)}$$

$$= \frac{1}{\sqrt{8\pi^3}} \iiint_{\mathbf{R}^3} j\boldsymbol{k}\, c(\boldsymbol{k})\, e^{j\boldsymbol{k}\cdot\boldsymbol{r}} dk_x\, dk_y\, dk_z\ . \quad \text{(A.31)}$$

For a real-vector-valued function $\boldsymbol{A}(\boldsymbol{r})$ we have

---

[12] Courant and John (1989), Vol. II, pp. 476–477.

[13] Sufficient conditions for A.22, A.25, and A.26 (Courant and John, 1989, Vol. II, pp. 476–477, 486–489): the three functions f, f', and f'' are everywhere defined and continuous, $\int_{-\infty}^{\infty} |f(t)|\, dt < \infty$, $\int_{-\infty}^{\infty} |f'(t)|\, dt < \infty$, and $\int_{-\infty}^{\infty} |f''(t)|\, dt < \infty$.

[14] Sufficient conditions for A.27–A.29 (Courant and John, 1989, Vol. II, pp. 491–492, 496): f and all its derivatives of order four or less are everywhere defined, and the integrals of their absolute values over all space are finite. Courant and John do not state conditions under which A.30 is true, as they do for the similar equations given above. We may presume that such conditions are fairly general.

$$\boldsymbol{A}(\boldsymbol{r}) = \frac{1}{\sqrt{8\pi^3}} \iiint_{\mathbf{R}^3} \mathbf{c}(\boldsymbol{k})\, e^{j\boldsymbol{k}\cdot\boldsymbol{r}}\, dk_x\, dk_y\, dk_z \tag{A.32}$$

[where $\mathbf{c}(-\boldsymbol{k}) = \mathbf{c}(\boldsymbol{k})^*$], and an equation similar to A.28–A.29 for norm($\boldsymbol{A}$). For the divergence and curl we have equations similar to A.31:

$$\boldsymbol{\nabla} \cdot \boldsymbol{A}(\boldsymbol{r}) = \frac{1}{\sqrt{8\pi^3}} \iiint_{\mathbf{R}^3} j\boldsymbol{k} \cdot \mathbf{c}(\boldsymbol{k})\, e^{j\boldsymbol{k}\cdot\boldsymbol{r}}\, dk_x\, dk_y\, dk_z\ , \tag{A.33}$$

$$\boldsymbol{\nabla} \times \boldsymbol{A}(\boldsymbol{r}) = \frac{1}{\sqrt{8\pi^3}} \iiint_{\mathbf{R}^3} j\boldsymbol{k} \times \mathbf{c}(\boldsymbol{k})\, e^{j\boldsymbol{k}\cdot\boldsymbol{r}}\, dk_x\, dk_y\, dk_z\ . \tag{A.34}$$

A.5 Characteristic Time $\mathcal{T}$ of a Function

As before, let t be the time. If $f(t)$ is periodic of period $T = 2\pi/\omega$ (or defined only on the interval from 0 to T), it can usually be expressed as a Fourier series A.21. The norm A.6 of f is then usually given by Parseval's equality A.23 and f can usually be differentiated simply as in A.24. Parseval's equality for f' then gives

$$\text{norm}^2(f') \equiv \frac{1}{T}\int_0^T |f'(t)|^2\, dt = \sum_{n=-\infty}^{\infty} |n\omega c_n|^2 = \sum_{n=-\infty}^{\infty} (n\omega)^2 |c_n|^2\ . \tag{A.35}$$

Let the **root-mean-square circular frequency** of f be the circular frequency $\omega_{\text{rms}}(f)$ calculated as follows:

$$\omega_{\text{rms}}(f) = \sqrt{\frac{\sum_{n=-\infty}^{\infty} (n\omega)^2 |c_n|^2}{\sum_{n=-\infty}^{\infty} |c_n|^2}} = \frac{\text{norm}(f')}{\text{norm}(f)}\ . \tag{A.36}$$

Let the **characteristic time** of f be

$$\mathcal{T}(f) = 1/\omega_{\text{rms}}(f)\ . \tag{A.37}$$

Then, tautologically,

$$\text{norm}(f') = \text{norm}(f)/\mathcal{T}(f)\ . \tag{A.38}$$

This equation may be a mathematical tautology, but it is a meaningful one, since $\omega_{\text{rms}}(f)$ is the rms circular frequency of the Fourier spectrum of f.

No simple and meaningful equation of this kind can be derived with amplitudes. *We must use norms!*

If $f(t)$ is nonperiodic and can be expressed as a Fourier integral (Eq. A.22), we define

$$\omega_{\text{rms}}(f) = \sqrt{\frac{\int_{-\infty}^{\infty} \omega^2 |c(\omega)|^2\, d\omega}{\int_{-\infty}^{\infty} |c(\omega)|^2\, d\omega}} = \frac{\text{norm}(f')}{\text{norm}(f)}\ . \tag{A.39}$$

With $\mathcal{T}(f) = 1/\omega_{\text{rms}}(f)$, we again obtain Eq. A.38.

We have the following result.

Result 1 (functions of t). *If the real-valued function $f(t)$ and the real-vector-valued function $\boldsymbol{A}(t)$ can be expressed as Fourier series or integrals, if f and $\boldsymbol{A}$ can be differentiated simply as in A.24 or A.26, and if f, f', $\boldsymbol{A}$, and $\boldsymbol{A}'$ all satisfy Parseval's equality, then* $\text{norm}(df/dt) = \text{norm}(f)/\mathcal{T}(f)$ *and* $\text{norm}(d\boldsymbol{A}/dt) = \text{norm}(\boldsymbol{A})/\mathcal{T}(\boldsymbol{A})$*, where $\mathcal{T}(f)$ and $\mathcal{T}(\boldsymbol{A})$ are the* characteristic times *of f and $\boldsymbol{A}$ (Eqs. A.36, A.39, A.37).*

Result 1 summarizes the method of characteristic *times*. Note the first assumption in this Result: that the function of t considered can be expressed as a Fourier series or integral. This condition is often satisfied and, if it is, then the other conditions mentioned are usually also satisfied.

Example 1. Consider the simple time-pulse

$$f(t) = e^{-t^2/a^2}\ , \quad \text{where } a > 0. \tag{A.40}$$

The inflection points of the curve of f occur at $t = \pm a/\sqrt{2}$. At these points, the value of f is approximately 61% of its peak value. Define the "width" of the pulse as the distance between the two inflection points, namely $a\sqrt{2}$. Using an integral table, we find that

$$\text{norm}^2(f) = a\sqrt{\frac{\pi}{2}}\ , \quad \text{norm}^2(f') = \frac{1}{a}\sqrt{\frac{\pi}{2}}\ , \tag{A.41}$$

$$\mathcal{T}^2(f) = \frac{\text{norm}^2(f)}{\text{norm}^2(f')} = a^2\ , \tag{A.42}$$

$$\mathcal{T}(f) = a\ . \tag{A.43}$$

The characteristic time $\mathcal{T}$ of the pulse is its width divided by $\sqrt{2}$.

Exercise 1. [15] Calculate the characteristic time of the pulse $g(t) = 1/(1 + t^2/a^2)$. Here, define the width of the pulse as the distance between the two values of t for which $g(t) = \frac{1}{2}\max g$.

Remember the following important point: Result 1 does not apply to all values of df/dt or $d\boldsymbol{A}/dt$, but only to the *norms* of these derivatives.

Section A.6 explains the method of characteristic *lengths*; the functions there are scalar or vector-valued functions of the variables x, y, and z. The method is based on similar assumptions.

A.6 Characteristic Length $\mathcal{L}$ of a Function

As before, let $\boldsymbol{r} = (x, y, z)$.

[15] Solutions to the exercises are given at the end of F. Lorrain (2003).

A.6.1 The Norm of $\boldsymbol{\nabla} f(\boldsymbol{r})$

Assume that $f(\boldsymbol{r})$ can be expanded as in A.27 and differentiated simply as in A.30–A.31. Let the **root-mean-square wave number** of f be the wave number $k_{\mathrm{rms}}(f)$ calculated as follows (with Parseval's equalities):

$$k_{\mathrm{rms}}(f) = \sqrt{\frac{\iiint_{\mathbf{R}^3} k^2\,|c(\boldsymbol{k})|^2\,dk_x\,dk_y\,dk_z}{\iiint_{\mathbf{R}^3} |c(\boldsymbol{k})|^2\,dk_x\,dk_y\,dk_z}} \tag{A.44}$$

$$= \frac{\mathrm{norm}(\boldsymbol{\nabla} f)}{\mathrm{norm}(f)}\,, \tag{A.45}$$

where $k = |\boldsymbol{k}|$. Let the **characteristic length** of f be

$$\mathcal{L}(f) = 1/k_{\mathrm{rms}}(f)\,. \tag{A.46}$$

This $\mathcal{L}(f)$ is a radian length, that is, a wavelength divided by 2π. Then $\mathrm{norm}(\boldsymbol{\nabla} f) = \mathrm{norm}(f)/\mathcal{L}(f)$. The meaning of this equation comes from the fact that $k_{\mathrm{rms}}(f)$ is the rms wave number of the Fourier spectrum of f. We have the following result.

Result 2 (gradient). *Assume that $f(\boldsymbol{r})$ can be expanded as a Fourier integral, that it can be differentiated simply as in A.30, and that f and $\boldsymbol{\nabla} f$ both satisfy Parseval's equality. Then* $\mathrm{norm}(\boldsymbol{\nabla} f) = \mathrm{norm}(f)/\mathcal{L}(f)$, *where $\mathcal{L}(f)$ is the* characteristic length *of f (Eqs. A.44 and A.46).*

A.6.2 Vectors with Complex Components

We shall soon calculate the norms of the divergence and curl of $\boldsymbol{A}$ given in Eqs. A.33–A.34. To do this, we need the moduli of $\boldsymbol{k}\cdot\boldsymbol{c}$ and of $\boldsymbol{k}\times\boldsymbol{c}$.

Using the definition A.5 of the modulus of a complex vector, we obtain the following proposition.

Proposition 1. *If $\boldsymbol{w} = \boldsymbol{a} + j\boldsymbol{b}$, where $\boldsymbol{a}$ and $\boldsymbol{b}$ are real vectors, then $|\boldsymbol{w}|^2 = |\boldsymbol{a}|^2 + |\boldsymbol{b}|^2$. And, if $\boldsymbol{k}$ is a real vector of magnitude k, then*

$$|\boldsymbol{k}\cdot\boldsymbol{w}|^2 + |\boldsymbol{k}\times\boldsymbol{w}|^2 = k^2|\boldsymbol{w}|^2\,, \tag{A.47}$$

so that $|\boldsymbol{k}\cdot\boldsymbol{w}|^2 \le k^2|\boldsymbol{w}|^2$ and $|\boldsymbol{k}\times\boldsymbol{w}|^2 \le k^2|\boldsymbol{w}|^2$.

A.6.3 The Norms of $\boldsymbol{\nabla}\cdot\boldsymbol{A}$ and of $\boldsymbol{\nabla}\times\boldsymbol{A}$

With the values given in A.32–A.34 for $\boldsymbol{A}$, $\boldsymbol{\nabla}\cdot\boldsymbol{A}$, and $\boldsymbol{\nabla}\times\boldsymbol{A}$, Parseval's equalities give

$$\mathrm{norm}^2(\boldsymbol{\nabla}\cdot\boldsymbol{A}) = \iiint_{\mathbf{R}^3} |\boldsymbol{k}\cdot[j\boldsymbol{c}(\boldsymbol{k})]|^2\,dk_x\,dk_y\,dk_z\,, \tag{A.48}$$

$$\mathrm{norm}^2(\boldsymbol{\nabla}\times\boldsymbol{A}) = \iiint_{\mathbf{R}^3} |\boldsymbol{k}\times[j\boldsymbol{c}(\boldsymbol{k})]|^2\,dk_x\,dk_y\,dk_z\,. \tag{A.49}$$

Adding these two equations and using Eq. A.47, we obtain

$$\mathrm{norm}^2(\boldsymbol{\nabla}\cdot\boldsymbol{A})+\mathrm{norm}^2(\boldsymbol{\nabla}\times\boldsymbol{A})=\iiint_{\mathbf{R}^3} k^2|\mathbf{c}(\boldsymbol{k})|^2\,dk_x\,dk_y\,dk_z\,. \quad (A.50)$$

Now let $k_{\mathrm{rms}}(\boldsymbol{A})$ be the **root-mean-square wave number** of $\boldsymbol{A}$, defined in a manner similar to Eq. A.44. Let the **characteristic length** of $\boldsymbol{A}$ be, as before,

$$\mathcal{L}(\boldsymbol{A})=1/k_{\mathrm{rms}}(\boldsymbol{A})\,. \quad (A.51)$$

This is again a wavelength divided by 2π. Equation A.50 can now be rewritten as

$$\mathrm{norm}^2(\boldsymbol{\nabla}\cdot\boldsymbol{A})+\mathrm{norm}^2(\boldsymbol{\nabla}\times\boldsymbol{A})=k_{\mathrm{rms}}^2(\boldsymbol{A})\,\mathrm{norm}^2(\boldsymbol{A})=\frac{\mathrm{norm}^2(\boldsymbol{A})}{\mathcal{L}^2(\boldsymbol{A})}\,. \quad (A.52)$$

This is a surprising equation. Note its similarity to Eq. A.47. Like A.38, it is a sort of mathematical tautology, but a meaningful one, since $k_{\mathrm{rms}}(\boldsymbol{A})$ is the rms wave number of the Fourier spectrum of $\boldsymbol{A}$.

We then have the inequalities

$$\mathrm{norm}(\boldsymbol{\nabla}\cdot\boldsymbol{A})\le\mathrm{norm}(\boldsymbol{A})/\mathcal{L}(\boldsymbol{A})\,, \quad (A.53)$$

$$\mathrm{norm}(\boldsymbol{\nabla}\times\boldsymbol{A})\le\mathrm{norm}(\boldsymbol{A})/\mathcal{L}(\boldsymbol{A})\,. \quad (A.54)$$

Result 3 (divergence, curl). *Let $\boldsymbol{A}(\boldsymbol{r})$ have the Fourier expansion A.32. Assume that $\boldsymbol{\nabla}\cdot\boldsymbol{A}$ and $\boldsymbol{\nabla}\times\boldsymbol{A}$ can be calculated simply as in A.33, and that $\boldsymbol{A}$, $\boldsymbol{\nabla}\cdot\boldsymbol{A}$ and $\boldsymbol{\nabla}\times\boldsymbol{A}$ all satisfy Parseval's equality. Then*

$$\mathrm{norm}^2(\boldsymbol{\nabla}\cdot\boldsymbol{A})+\mathrm{norm}^2(\boldsymbol{\nabla}\times\boldsymbol{A})=\frac{\mathrm{norm}^2(\boldsymbol{A})}{\mathcal{L}^2(\boldsymbol{A})}\,, \quad (A.55)$$

so that $\mathrm{norm}(\boldsymbol{\nabla}\cdot\boldsymbol{A})\le\mathrm{norm}(\boldsymbol{A})/\mathcal{L}(\boldsymbol{A})$ *and* $\mathrm{norm}(\boldsymbol{\nabla}\times\boldsymbol{A})\le\mathrm{norm}(\boldsymbol{A})/\mathcal{L}(\boldsymbol{A})$, *where $\mathcal{L}(\boldsymbol{A})$ is the* characteristic length *of $\boldsymbol{A}$ (Eq. A.51).*

Note a most important fact: we have *inequalities* here, not only equalities! These inequalities reflect the elementary vector inequalities of Proposition 1, which are intrinsic to the dot and cross products.

Inequalities – instead of equalities – can sometimes be a *handicap*. For example, if $a\le b$ and $a\le c$, then no conclusion can be drawn as to the relative sizes of b and c.

But inequalities are also an *advantage* here, because we want our results A.53–A.54 to be applicable to nonzero fields $\boldsymbol{A}$ that have zero divergence or zero curl.

These results can be considerably refined, however. We shall use the following definition.

Definition 3. *Let a field $\boldsymbol{A}(\boldsymbol{r})$ have the Fourier expansion A.32. Call this field **transversal** if its Fourier coefficients $\mathbf{c}(\boldsymbol{k})$ are such that $\boldsymbol{k}\cdot\mathbf{c}(\boldsymbol{k})=0$ for almost all $\boldsymbol{k}$'s. Call this field **longitudinal** if its Fourier coefficients $\mathbf{c}(\boldsymbol{k})$ are such that $\boldsymbol{k}\times\mathbf{c}(\boldsymbol{k})=\mathbf{0}$ for almost all $\boldsymbol{k}$'s.*

Result 3 notwithstanding, under certain conditions spelled out in the following Result, our inequalities can also be replaced by approximate equalities as in A.20 or even strict equalities. This follows from A.48, A.49, and A.52.

Result 4 (transversality, longitudinality). *Let $\boldsymbol{A}(\boldsymbol{r})$ have the Fourier expansion A.32 and satisfy Parseval's equality. Assume that $\nabla \cdot \boldsymbol{A}$ and $\nabla \times \boldsymbol{A}$ can be calculated simply as in A.33 and that they satisfy Parseval's equalities A.48 and A.49.*

(a) Then $\boldsymbol{A}$ is transversal if, and only if, $\nabla \cdot \boldsymbol{A} = 0$ almost everywhere. In such a case $\mathrm{norm}(\nabla \times \boldsymbol{A}) = \mathrm{norm}(\boldsymbol{A})/\mathcal{L}(\boldsymbol{A})$.

(b) If $\boldsymbol{A}$ is not transversal, but also not too longitudinal, then we have both

$$\mathrm{norm}(\nabla \times \boldsymbol{A}) < \frac{\mathrm{norm}(\boldsymbol{A})}{\mathcal{L}(\boldsymbol{A})} \quad \textit{and} \quad \mathrm{norm}(\nabla \times \boldsymbol{A}) \sim \frac{\mathrm{norm}(\boldsymbol{A})}{\mathcal{L}(\boldsymbol{A})} . \tag{A.56}$$

(c) Similarly, $\boldsymbol{A}$ is longitudinal if, and only if, $\nabla \times \boldsymbol{A} = \boldsymbol{0}$ almost everywhere. In such a case $\mathrm{norm}(\nabla \cdot \boldsymbol{A}) = \mathrm{norm}(\boldsymbol{A})/\mathcal{L}(\boldsymbol{A})$.

(d) If $\boldsymbol{A}$ is not longitudinal, but also not too transversal, then we have both

$$\mathrm{norm}(\nabla \cdot \boldsymbol{A}) < \frac{\mathrm{norm}(\boldsymbol{A})}{\mathcal{L}(\boldsymbol{A})} \quad \textit{and} \quad \mathrm{norm}(\nabla \cdot \boldsymbol{A}) \sim \frac{\mathrm{norm}(\boldsymbol{A})}{\mathcal{L}(\boldsymbol{A})} . \tag{A.57}$$

Zero-divergence fields such as the magnetic flux density $\boldsymbol{B}$ are transversal; this immediately implies that $\mathrm{norm}(\nabla \times \boldsymbol{B}) = \mathrm{norm}(\boldsymbol{B})/\mathcal{L}(\boldsymbol{B})$.

Zero-curl fields, such as the electric field intensity $\boldsymbol{E}$ when $\boldsymbol{B}$ is static, are longitudinal. For such an $\boldsymbol{E}$, therefore, $\mathrm{norm}(\nabla \cdot \boldsymbol{E}) = \mathrm{norm}(\boldsymbol{E})/\mathcal{L}(\boldsymbol{E})$.

Example 2. Consider the field $\boldsymbol{E}$ of a charge Q uniformly distributed inside a ball of radius a. A straightforward computation gives

$$\mathrm{norm}^2(\boldsymbol{E}) = \frac{3Q^2}{10\pi\epsilon_0^2 a} , \quad \mathrm{norm}^2(\nabla \cdot \boldsymbol{E}) = \frac{3Q^2}{4\pi\epsilon_0^2 a^3} , \tag{A.58}$$

$$\mathcal{L}^2(\boldsymbol{E}) = \frac{\mathrm{norm}^2(\boldsymbol{E})}{\mathrm{norm}^2(\nabla \cdot \boldsymbol{E})} = \frac{2a^2}{5} , \tag{A.59}$$

$$\mathcal{L}(\boldsymbol{E}) = a\sqrt{2/5} \approx 0.63\, a . \tag{A.60}$$

The characteristic length is a function of the radius a only and is proportional to a.

Exercise 2. [16] Calculate the characteristic length of the zero-curl field $\boldsymbol{E}$ of a charge Q uniformly distributed inside a spherical shell of external radius a and internal radius $b < a$. Express your final result in the following form:

[16] Solutions to the exercises are given at the end of F. Lorrain (2003).

$$\mathcal{L}^2(\boldsymbol{E}) = \frac{1}{3}\left\{\frac{a^3-b^3}{a} + \frac{1}{a^3-b^3}\left[\frac{a^5-b^5}{5} - b^3(a^2-b^2) + b^5\,\frac{a-b}{a}\right]\right\} . \tag{A.61}$$

Verify first that, when $b \to 0$, the $\mathcal{L}$ obtained is that of Eq. A.60. Then see what happens to $\mathcal{L}$ when the shell becomes infinitely thin ($b \to a$); in order to calculate this limit, however, you must first divide the denominator $a^3 - b^3$ in the above equation, as well as the numerators $a^5 - b^5$, $a^2 - b^2$, and $a - b$, by the common factor $a - b$. If you wish to be surprised, calculate the limit before reading any further!
(At the limit $b \to a$, the characteristic length vanishes. Recall our first rough description of $\mathcal{L}$ as a distance over which the field changes by an appreciable fraction.)

Some fields have zero divergence and zero curl. The field $\boldsymbol{E} = -\boldsymbol{\nabla} V$ with $\boldsymbol{\nabla}^2 V = 0$ is an example. The simplest case is that of a uniform $\boldsymbol{E}$. But a non-zero field cannot be at the same time transversal and longitudinal. The existence of nonzero fields with zero divergence and curl shows the importance of being careful with one's assumptions. Such a field does not satisfy all the conditions behind Eqs. A.48 and A.49: that it have a Fourier expansion, and so on. Moreover, we also implicitly assume that the field has a finite norm, which means that it is relatively localized. By contrast, non-zero fields with zero divergence and curl *over all space* have infinite norms (an $\boldsymbol{E}$ field such as the one just described has an infinite energy[17]).

A.6.4 The Norms of $\boldsymbol{\nabla}^2 f$, $\boldsymbol{\nabla}^2 \boldsymbol{A}$, and $\boldsymbol{\nabla} \times \boldsymbol{\nabla} \times \boldsymbol{A}$

We have $\boldsymbol{\nabla}^2 f = \boldsymbol{\nabla} \cdot \boldsymbol{\nabla} f$. Now $\boldsymbol{\nabla} f$ is longitudinal, since its curl is zero (Eq. A.49). Therefore, by Results 4(c) and 2 (under the usual conditions),

$$\mathrm{norm}(\boldsymbol{\nabla} \cdot \boldsymbol{\nabla} f) = \frac{\mathrm{norm}(\boldsymbol{\nabla} f)}{\mathcal{L}(\boldsymbol{\nabla} f)} = \frac{\mathrm{norm}(f)}{\mathcal{L}(\boldsymbol{\nabla} f)\mathcal{L}(f)} . \tag{A.62}$$

Similarly, $\boldsymbol{\nabla} \times \boldsymbol{A}$ is transversal, since its divergence is zero (Eq. A.48). Therefore, by Results 4(a) and 3 (under the usual conditions),

$$\mathrm{norm}(\boldsymbol{\nabla} \times \boldsymbol{\nabla} \times \boldsymbol{A}) = \frac{\mathrm{norm}(\boldsymbol{\nabla} \times \boldsymbol{A})}{\mathcal{L}(\boldsymbol{\nabla} \times \boldsymbol{A})} \leq \frac{\mathrm{norm}(\boldsymbol{A})}{\mathcal{L}(\boldsymbol{\nabla} \times \boldsymbol{A})\mathcal{L}(\boldsymbol{A})} . \tag{A.63}$$

If $\boldsymbol{A}$ is transversal, then, by Result 4(a), the $\leq$ sign above can be replaced by an equal sign.

The case of $\boldsymbol{\nabla}^2 \boldsymbol{A}$ is more delicate. We could look at the norms of its components $\boldsymbol{\nabla}^2 A_x$, $\boldsymbol{\nabla}^2 A_y$, and $\boldsymbol{\nabla}^2 A_z$, but these are relative to a particular

[17] The potential of such an $\boldsymbol{E}$ is not constant and it is a harmonic function. Liouville's theorem states that all nonconstant harmonic functions are unbounded; see Axler et al. (2001), p. 31. The potential differences across this field are infinite and the field's energy is infinite.

set of axes. It might be best to use the following invariant definition of the Laplacian of $\boldsymbol{A}$:

$$\boldsymbol{\nabla}^2 \boldsymbol{A} \equiv \boldsymbol{\nabla}(\boldsymbol{\nabla} \cdot \boldsymbol{A}) - \boldsymbol{\nabla} \times \boldsymbol{\nabla} \times \boldsymbol{A} \,. \tag{A.64}$$

A bound for $\boldsymbol{\nabla}^2 \boldsymbol{A}$ can then be obtained from the triangle inequality A.8, and Results 2 and 3.

Result 5 (Laplacian, curl of the curl). *Under our usual conditions (see Results 2 and 3),*

$$\mathrm{norm}(\boldsymbol{\nabla}^2 f) = \frac{\mathrm{norm}(f)}{\mathcal{L}(f)\mathcal{L}(\boldsymbol{\nabla} f)} \,, \tag{A.65}$$

$$\mathrm{norm}(\boldsymbol{\nabla} \times \boldsymbol{\nabla} \times \boldsymbol{A}) \leq \frac{\mathrm{norm}(\boldsymbol{A})}{\mathcal{L}(\boldsymbol{A})\mathcal{L}(\boldsymbol{\nabla} \times \boldsymbol{A})} \,, \tag{A.66}$$

$$\mathrm{norm}(\boldsymbol{\nabla}^2 \boldsymbol{A}) \leq \frac{\mathrm{norm}(\boldsymbol{A})}{\mathcal{L}(\boldsymbol{A})} \left[\frac{1}{\mathcal{L}(\boldsymbol{\nabla} \cdot \boldsymbol{A})} + \frac{1}{\mathcal{L}(\boldsymbol{\nabla} \times \boldsymbol{A})} \right] . \tag{A.67}$$

In A.66 we have an equal sign when $\boldsymbol{A}$ *is transversal, a $<$ sign when* $\boldsymbol{A}$ *is not transversal, and, if* $\boldsymbol{A}$ *is not transversal and at the same time not too longitudinal, the $<$ and $\sim$ signs are both valid.*

Results 1 to 5 summarize the method of characteristic times and lengths that was announced at the end of Sects. A.3 and A.5.

A.7 Products of Functions

One often encounters products of functions such as fg, $f\boldsymbol{A}$, $\boldsymbol{A} \cdot \boldsymbol{B}$, or $\boldsymbol{A} \times \boldsymbol{B}$. In electromagnetism, for example, $\rho \boldsymbol{v}$ and $\boldsymbol{v} \times \boldsymbol{B}$ appear in many equations.

Take the simple product $f(t)g(t)$. Is there some sort of general inequality that compares the norm of fg to the product $\mathrm{norm}(f)\,\mathrm{norm}(g)$? It turns out that, in certain cases, $\mathrm{norm}(fg)$ is larger, or even much larger, than $\mathrm{norm}(f)\,\mathrm{norm}(g)$, and that, in other cases, the reverse is true! Much can be said on this matter.[18] For instance, it can be shown that, for periodic functions f and g, the two quantities $\mathrm{norm}(fg)$ and $\mathrm{norm}(f)\,\mathrm{norm}(g)$ are never smaller than the average value of $|fg|$.[19] But, further than that, it is difficult to relate the norm of fg with the product $\mathrm{norm}(f)\,\mathrm{norm}(g)$.

Let us be more specific and consider the classic example of a sinusoidal voltage V and the corresponding current I in a simple RCL circuit. A short calculation shows that the root-mean-square power dissipated $(VI)_{\mathrm{rms}}$ (the norm of VI) is sometimes larger, sometimes smaller than $V_{\mathrm{rms}}\, I_{\mathrm{rms}}$ [the product $\mathrm{norm}(V)\,\mathrm{norm}(I)$], but that, whatever the phase difference between V

[18] The question is discussed in more detail, with examples, in F. Lorrain (2003), Sect. 10.

[19] F. Lorrain (2003), Sec. 10.3.

and I, $(VI)_{\text{rms}}$ is always of the same order of magnitude as $V_{\text{rms}}\, I_{\text{rms}}$. Why is that?

A sinusoidal function has the following interesting property:

$$\textit{it is "most of the time" "of the same order" as its amplitude.} \tag{A.68}$$

Although the expressions "most of the time" and "of the same order" have not been given precise meanings, let us nevertheless speculate. If f and g are periodic and have property A.68, then it is likely that

$$\text{norm}(f) \sim \text{amplitude}(f) \,, \tag{A.69}$$

$$\text{norm}(g) \sim \text{amplitude}(g) \,, \tag{A.70}$$

$$\text{norm}(fg) \sim \text{norm}(f)\,\text{norm}(g) \,. \tag{A.71}$$

If these equations could be proven, they would be very useful. Before attempting such a proof, however, property A.68 must first be expressed in rigorous mathematical terms. This can be done, and it is then easy to show that the first two equations, A.69–A.70, are true "in most cases".[20] Equation A.71, however, is much more difficult to prove; I was able to show that it is valid in most cases (as the two other equations), but under conditions that are not quite as specific as I should wish.[21] Unfortunately my demonstration is long and tedious and so cannot be included here. The matter requires further study. For the purposes of the present appendix, the reader's intuitive appreciation of property A.68 and of the likeliness of Eqs. A.69–A.71 will be a sufficient foundation for the considerations that follow.

The above approximate equalities are for periodic functions, whose norm A.6 is the rms value of the function. What about nonperiodic functions, whose norm A.7 is not an rms value? Let $F(t)$ and $G(t)$ be two such functions. The very fact that they have finite norms implies that *the integrals of F^2 and G^2 are both negligible outside a given interval I of finite length T*. Then we can use the norm A.6 over I, instead of the norm A.7, so that norm(F) is the rms value of F over I, and the same for G. If F and G have property A.68 over I, then they very likely obey approximate equalities similar to A.69–A.71.

Finally we have the following tentative result.[22]

Result 6 (tentative). *Let $f(t)$, $g(t)$, $\boldsymbol{A}(t)$, and $\boldsymbol{B}(t)$ be periodic functions of period T, or let them be functions that are negligible outside a given interval I of length T. Let us then use the norm A.6 over I for all these functions. Assume that they have finite amplitudes and that they have property A.68 over I. Let their norms, and the norms of the products fg, $f\boldsymbol{A}$, $\boldsymbol{A} \cdot \boldsymbol{B}$,*

[20] F. Lorrain (2003), Sect. 10.4.2.

[21] F. Lorrain (2003), Sect. 10.4.11.

[22] I present this result as tentative, because, as mentioned earlier, the conditions under which I have been able to prove it are too general. See F. Lorrain (2003), Sect. 10.4.11.

$\boldsymbol{A} \times \boldsymbol{B}$, and $|\boldsymbol{A}||\boldsymbol{B}|$, be well defined. Then, in most cases, the norms of these functions are of the same order of magnitude as their amplitudes and

$$\operatorname{norm}(fg) \sim \operatorname{norm}(f)\operatorname{norm}(g) , \tag{A.72}$$
$$\operatorname{norm}(f\boldsymbol{A}) \sim \operatorname{norm}(f)\operatorname{norm}(\boldsymbol{A}) , \tag{A.73}$$
$$\operatorname{norm}(|\boldsymbol{A}||\boldsymbol{B}|) \sim \operatorname{norm}(\boldsymbol{A})\operatorname{norm}(\boldsymbol{B}) . \tag{A.74}$$

Then, if $\boldsymbol{A}$ and $\boldsymbol{B}$ are not too perpendicular (that is, if $|\boldsymbol{A} \cdot \boldsymbol{B}| \sim |\boldsymbol{A}||\boldsymbol{B}|$), we have[23]

$$\operatorname{norm}(\boldsymbol{A} \cdot \boldsymbol{B}) \sim \operatorname{norm}(|\boldsymbol{A}||\boldsymbol{B}|) \sim \operatorname{norm}(\boldsymbol{A})\operatorname{norm}(\boldsymbol{B}) . \tag{A.75}$$

And, if $\boldsymbol{A}$ and $\boldsymbol{B}$ are not too parallel (that is, if $|\boldsymbol{A} \times \boldsymbol{B}| \sim |\boldsymbol{A}||\boldsymbol{B}|$), we have

$$\operatorname{norm}(\boldsymbol{A} \times \boldsymbol{B}) \sim \operatorname{norm}(|\boldsymbol{A}||\boldsymbol{B}|) \sim \operatorname{norm}(\boldsymbol{A})\operatorname{norm}(\boldsymbol{B}) . \tag{A.76}$$

Similar tentative results of course apply to functions of x, y, and z.

This tentative result could be useful when used in conjunction with our previous results.[24]

A.8 Summary and Discussion

Results 1 to 6 are similar to the informal procedures that are commonly used in the literature – except for the divergence, curl, and Laplacian, for which *inequalities* must often be used, instead of equalities or even approximate equalities.

In practice, of course, we might be allowed to use our method in Physics without paying too much attention to the detailed mathematical conditions under which we have seen that it is valid.

Consider a physical phenomenon that is sufficiently localized, so that we can use the norm A.6 over a finite interval of time or finite region of space. Assume that all the functions we encounter have Fourier expansions. Assume also that the norms (rms values) of these functions are typical absolute values of the same.

Our characteristic length $\mathcal{L}$ is a typical value among the radian lengths that underlie the phenomenon studied. Our characteristic time $\mathcal{T}$ is a typical

[23] See Footnote 9.

[24] One application would be to the calculation of a bound for the norm of the quantity $(\boldsymbol{A} \cdot \boldsymbol{\nabla})\boldsymbol{A}$, which is often encountered in Fluid- and Magneto-Fluid-Dynamics. A bound could be obtained by applying the triangle inequality and our previous results, including Tentative Result 6, to the right side of the identity $(\boldsymbol{A} \cdot \boldsymbol{\nabla})\boldsymbol{A} \equiv \boldsymbol{\nabla}(A^2)/2 + \boldsymbol{A} \times (\boldsymbol{\nabla} \times \boldsymbol{A})$. See F. Lorrain (2003), Sect. 10.5, Tentative Result 13.

value among the $T/2\pi$'s that underlie the same phenomenon (where the symbol T designates the period of a periodic process). In principle every function has its own particular $\mathcal{T}$ or $\mathcal{L}$. Assume that all the characteristic times involved are of the same order of magnitude, and similarly for the characteristic lengths.

Suppose finally that products of functions can all be treated simply as in Tentative Result 6 above.

Let $|f|_{\text{typical}}$ and A_{typical} be "typical values" of $|f|$ and $|\boldsymbol{A}|$ respectively. Let the symbol $\sim$ mean, as before, "is of the same order of magnitude as". Then our results can be summarized simply, as follows.

THE METHOD OF CHARACTERISTIC TIMES AND LENGTHS – *simplified version*

$$|df/dt|_{\text{typical}} \sim |f|_{\text{typical}}/\mathcal{T}\,, \tag{A.77}$$

$$|d\boldsymbol{A}/dt|_{\text{typical}} \sim A_{\text{typical}}/\mathcal{T}\,, \tag{A.78}$$

$$|\boldsymbol{\nabla} f|_{\text{typical}} \sim |f|_{\text{typical}}/\mathcal{L}\,, \tag{A.79}$$

$$|\boldsymbol{\nabla}\cdot\boldsymbol{A}|^2_{\text{typical}} + |\boldsymbol{\nabla}\times\boldsymbol{A}|^2_{\text{typical}} \sim A^2_{\text{typical}}/\mathcal{L}^2 \quad \textit{(all } \boldsymbol{A}\textit{'s)}, \tag{A.80}$$

$$|\boldsymbol{\nabla}\cdot\boldsymbol{A}|_{\text{typical}} \lesssim A_{\text{typical}}/\mathcal{L} \quad \textit{(all } \boldsymbol{A}\textit{'s)}, \tag{A.81}$$

$$|\boldsymbol{\nabla}\cdot\boldsymbol{A}|_{\text{typical}} \sim A_{\text{typical}}/\mathcal{L} \quad \textit{(if } \boldsymbol{A} \textit{ is not too transversal)}, \tag{A.82}$$

$$|\boldsymbol{\nabla}\times\boldsymbol{A}|_{\text{typical}} \lesssim A_{\text{typical}}/\mathcal{L} \quad \textit{(all } \boldsymbol{A}\textit{'s)}, \tag{A.83}$$

$$|\boldsymbol{\nabla}\times\boldsymbol{A}|_{\text{typical}} \sim A_{\text{typical}}/\mathcal{L} \quad \textit{(if } \boldsymbol{A} \textit{ is not too longitudinal)}, \tag{A.84}$$

$$|\boldsymbol{\nabla}^2 f|_{\text{typical}} \sim |f|_{\text{typical}}/\mathcal{L}^2\,, \tag{A.85}$$

$$|\boldsymbol{\nabla}\times\boldsymbol{\nabla}\times\boldsymbol{A}|_{\text{typical}} \lesssim A_{\text{typical}}/\mathcal{L}^2 \quad \textit{(all } \boldsymbol{A}\textit{'s)}, \tag{A.86}$$

$$|\boldsymbol{\nabla}\times\boldsymbol{\nabla}\times\boldsymbol{A}|_{\text{typical}} \sim A_{\text{typical}}/\mathcal{L}^2 \quad \textit{(if } \boldsymbol{A} \textit{ is not too longitudinal)}, \tag{A.87}$$

$$|\boldsymbol{\nabla}^2 \boldsymbol{A}|_{\text{typical}} \lesssim 2\,A_{\text{typical}}/\mathcal{L}^2\,, \tag{A.88}$$

where $\mathcal{L}$ is a typical radian length and $\mathcal{T}$ is a typical $T/2\pi$ of the phenomenon studied.

(For more information on the subject of transversality and longitudinality, see Def. 3 and Result 4.)

Note the factor of 2 in Eq. A.88; its origin lies in Eq. A.67.

Still, even with this simplified version of the method, a certain amount of care must be exercised. Obeying the following three rules is paramount.

Rule 1. *The above equalities and inequalities apply only to* typical *values of the functions and of their derivatives (or gradients, divergences, curls, and so on).*

Rule 2. *The nature of the characteristic time $\mathcal{T}$ (a typical $T/2\pi$) and of the characteristic length $\mathcal{L}$ (a typical radian length) must be well understood and their values chosen accordingly.*

The factor of 2π represents almost one order of magnitude. It can often be neglected, but in tighter cases it might be necessary to take it into account.

Example 3. If one studies the magnetic field of the Earth, an $\mathcal{L}$ equal to the Earth's radius R might be too large. With the factor of 2π, an $\mathcal{L}$ equal to $R/6$ might be more appropriate. Recall the characteristic length of the field $\boldsymbol{E}$ of a spherical charge (Eq. A.60): it is about six tenths of the radius. For a roughly dipolar magnetic field, which is not spherically symmetrical and decreases more quickly with distance than this $\boldsymbol{E}$, one expects a characteristic length appreciably smaller than $0.6\,R$. Much depends, of course, on the scale of the particular aspect of the Earth's magnetic field under study. Needless to say, different values of $\mathcal{L}$ give rise to different estimations for $|\boldsymbol{\nabla} f|_{\text{typical}}$, $|\boldsymbol{\nabla}\cdot\boldsymbol{A}|_{\text{typical}}$, etc.

Rule 3. *For the divergence, always use the inequality A.81 if the field is, or could easily be, transversal; for fields that are longitudinal, or not too transversal, the approximate equality A.82 may be used.*

For the curl, and for the curl of the curl, always use the inequalities A.83 and A.86 if the field is, or could easily be, longitudinal; for fields that are transversal, or not too longitudinal, the approximate equalities A.84 and A.87 may be used.

*For the Laplacian of the vector field **A**, the inequality A.88 should almost always be used. Only under exceptional circumstances could we have an equality here.*

(For more information on the subject of transversality and longitudinality, see Def. 3 and Result 4.)

Unfortunately, the direction of these unavoidable inequalities imposes severe restrictions on the conclusions that can be drawn from them.

Example 4. Consider the classic case of a highly conducting moving plasma in which the free charge density ρ_{f} continually adjusts so that $\boldsymbol{E}+\boldsymbol{v}\times\boldsymbol{B}_{\text{net}}=\boldsymbol{0}$, where $\boldsymbol{B}_{\text{net}}$ is the net magnetic flux density from all sources.[25] Let $\boldsymbol{B}_{\text{conv}}$ be the magnetic flux density due to the convection current $\rho_{\text{f}}\boldsymbol{v}$. We wish to evaluate the ratio of $B_{\text{conv typical}}$ to $B_{\text{net typical}}$.

We have

$$\boldsymbol{\nabla}\times\boldsymbol{B}_{\text{conv}}=\mu_0\rho_{\text{f}}\boldsymbol{v}\,, \tag{A.89}$$

$$\boldsymbol{\nabla}\cdot(\boldsymbol{v}\times\boldsymbol{B}_{\text{net}})=-\rho_{\text{f}}/\epsilon_0\,. \tag{A.90}$$

The latter equation is an immediate consequence of our hypothesis that $\boldsymbol{E}+\boldsymbol{v}\times\boldsymbol{B}_{\text{net}}=\boldsymbol{0}$. Equation A.90 was also be obtained earlier in the book

[25] Davidson (2001), pp. 29–30.

(Eq. 7.10) from a different hypothesis: steady conditions and a uniform σ – which entails that $\nabla \cdot (\boldsymbol{E} + \boldsymbol{v} \times \boldsymbol{B}_{\mathrm{net}}) = 0$ (see Sect. 7.2).

Under our simplified version of the method, Eq. A.89 gives us

$$\mu_0 |\rho_{\mathrm{f}}|_{\mathrm{typical}}\, v_{\mathrm{typical}} \lesssim \frac{B_{\mathrm{conv\ typical}}}{\mathcal{L}} , \tag{A.91}$$

$$|\rho_{\mathrm{f}}|_{\mathrm{typical}} \lesssim \frac{B_{\mathrm{conv\ typical}}}{\mu_0 \mathcal{L} v_{\mathrm{typical}}} , \tag{A.92}$$

while Eq. A.90 gives

$$|\rho_{\mathrm{f}}|_{\mathrm{typical}} \lesssim \frac{\epsilon_0 v_{\mathrm{typical}}\, B_{\mathrm{net\ typical}}}{\mathcal{L}} . \tag{A.93}$$

From the last two inequalities, no conclusion can be drawn as to the ratio of $B_{\mathrm{conv\ typical}}$ to $B_{\mathrm{net\ typical}}$.

If we had approximate equalities instead of inequalities, however, the ratio of these two quantities could be evaluated. Are Eqs. A.82 and A.84 applicable here? This requires some thought.

First of all, recall that, since $\nabla \cdot \boldsymbol{B}_{\mathrm{conv}} = 0$, the field $\boldsymbol{B}_{\mathrm{conv}}$ is transversal [Result 4(a)]. So are all magnetic flux density fields, of course. According to Rule 3, then, the $\lesssim$ sign in A.92 can be replaced by a $\sim$ sign:

$$|\rho_{\mathrm{f}}|_{\mathrm{typical}} \sim \frac{B_{\mathrm{conv\ typical}}}{\mu_0 \mathcal{L} v_{\mathrm{typical}}} . \tag{A.94}$$

We could now ask ourselves if the field $\boldsymbol{v} \times \boldsymbol{B}_{\mathrm{net}}$ is transversal or longitudinal. We might also discuss to what degree $\boldsymbol{v}$ may be parallel or perpendicular to $\boldsymbol{B}_{\mathrm{net}}$ (see Result 6). But we need not do any of this here, because, putting together A.93 and A.94, we find that, roughly,

$$\frac{B_{\mathrm{conv\ typical}}}{\mu_0 \mathcal{L} v_{\mathrm{typical}}} \lesssim \frac{\epsilon_0 v_{\mathrm{typical}}\, B_{\mathrm{net\ typical}}}{\mathcal{L}} , \tag{A.95}$$

$$\frac{B_{\mathrm{conv\ typical}}}{B_{\mathrm{net\ typical}}} \lesssim \epsilon_0 \mu_0 v_{\mathrm{typical}}^2 = \frac{v_{\mathrm{typical}}^2}{c^2} . \tag{A.96}$$

Since v is normally very small, compared to the velocity of light c, we conclude (as in Sect. 7.2.3 above) that $\boldsymbol{B}_{\mathrm{conv}}$ is negligible, compared to $\boldsymbol{B}_{\mathrm{net}}$.

Very well, but notice that, in order to find our way out of the blind alley of the inequalities A.92–A.93, from which nothing could be concluded, we had to take a good look at the fields involved.

Exercise 3. [26] Consider the same highly conducting moving plasma, in which the $\boldsymbol{E}$ and $\boldsymbol{v} \times \boldsymbol{B}$ fields practically cancel each other. Find conditions under which the displacement current density $\boldsymbol{J}_{\mathrm{d}} = \epsilon_0 \partial \boldsymbol{E}/\partial t$ is negligible, compared to the sum

[26] The solution to this exercise is given below.

$$\boldsymbol{J}_{\mathrm{d}} + \boldsymbol{J}_{\mathrm{ch}} = \frac{1}{\mu_0} \nabla \times \boldsymbol{B} , \tag{A.97}$$

where $\boldsymbol{J}_{\mathrm{ch}}$ is the net current density of all moving charges. Start with

$$E_{\mathrm{typical}} \sim |\boldsymbol{v} \times \boldsymbol{B}|_{\mathrm{typical}} \leq (vB)_{\mathrm{typical}} \sim v_{\mathrm{typical}}\, B_{\mathrm{typical}} \,. \tag{A.98}$$

Interpret physically the mathematical condition finally obtained.

Solution of Exercise 3. We have

$$J_{\mathrm{d\,typical}} = \epsilon_0 \left| \frac{\partial \boldsymbol{E}}{\partial t} \right|_{\mathrm{typical}} \sim \epsilon_0 \frac{E_{\mathrm{typical}}}{\mathcal{T}} \tag{A.99}$$

$$\lesssim \epsilon_0 \frac{v_{\mathrm{typical}}\, B_{\mathrm{typical}}}{\mathcal{T}} , \tag{A.100}$$

$$|\boldsymbol{J}_{\mathrm{d}} + \boldsymbol{J}_{\mathrm{ch}}|_{\mathrm{typical}} = \frac{|\nabla \times \boldsymbol{B}|_{\mathrm{typical}}}{\mu_0} = \frac{B_{\mathrm{typical}}}{\mu_0 \mathcal{L}} . \tag{A.101}$$

Therefore

$$\frac{J_{\mathrm{d\,typical}}}{|\boldsymbol{J}_{\mathrm{d}} + \boldsymbol{J}_{\mathrm{ch}}|_{\mathrm{typical}}} \lesssim \frac{\epsilon_0 \mu_0 v_{\mathrm{typical}}\, \mathcal{L}}{\mathcal{T}} = \frac{v_{\mathrm{typical}}\, \mathcal{L}}{c^2 \mathcal{T}} . \tag{A.102}$$

The displacement current $\boldsymbol{J}_{\mathrm{d}}$ will be negligible, relative to the net current $\boldsymbol{J}_{\mathrm{ch}}$ of all moving charges, if

$$\mathcal{T} \gg \frac{v_{\mathrm{typical}}}{c} \frac{\mathcal{L}}{c} . \tag{A.103}$$

This condition will be satisfied if, for example,

$$\frac{v_{\mathrm{typical}}}{c} \ll 1 \ \text{ and } \ \frac{\mathcal{L}}{c} < \mathcal{T} , \tag{A.104}$$

or if

$$\frac{v_{\mathrm{typical}}}{c} < 1 \ \text{ and } \ \frac{\mathcal{L}}{c} \ll \mathcal{T} . \tag{A.105}$$

The plasma velocity v is usually orders of magnitude smaller than c. As to $\mathcal{L}/c$, it is the time required for light to travel a distance equal to the characteristic length; Equations A.104–A.105 require that this time be smaller or much smaller than the characteristic time $\mathcal{T}$ of the system.[27]

The above exercise concerns *moving* media. Section 2.4 shows that the displacement current is also negligible in ordinary *stationary* conducting media.

It is often difficult to evaluate the orders of magnitude of the various quantities that appear in physical equations. Hopefully the approach presented here – in particular the above three rules – will help us attain greater certainty and precision in this task. For other examples where our method is used, see Sects. 2.5, 6.6, and 9.2.

[27] See Ferraro and Plumpton (1966), p. 16.

B SI prefixes

Factor	Prefix	Symbol
10^{24}	yotta	Y
10^{21}	zetta	Z
10^{18}	exa	E
10^{15}	peta	P
10^{12}	tera	T
10^{9}	giga	G
10^{6}	mega	M
10^{3}	kilo	k
10^{2}	hecto	h
10^{1}	deka	da
10^{-1}	deci	d
10^{-2}	centi	c
10^{-3}	milli	m
10^{-6}	micro	μ
10^{-9}	nano	n
10^{-12}	pico	p
10^{-15}	femto	f
10^{-18}	atto	a
10^{-21}	zepto	z
10^{-24}	yocto	y

References

Abbreviations

A&A: Astronomy and Astrophysics
ApJ: Astrophysical Journal

1. V.E. Abramov-Maksimov, G.F. Vyalshin, G.B. Gelfreikh, V.T. Shatilov: Solar Phys. **164**, 333 (1996)
2. A. Adjabshirizadeh, S. Koutchmy: A&A **122**, 1 (1983)
3. I. Alexeff: IEEE Trans. on Plasma Science **17**, 282 (1989)
4. I. Alexeff, J.R. Roth: IEEE Trans. on Plasma Science **PS-14**, 862 (1986)
5. H. Alfvén: Arkiv för Mat. Astron. o Fysik **29B** No 2 (1943)
6. H. Alfvén: Electric current structure of the magnetosphere. In: *Physics of the Hot Plasma in the Magnetosphere*, ed by B. Hultqvist, L. Stenflo (Plenum, New York 1975)
7. H. Alfvén: *Cosmic Plasma* (D. Reidel, Dordrecht 1981)
8. H. Alfvén, C.G. Fälthammar: *Cosmical Electrodynamics* (Clarendon Press, Oxford 1963)
9. C.E. Alissandrakis, D. Dialetis, G. Tsiropoulos: A&A **174**, 275 (1987)
10. C.E. Alissandrakis, H.C. Dara, S. Koutchmy: A&A **249**, 533 (1991)
11. C.W. Allen: *Astrophysical Quantities* (Athlone Press, London 1973)
12. M.D. Altschuler: IAU Symposium **57** (1974)
13. A.S. Andreev, A.G. Kosovichev. In: *Second SoHO Workshop on Mass Supply and Flows in the Solar Corona*, ed by B. Fleck, G. Noci, G. Poletto (Kluwer Academic, Newell (MA) 1994)
14. P. Arena, E. Landi Degl'Innocenti, G. Noci: Solar Phys. **129**, 259 (1990)
15. Brother Arnold, Brother Potamian: *The Letter of Petrus Peregrinus on the Magnet, A.D. 1269*, transl by Brother Arnold, with introductory notice by Brother Potamian (McGraw, New York 1904)
16. M.J. Aschwanden, L. Fletcher, C.J. Schrijver, D. Alexander: ApJ **520**, 880 (1999)
17. M.J. Aschwanden, T. Kosugi, Y. Hanaoka, M. Nishio, D.B. Melrose: ApJ **526**, 1026 (1999)
18. M.J. Aschwanden, J.S. Newmark, J.P. Delaboudinière, W.M. Neupert, J.A. Klimchuk, G.A. Gary, F. Portier-Fozzani, A. Zucker: ApJ **515**, 842 (1999)
19. M.J. Aschwanden, C.J. Schrijver, D. Alexander: ApJ **550**, 1036 (2001)
20. S. Axler et al: *Harmonic Function Theory*, 2nd edn (Springer, New York 2001)

21. G. Backus, R. Parker, C. Constable: *Foundations of Geomagnetism* (Cambridge University Press, Cambridge, UK 1996)
22. J.N. Bahcall, M.H. Pinsonneau: Rev. Mod. Phys. **64**, 885 (1992)
23. J.N. Bahcall, R.K. Ulrich: Rev. Mod. Phys. **60**, 297 (1988)
24. H. Balthasar, H. Schleicher, C. Bendlim, R. Volkmer: A&A **315**, 603 (1996)
25. F. Beaudin, E. Bocchialini, S. Koutchmy: A&A **314**, L9 (1996)
26. F. Beaudin, R. Molowny-Horas, S. Koutchmy: A&A **326**, 842 (1997)
27. G. Beaujouan: La science dans l'Occident médiéval chrétien. In: *Histoire générale des sciences*, vol 1 (Presses Universitaires de France, Paris 1957)
28. J.M. Beckers: Ann. Rev. Astron. and Astrophys. **10**, 73 (1972)
29. J.M. Beckers, E.H. Schröter: Solar Phys. **125**, 37 (1989)
30. G. Bekefi, B.T. Feld, J. Parmentola, K. Tsipis: Nature **284**, 219 (1980)
31. W.H. Bennett: Phys. Rev. **45**, 890 (1934)
32. A. Berkovic, E. Landi, M. Landini, I. Ruedi, S.K. Solanki: A&A **383**, 661 (2002)
33. D. Bercovici, G. Schubert, G.A. Glatzmaier: Geophys. Res. Letters **16**, 617 (1989)
34. H.A. Bethe, J. Ashkin. In: *Experimental Nuclear Phys.*, vol 1, ed by E. Segrè (Wiley, New York; Chapman and Hall, London 1953)
35. J.W. Bieber, D.M. Rust: ApJ **453**, 911 (1995)
36. F.H. Bigelow: *The Solar Corona* (US Weather Bureau Bulletin 21, Washington 1898)
37. C.H. Blanchard: National Bureau of Standards Circular **527**, 9 (1954)
38. K. Bocchialini, S. Koutchmy, J.C. Vial: Space Science Rev. **70**, 57 (1994)
39. K. Bocchialini, J.C. Vial, S. Koutchmy: ApJ **423**, L67 (1994)
40. T.J. Bogdan, D.C. Braun. In: *Fourth SoHO Workshop on Helioseismology*, vol 1, ESA SP-376 (1995)
41. T.J. Bogdan, D.C. Braun, B.W. Lites, J.H. Thomas: ApJ **492**, 369 (1998)
42. A.M. Bork: Amer. Jour. Phys. **31**, 854 (1963)
43. T.J.M. Boyd, J.J. Sanderson: *Plasma Dynamics* (Nelson, London, UK 1969)
44. S.I. Braginskii: Rev. Plasma Phys. (Consultants Bureau, New York) **1**, 205 (1965)
45. P.N. Brandt, G.B. Scharmer, S. Ferguson et al: Nature **335**, 238 (1988)
46. D.C. Braun: Astronomical Society of the Pacific Conference Series **76** (1995)
47. D.C. Braun, Y. Fan, C. Lindsey, S.M. Jefferies. Preprint, National Solar Observatory, Tucson (AZ), USA (1996)
48. R.J. Bray, R.E. Loughhead: Solar Phys. **85**, 131 (1983)
49. R.J. Bray, R.E. Loughhead: A&A **142**, 199 (1985)
50. R.J. Bray, L.E. Cram, C.J. Durrant, R.E. Loughhead: *Plasma Loops in the Solar Corona* (Cambridge University Press, Cambridge, UK 1991)
51. P. Brekke, O. Kjeldseth-Moe, J.D. Bartoe, G.E. Brueckner: ESA **SP-270** (1987)
52. P. Brekke, O. Kjeldseth-Moe, G.E. Brueckner: ESA **SP-270**, 341 (1987)
53. P. Brekke, O. Kjeldseth-Moe, G.E. Brueckner: Astrophys. and Space Science **170**, 135 (1990)
54. J.C. Bremaeker: *Geophysics: the Earth's Interior* (Wiley, New York 1985)
55. R.J. Briggs, J. Clark, T. Fessenden, E. Lee, E. Lauer: Lawrence Livermore Laboratory Report **UCID-17516** (1977)

56. G.E. Brueckner. In: *Solar Active Regions*, ed by F. Orall (Colorado Associated University Press 1981)
57. G.E. Brueckner: ApJ **272**, 329 (1983)
58. G.E. Brueckner: Solar Phys. **123**, 41 (1989)
59. G.E. Brueckner, J.D.F. Bartoe: ApJ **272**, 329 (1983)
60. E. Bullard: Proc. Cambridge Philos. Soc. **51**, 744 (1955)
61. E. Bullard: AIP Conference Proceedings **46**, 373 (1978)

62. A.B. Cambel: *Plasma Physics and Magnetofluidmechanics* (McGraw–Hill, New York 1963)
63. P.J. Cargill, J. Chen, D.A. Garren: ApJ **423**, 854 (1994)
64. G.A. Chapman: Nature **308**, 252 (1984)
65. F.F. Chen: *Introduction to Plasma Physics and Controlled Fusion* (Springer, New York 1984)
66. J. Chen: ApJ **338**, 453 (1989)
67. J. Chen: Dynamics, catastrophe and magnetic energy release of toroidal solar current loops. In: *Physics of Magnetic Flux Ropes*, ed by C.T. Russell et al (Geophysical Monograph 58, American Geophysical Union, 1990)
68. A.R. Choudhuri: *The Physics of Fluids and Plasmas* (Cambridge University Press, Cambridge, UK 1998)
69. E.R. Cohen, B.N. Taylor: Phys. Today, Buyer's Guide (August 2001)
70. A.H. Cook: *Physics of the Earth and Planets* (Wiley, New York 1973)
71. D.R. Corson: Amer. Jour. Phys. **24**, 126 (1956)
72. R. Courant, F. John: *Introduction to Calculus and Analysis* (Springer, New York 1989)
73. T.G. Cowling: Monthly Notices of the Royal Astronomical Society (GB) **89**, 140 (1929)
74. T.G. Cowling: Monthly Notices of the Royal Astronomical Society (GB) **94**, 39 (1934)
75. T.G. Cowling: Monthly Notices of the Royal Astronomical Society (GB) **105**, 166 (1945)
76. T.G. Cowling: Quart. Jour. Mech. Appl. Math. **10**, 129 (1957)
77. T.G. Cowling: Monthly Notices of the Royal Astronomical Society (GB) **140**, 547 (1968)
78. T.G. Cowling: *Magnetohydrodynamics* (Adam Hilger, Bristol, UK 1976)
79. M.J. Crooks, D.B. Litvin, P.W. Matthews, R. Macaulay, J. Shaw: Amer. Jour. Phys. **46**, 729 (1978)

80. H.C. Dara, C.E. Alissandrakis, S. Koutchmy: Solar Phys. **109**, 19 (1987)
81. H.C. Dara, C.E. Alissandrakis, S. Koutchmy: Solar Phys. **127**, 403 (1990)
82. H.C. Dara, S. Koutchmy, Y. Suematsu: *Proceedings of an International Meeting Held in Guadeloupe, DOM, France* (1998) p 255
83. P.A. Davidson: *An Introduction to Magnetohydrodynamics* (Cambridge University Press, Cambridge, UK 2001)
84. J.M. Davila: ApJ **317**, 514 (1987)
85. J.C. De Bremaeker: *Geophysics, the Earth's Interior* (Wiley, New York 1985)
86. W. Deinzer, M. Knölker, H.H. Voigt: *Small-Scale Magnetic Flux Concentrations in the Solar Photosphere* (Vandenhoeck & Ruprecht, Göttingen 1985)

87. J.C. del Toro Iniesta, M. Collados, Sánchez et al: A&A **233**, 570 (1990)
88. P. Demoulin, K.L. Klein: Transport and energy conversion in the heliosphere. In: *Lectures Given at the CNRS Summer School on Solar Astrophysics* (2000) p 99
89. K.P. Dere: Solar Phys. **77**, 77 (1982)
90. K.P. Dere, J.D.F. Bartoe, G.E. Brueckner: Solar Phys. **123**, 41 (1989)
91. C. Domb: Nature **282**, 235 (1979)
92. G.A. Doschek, U. Feldman: ApJ **459**, 773 (1996)
93. G.A. Doschek, K.T. Strong, S. Tsuneta: ApJ **440**, 370 (1995)
94. J.J. Drake, G. Peres, S. Orlando, J.M. Laming, A. Maggio: ApJ **545**, 1074 (2000)
95. S. D'Silva: ApJ **407**, 385 (1993)
96. M.A. Dubois, F. Bely-Duban, D. Grisillon (eds): *Plasma Phenomena in the Solar Atmosphere* (Les Éditions de Physique, Les Ulis, France 1990)
97. S.V. Dudin, A.V. Zykov, V.I. Farenik: Rev. Sci. Instr. **65**, 1451 (1994)
98. R.B. Dunn. Thesis, Harvard University, Cambridge (MA) (1960)
99. R.B. Dunn. Private communication (1976)
100. R.B. Dunn, J.B. Zirker: Solar Phys. **33**, 281 (1973)
101. E. Durand: *Magnétostatique* (Masson, Paris 1968)
102. T.L. Duvall: Astronomical Society of the Pacific Conference Series **76** (1995)
103. T.L. Duvall, S.M. Jefferies, J.W. Harvey, M.A. Pomerantz: Nature **362**, 430 (1993)
104. T.L. Duvall Jr, S. D'Silva, S.M. Jefferies, J.W. Harvey, J. Schou: Nature **379**, 235 (1996)

105. J. Evershed: Monthly Notices of the Royal Astronomical Society (GB) **69**, 454 (1909)
106. J. Evershed: Monthly Notices of the Royal Astronomical Society (GB) **70**, 217 (1910)

107. M. Faraday: Phil. Trans. Roy. Soc. **25** (1852)
108. M. Faraday: *Faraday's Diary*, vol I (G. Bell and Sons, London 1932) paragraphs 255–257
109. M. Faraday: *Faraday's Diary*, vol V (Martin T. Bell, London 1852) p 403
110. M. Faraday: *Experimental Researches in Electricity*, vol 1 (New York, Dover 1965) paragraph 220
111. D.R. Fearn, P.H. Roberts, A.M. Soward: Convection, stability, and the dynamo. In: *Energy Stability and Convection*, ed by G.P. Galdi, B. Straughan (Longman, Harlow (Essex), UK 1988)
112. R.F. Fernsler, S.P. Slinker, R.F. Hubbard: Phys. of Fluids **B3**, 2696 (1991)
113. V.C.A. Ferraro: Monthly Notices of the Royal Astronomical Society (GB) **97**, 458 (1937)
114. V.C.A. Ferraro, C. Plumpton: *An Introduction to Magneto-fluid Mechanics*, 2nd edn (Clarendon Press, Oxford 1966)
115. J.M. Ferreira, C.A. Mendoza-Briceno: A&A **327**, 252 (1997)
116. R.P. Feynman: *The Feynman Lectures on Physics*, vol 2 (Addison–Wesley, Reading (MA) 1964)
117. C.E. Fichtel, F.B. McDonald: Annual Rev. of Astron. and Astrophys. **5**, 351 (1967)

118. B. Fleck: Rev. Mod. Astron. **4**, 90 (1991)
119. B. Fleck, G. Noci, G. Poletto (eds): *Second SoHO Workshop on Mass Supply and Flows in the Solar Corona* (Kluwer Academic, Newell (MA) 1994)
120. J.A. Fleming: The Earth's magnetism and magnetic surveys. In: *Terrestrial Magnetism and Electricity* (McGraw–Hill, New York 1939)
121. P. Foukal: ApJ **223**, 1046 (1978)
122. P. Foukal: *Solar Astrophysics* (John Wiley, New York 1990)
123. P. Foukal: Nature **358**, 285 (1992)
124. P.V. Foukal, B.B. Behr. In: IAU Colloquium 144, *Solar Coronal Structures*, ed by V. Rušin, P. Heinzel, J.C. Vial (VEDA Publishing Company, Bratislava, Slovakia 1994)
125. R.L. Freeman, E.M. Jones: *Atomic Collision Processes in Physics Plasma Experiments* (UKAEA Research Group, Culham Laboratory, Abingdon (Berkshire), UK 1974)
126. A.P. French: Amer. Jour. Phys. **62**, 972 (1994)

127. D.E. Gary, H. Wang, N. Nitta, T. Kosugi: ApJ **464**, 965 (1996)
128. G.A. Gary, F. Portier-Fozzani, A. Zucker: ApJ **515**, 842 (1999)
129. C.C. Gillispie (ed): *Dictionary of Scientific Biography* (Charles Scribner, New York 1970–1978)
130. R.G. Giovanelli. In: *Solar Physics*, ed by J.N. Xanthikis (Interscience, London 1967)
131. G.A. Glatzmaier, P.H. Roberts: Nature **377**, 203 (1995)
132. L. Golub. In: *Mechanisms of Chromospheric and Coronal Heating*, ed by P. Ulmschneider, E.R. Priest, R. Rosner (Springer, Berlin 1991) p 115
133. M.L. Goodman: ApJ **443**, 450 (1995)
134. M.L. Goodman: ApJ **463**, 784 (1996)
135. G. Gouy: Comptes rendus hebdomadaires des séances de l'Académie des sciences (Gauthier-Villars, Paris) **155**, 608 (1912)
136. D. Gubbins, P.H. Roberts: Magnetohydrodynamics of the Earth's core. In: *Geomagnetism*, vol 2, ed by J.A. Jacobs (Academic Press, San Diego (CA) 1987)

137. G. Haerendel: Nature **360**, 241 (1992)
138. G.E. Hale: ApJ **28**, 100, 315 (1908)
139. Y. Hanoaka: Solar Phys. **165**, 275 (1996)
140. P.M. Harman: *The Natural Philosophy of James Clerk Maxwell* (Cambridge University Press, Cambridge, UK 1998)
141. R. Harmon, R. Rosner, H. Zirin, E. Spiller, L. Golub: ApJ **417**, L83 (1993)
142. L.K. Harra-Murnion, S.A. Matthews, H. Hara, K. Ichimoto: A&A **345**, 1011 (1999)
143. A. Haudricourt, J. Needham: Les sciences en Chine médiévale. In: *Histoire générale des sciences*, vol 1 (Presses Universitaires de France, Paris 1957)
144. O. Heaviside: *Electromagnetic Theory*, with a critical and historical introduction by E. Weber (Dover, New York 1950)
145. A. Hellemans, B. Bunch: *The Timetables of Science* (Simon and Schuster, New York 1988)
146. A.W. Hendry: Amer. Jour. Phys. **61**, 906 (1993)

147. J.C. Hénoux, B.V. Somov: A&A **185**, 306 (1987)
148. D. Heristchi, Z. Mouradian: Solar Phys. **142**, 21 (1992)
149. J. Heyvaerts, M.J. Hagyard: Report of Working Group 1 on the energy storage in solar flares. In: *Flares 22 Workshop, Dynamics of Solar Flares*, ed by B. Schmeider, E. Priest (Observatoire de Paris, Paris 1991)
150. E. Hiei. In: *Second SoHO Workshop on Mass Supply and Flows in the Solar Corona*, ed by B. Fleck, G. Noci, G. Poletto (Kluwer Academic, Newell (MA) 1994) p 189
151. E. Hiei. In: *Solar Coronal Structures* (VEDA Publishing Company, Bratislava, Slovakia 1994)
152. T. Hirayama: Solar Phys. **137**, 33 (1991)
153. T. Hirayama: Solar Phys. **137**, 33 (1992)
154. J.V. Holweg: ApJ **320**, 875 (1987)
155. R.F. Howard: Solar Phys. **167**, 95 (1996)
156. T.P. Hughes, B.B. Godfrey: Phys. of Fluids **27**, 1531 (1984)
157. B.J. Hunt: *The Maxwellians* (Cornell University Press, Ithaca (NY) 1991)

158. K. Ichimoto, H. Hara, A. Takeda, K. Kumagai, T. Sakurai, T. Shimizu, H.S. Hudson: ApJ **445**, 978 (1995)
159. R.L. Ingraham: Physica Scripta **52**, 60 (1995)
160. R.L. Ingraham: Physica Scripta **53**, 700 (1996)
161. R.L. Ingraham, T. Vulcan: Phys. Plasmas **4**, 3173 (1997)
162. J.A. Ionson: ApJ **226**, 650 (1978)
163. J.A. Ionson: ApJ **254**, 318 (1982)
164. J.A. Ionson: ApJ **271**, 778 (1983)

165. J.A. Jacobs: *Geomagnetism* (Academic Press, London 1987)
166. K. Jahn: A&A **222**, 264 (1983)
167. D. Jault, C. Gire, J.L. LeMouël: Nature **333**, 353 (1988)
168. J.T. Jefferies, F.Q. Orrall: ApJ **141**, 505 and 519 (1965)

169. R. Kano, S. Tsuneta: ApJ **454**, 934 (1995)
170. M. Karlicky, G. Mann, H. Aurass: A&A **314**, 303 (1996)
171. H. Kawakami: PASJ **35**, 459 (1983)
172. R. Keppens, V. Martínez Pillet: A&A **316**, 229 (1996)
173. J.I. Khan, L.K. Harra-Murnion, H.S. Hudson, J.R. Lemen, A.C. Sterling: ApJ **452**, L153 (1995)
174. O. Kjeldseth, P. Brekke, S.V.H. Haugan: Proceedings of the Conference *A Crossroads for European Solar and Heliospheric Physics. Recent Achievements and Future Mission Possibilities* **SP-417**, 153 (1998)
175. J.A. Klimchuk, L.J. Porter: Nature **337**, 131 (1995)
176. J.A. Klimchuk, J.R. Leman, U. Feldman, S. Tsuneta, Y. Uchida: Publ. Astron. Soc. Japan **44**, L181 (1992)
177. J.A. Klimchuk, S.K. Antiochos, D. Norton: ApJ **542**, 504 (2000)
178. H. Knoepfel: *Pulsed High Magnetic Fields* (North Holland 1970)
179. H.H. Kolm, O.K. Mawardi: Jour. Appl. Phys. **32**, 1296 (1961)
180. M. Kopecký, G.V. Kuklin: Bull. Astron. Inst. Czech. **17**, 45 (1966)
181. M. Kopecký, E. Soytürk: Bull. Astron. Inst. Czech. **22**, 154 (1971)

182. A. Kosovichev: ApJ **461**, L55 (1996)
183. S. Koutchmy. In: *Solar Polarimetry, Proceedings of the 11th NSO/SPO Summer Workshop*, ed by L.J. November (1991) p 237
184. S. Koutchmy, A. Adjabshirizadeh: A&A **99**, 111 (1981)
185. S. Koutchmy, M. Livshits: Space Science Rev. **61**, 393 (1992)
186. S. Koutchmy, P. Lorrain. In: *Recent Observations on Solar Spicules; a Preliminary Model*, Sumer/CDS meeting, Orsay (France), ed by J.C. Vial (Institut d'astrophysique spatiale, Université de Paris XI, Orsay, France 1992)
187. S. Koutchmy, M.L. Loucif: Properties of impulsive events in a polar coronal hole. In: *Mechanisms of Chromospheric and Coronal Heating*, ed by P. Ulmschneider, E.R. Priest, R. Rosner (Springer, Berlin 1991)
188. S. Koutchmy, G. Stellmacher: Solar Phys. **49**, 253 (1976)
189. S. Koutchmy, G. Stellmacher: A&A **67**, 93 (1978)
190. S. Koutchmy, V.A. Koutvitsky, M.M. Molodensky, L.S. Solov'iev, O. Koutchmy: Space Science Rev. **70**, 283 (1994)
191. S. Koutchmy, H. Hara, K. Shibata, Y. Suematsu, K. Reardon. In: *Observations of Plasma Astrophysics, Five Years of Yohkoh and Beyond, Proceedings of the Fifth Annual Symposium*, ed by T. Watanabe, T. Kosugi, A.C. Sterling (1997)
192. V. Krishan: Jour. Plasma Phys. **56**, 427 (1996)

193. B. Ladbury: Phys. Today (April 1996) p 20
194. L.D. Landau, E.M. Lifshitz: *Fluid Mechanics*, 2nd edn (Butterworth–Heinemann, Oxford 1987)
195. M. Landini, L. Landini, E. Landini: A&A **383**, 653 (2002)
196. K.R. Lang In: IAU Colloquium 144, *Solar Coronal Structures*, ed by V. Rušin, P. Heinzel, J.C. Vial (VEDA Publishing, Bratislava, Slovakia 1994) p 195
197. J. Larmor: *Report of the British Association for the Advancement of Science*, p 159 (1919)
198. J. Larmor: *Mathematical and Physical Papers*, vol 2 (Cambridge University Press, Cambridge, UK 1929) p 611, 615
199. J. Larmor: Monthly Notices of the Royal Astronomical Society (GB) **94**, 469 (1934)
200. J.D. Lawson: *The Physics of Charged-Particle Beams* (Clarendon, Oxford 1977)
201. R. Lee, M. Lampe: Phys. Rev. Letters **31**, 1390 (1973)
202. K.D. Leha, R.C. Canfield, A.N. McClymont, L. Van Driel-Gesztelyi: ApJ **462**, 547 (1996)
203. R. Lenoble: Le magnétisme et l'électricité. In: *Histoire générale des sciences*, vol 2 (Presses Universitaires de France, Paris 1958)
204. D.D. Lenz: ApJ **517**, 497 (1999)
205. G. Léonard: L'Italie médiévale. In: *Encyclopédie de la Pléiade, Histoire Universelle*, vol 2 (Gallimard, Paris 1957)
206. C. Lindsey, J.T. Jefferies, T.A. Clark et al: Nature **358**, 308 (1992)
207. C. Lindsey, D.C. Braun, S.M. Jefferies, M. Woodward, Y. Fan, Y. Gu, S. Redfield. Preprint, Solar Physics Research Corporation, Tucson (AZ), USA 1996
208. B.W. Lites, A. Skumanich: ApJ **348**, 747 (1990)

209. B.W. Lites, D.F. Elmore, P. Seagraves, A.P. Skumanich: ApJ **418**, 928 (1990)
210. W. Livingston: Nature **350**, 45 (1991)
211. F. Lorrain: Using characteristic lengths and times to estimate the order of magnitude of derivatives in physics. Unpublished document (2003) available via the EPAPS homepage (www.aip.org/pubservs/epaps.html – click on "Retrieve files") or from ftp://ftp.aip.org/epaps/ in the directory am_j_phys, document number E-AJPIAS-72-027401
212. F. Lorrain: Amer. Jour. Phys. **72**, 683 (2004)
213. P. Lorrain: Eur. Jour. Phys. **11**, 94 (1990)
214. P. Lorrain: IEEE Trans. Plasma Sci. **19**, 862 (1991)
215. P. Lorrain: Physica Scripta **47**, 461 (1993a)
216. P. Lorrain: Amer. Jour. Phys. **61**, 811 (1993b)
217. P. Lorrain: Physica Scripta **52**, 349 (1995)
218. P. Lorrain: Eur. Jour. Phys. **22**, L3 (2001)
219. P. Lorrain, Corson D.R.: *Electromagnetism: Principles and Applications*, 2nd edn (W.H. Freeman, New York 1990)
220. P. Lorrain, O. Koutchmy: A&A **339**, 610 (1998)
221. P. Lorrain, S. Koutchmy: A&A **269**, 518 (1993)
222. P. Lorrain, S. Koutchmy: Solar Phys. **165**, 115 (1996)
223. P. Lorrain, S. Koutchmy: Solar Phys. **178**, 39 (1998)
224. P. Lorrain, N. Salingaros: Amer. Jour. Phys. **61**, 811 (1993)
225. P. Lorrain, D.R. Corson, F. Lorrain: *Electromagnetic Fields and Waves*, 3rd edn (W.H. Freeman, New York 1988)
226. P. Lorrain, D.R. Corson, F. Lorrain: *Fundamentals of Electromagnetic Phenomena* (W.H. Freeman, New York 2000)
227. P. Lorrain, J. McTavish, F. Lorrain: Eur. Jour. Phys. **19**, 451 (1998)
228. P. Lorrain, D. Redžić, S. Koutchmy, J. McTavish: accepted for publication in A&A (2004)
229. P. Lorrain, D. Redžić, S. Koutchmy, J. McTavish, O. Koutchmy: Solar Phys., submitted (2005)
230. S. Lundquist: Arkiv för Fysik **5**, 297 (1952)

231. A.W. Macdonald: La Chine. In: *Encyclopédie de la Pléiade, Histoire Universelle*, vol 2 (Gallimard, Paris 1957)
232. S. Malin: Historical introduction to geomagnetism. In: *Geomagnetism*, ed by J.A. Jacobs (Academic Press, San Diego (CA) 1989)
233. J.T. Mariska: *The Solar Transition Region* (Cambridge Astrophysics Series No 23, Cambridge University Press, Cambridge, UK 1992)
234. P.C.H. Martens, N.E. Hurlburt, A.M. Title, L.W. Acton: ApJ **463**, 372 (1996)
235. M. Mason, W. Weaver: *The Electromagnetic Field* (Dover 1929)
236. S. Masuda, T. Kosugi, H. Hara, S. Tsuneta, Y. Ogawara: Nature **371**, 495 (1994)
237. J.C. Maxwell: *A Treatise on Electricity and Magnetism* (Clarendon, Oxford 1873)
238. A.N. McClymont, Z. Mikic: ApJ **422**, 899 (1994)
239. J.E. McCune, W.R. Sears: Jour. of the Aero/Space Sciences **26**, 674 (1959)
240. J.P. Mehltretter: Solar Phys. **38**, 43 (1974)
241. D.B. Melrose: ApJ **381**, 306 (1991)

242. R.T. Merrill, M.W. McElhinny: *The Earth's Magnetic Field* (Academic Press, London 1983)
243. J.P. Meyer: Adv. Space Res. **11**, 269 (1991)
244. R.B. Miller: *Introduction to the Physics of Intense Charged-Particle Beams* (Plenum, New York 1982)
245. K. Miyamoto: *Plasma Physics for Nuclear Fusion* (MIT Press, Cambridge (MA) 1989)
246. H.K. Moffatt: *Magnetic Field Generation in Electrically Conducting Fluids* (Cambridge University Press, Cambridge, UK 1978)
247. Y. Mok, Z. Mikic, J. Linker: ApJ **555**, 440 (2001)
248. C. Møller: *The Theory of Relativity* (Oxford university Press, London 1974)
249. M.M. Molodenskii, L.S. Solov'ev: Astron. Reports **37**, 83 (1993)
250. P.K. Monoharan, L. Van Driel-Gesztelyi, M. Pick, P. Demoulin: ApJ **468**, L73 (1996)
251. R. Moreau: *Magnetohydrodynamics* (Kluwer Academic, Newell (MA) 1990)
252. D. Moss, A. Brandenburg: A&A **346**, 1009 (1999)
253. R. Muller, Th. Roudier, J.C. Hulot: Solar Phys. **119**, 229 (1989)

254. V.M. Nakariakov, L. Ofman, E.E. DeLuca, B. Roberts, J.M. Davila: Science **285**, 862 (1999)
255. J. Needham, W. Ling: *Science and Civilization in Ancient China* (Cambridge University Press, Cambridge, UK 1954–1956)
256. J. Needham, W. Ling, K.G. Robinson: *Science and Civilization in Ancient China* vol 4, part 1 (Cambridge University Press, Cambridge, UK 1962)
257. W.A. Newcomb: Annals of Phys. **3**, 347 (1958)
258. K.P. Nicolas, O. Kjeldseth-Moe, J.D. Bartoe, G.E. Brueckner: Solar Phys. **81**, 253 (1982)
259. K. Niu: American Institute of Physics Conference Proceedings **318**, 421 (1994)
260. L.J. November, S. Koutchmy: ApJ **466**, 512 (1996)
261. L.J. November, J. Toomre, K.B. Gebbie et al: ApJ **245**, L123 (1981)

262. H.M. Oluseyi, A.B.C. Walker II, D.I. Santiago, R.B. Hoover, T.W. Barbee Jr: ApJ **527**, 992 (1999)
263. V.A. Osherovich, H.A. Garcia: Geophys. Res. Letters **17**, 2273 (1990)
264. L. Oster: Solar Phys. **3**, 543 (1968)
265. G.E. Owen: *Electromagnetic Theory* (Allyn and Bacon, Boston 1963)

266. F. Packer: ApJ **381**, 306 (1991)
267. E.N. Parker: *Comical Magnetic Fields* (Clarendon, Oxford 1979)
268. E.N. Parker: Solar Phys. **176**, 219 (1997)
269. P. Penfield Jr, H. Haus: *Electrodynamics of Moving Media* (Research Monograph No 40, MIT, Cambridge (MA) 1967)
270. A.L. Peratt: IRE Trans. on Plasma Science **PS-14**, 639, 763 (1986)
271. A.L. Peratt: *Physics of the Plasma Universe* (Springer, New York 1991)
272. V.J. Pizzo: ApJ **365**, 764 (1990)
273. Pliny the Elder: *Naturalis Historia – the Natural History of Pliny* (H.G. Bohn, London, UK 1855–1857)

274. L.J. Porter, J.A. Klimchuk: ApJ **454**, 499 (1995)
275. K. Prelec (ed): Proceedings of the Symposium on *Production and Neutralization of Negative Hydrogen Ions and Beams*, Brookhaven National Laboratory Report 50727 UC-34a (1977)
276. E.R. Priest. In: *Solar Flare MHD*, vol 1 (Gordon and Breach, New York 1981)
277. E.R. Priest: *Solar Magneto-Hydrodynamics* (Geophysics and Astrophysics Monographs 21, Kluwer, Dordrecht 1982)
278. E.R. Priest. In: *Second SoHO Workshop on Mass Supply and Flows in the Solar Corona*, ed by B. Fleck, G. Noci, G. Poletto (Kluwer Academic, Newell (MA) 1994) p 224
279. E.R. Priest, A.W. Hood (eds): *Advances in Solar System Magnetohydrodynamics* (Cambridge University Press, Cambridge, UK 1991)

280. F. Reale: Memorie della Società Astronomica Italiana **69**, 669 (1998)
281. F. Reale, G. Peres, S. Serio, E.E. DeLuca, L. Golub: ApJ **535**, 412 (2000)
282. D. Redžić: Eur. Jour. Phys. **22**, L1 (2001)
283. M. Reises: *Theory and Design of Charged-Particle Beams* (Wiley, New York 1994)
284. T. Rikitake: *Electromagnetism and the Earth's Interior* (Elsevier, Amsterdam 1966)
285. E. Robbrecht, E. Verwichte, D. Berghmans, J.F. Hochedez, S. Poedts: A&A **370**, 591 (2001)
286. B. Rompolt: Solar Phys. **141**, 1 (1993)
287. W.G.V. Rosser: *An Introduction to the Theory of Relativity* (Butterworth, London 1964)
288. V. Rušin, P. Heinzel, J.C. Vial (eds): IAU Colloquium 144, *Solar Coronal Structures* (VEDA Publishing Company, Bratislava, Slovakia 1994)
289. C.T. Russell, E.R. Priest, L.C. Lee: *Physics of Magnetic Flux Ropes* (Geophysical Monograph 58, American Geophysical Union, Washington (DC) 1990)
290. D.E. Rutherford: *Fluid Dynamics* (Oliver and Boyd, Edinburgh 1959)
291. A. Ruzmaikin: A&A **319**, L13 (1997)

292. N. Salingaros: Controlled Fusion **34**, 191 (1992)
293. N. Salingaros: Europhys. Letters **24**, 467 (1993)
294. B.J. Sams III, L. Golub, N.O. Weiss: ApJ **399**, 313 (1992)
295. G. Sarton: *Introduction to the History of Science* (William and Wilkins, Baltimore 1927–1948)
296. K.H. Schatten, H.G. Mayr, H. Omidvar: ApJ **311**, 60 (1986)
297. B. Schmieder, P. Heinzel, J.E. Wiik, J. Lemen, B. Anwar, P. Kotrc, E. Hiei: Solar Phys. **156**, 337 (1995)
298. C.J. Schrijver: Solar Phys. **198**, 325 (2001)
299. G. Schubert, D. Bercovici, G.A. Glatzmaier: Jour. Geophys. Res. **95**, 14, 105, 129 (1990)
300. P.A. Secchi: *Le soleil* (Gauthier-Villars, Paris 1877)
301. J.A. Shercliff: *A Textbook on Magnetohydrodynamics* (Pergamon, Oxford 1965)
302. I.P. Shkarofsky: Can. Jour. Phys. **39**, 1619 (1960)

303. G.W. Simon, A.M. Title, K.P. Topka et al: ApJ **327**, 964 (1988)
304. J. Slepian: Amer. Jour. Phys. **19**, 87 (1951)
305. S.K. Solanki: Empirical models of flux tubes. In: *Solar Photosphere: Structure, Convection, and Magnetic Fields*, ed by Stenflo (IAU 1990)
306. S.K. Solanki, H.U. Schmidt: A&A **267**, 287 (1993)
307. S.K. Solanki, I. Rüedi, W. Livingston: A&A **263**, 339 (1992)
308. A.A. Solov'ev: Astronomy Reports **19**, 1101 (1993)
309. A.A. Solov'ev: Astronomy Reports **41**, 121 (1997)
310. B.V. Somov: *Physical Processes in Solar Flares* (Kluwer Academic, Newell (MA) 1992)
311. L. Spitzer Jr: *The Physics of Fully Ionized Gases*, 1st edn (Interscience, New York 1956)
312. F.D. Stacey: *Physics of the Earth*, 3rd edn (Brookfield Press, Brisbane, Australia 1992)
313. D.C.H. Stanchfield, J.H. Thomas, B.W. Lites: ApJ **477**, 485 (1997)
314. O. Steiner, U. Grossmann-Doerth, M. Knölker, M. Schlüsser: ApJ **495**, 468 (1998)
315. J.O. Stenflo: Astron. Astrophys. Rev. **1**, 3 (1989)
316. K.T. Strong: Space Science Rev. **70**, 133 (1994)
317. P. Sturrock (ed): The solar atmosphere. In: *Physics of the Sun*, vol II (D. Reidel, Dordrecht 1986)
318. Y. Suematsu, H. Wang, H. Zirin: ApJ (1994)
319. Y. Suematsu: *Proceedings of an International Meeting Held in Guadeloupe, DOM, France* (1998)
320. M.T. Sun, D.Y. Chou, C.H. Lin: Solar Phys. **176**, 59 (1997)
321. P. Sutterlin, E.H. Schroter, K. Muglach: Solar Phys. **164**, 311 (1996)
322. G.W. Sutton, A. Sherman: *Engineering Magnetohydrodynamics* (McGraw–Hill, New York 1965)
323. D.R. Sweetman: Culham Laboratory Report No R112 (Abingdon, UK 1971)
324. E. Tandberg-Hanssen: *Solar Prominences* (Geophysics and Astrophysics Monographs, Reidel, Dordrecht 1974)
325. T. Tanimoto: Jour. Geophys. Res. **94**, **B3**, 3030 (1989)
326. E.F. Taylor, J.A. Wheeler: *Spacetime Physics: Introduction to Special Relativity* (W.H. Freeman, New York 1992)
327. J.J. Thomson: *Notes on Recent Researches on Electricity and Magnetism* (Clarendon, Oxford 1893)
328. A.M. Title, Z.A. Frank, R.A. Shine, T.D. Tarbell, K.P. Topka: ApJ **403**, 780 (1993)
329. D.J. Tritton: *Physical Fluid Dynamics*, 2nd edn (Clarendon Press, Oxford 1988)
330. S. Tsuneta: Publ. Astron. Soc. Japan **44**, L147 (1992)
331. P. Ulmschneider, E.R. Priest, R. Rosner: *Mechanisms of Chromo-spheric and Coronal Heating* (Springer, Berlin 1991)
332. J. van Bladel: *Relativity and Engineering* (Springer, Berlin 1984)
333. J. van Bladel: *Electromagnetic Fields* (Hemisphere, Washington 1985)

334. J.E. Vernazza, E.H. Avrett, R. Loeser: ApJ, Supplement Series **30**, 1 (1976)
335. J.E. Vernazza, E.H. Avrett, R. Loeser: ApJ, Supplement Series **45**, 635 (1981)
336. P. Veselovsky, L. Triskova, S. Koutchmy: Solar Phys. **153**, 237 (1994)

337. A. Walstad: Amer. Jour. Phys. **65**, 462 (1997)
338. H. Wang, H. Zirin: Solar Phys. **140**, 41 (1992)
339. J.A. Watko, J.A. Klimchuk: Solar Phys. **193**, 77 (2000)
340. E.S. Weibel: Phys. Rev. Letters **2**, 83 (1959)
341. N.O. Weiss. In: *Basic Plasma Processes in the Sun*, ed by E.R. Priest, V. Krishan (1990)
342. E.J. Williams: Proc. Roy. Soc. **A169**, 542 (1939)
343. R.F. Wilson, K.R. Lang, D.E. Gary: ApJ **418**, 490 (1993)
344. A.D. Wittmann: Solar Phys. **36**, 29 (1974)
345. B.E. Wood, M. Karovska, J.W. Cook, G.E. Brueckner, R.A. Howard, C.M. Korendyke, D.G. Socker: ApJ **505**, 432 (1998)

346. V.V. Zaitsev, A.V. Stepanov, S. Urpo, S. Pohjolainen: Astronomicheskii Zhurnal **75**, 455 (1998)
347. V.V. Zaitsev, A.V. Stepanov, S. Urpo, S. Pohjolainen: A&A **337**, 887 (1998)
348. P. Zeeman: Philosophical Magazine, 5th series **43**, 226 (March 1897); and ApJ **5**, 332 (May 1897)
349. H. Zirin: IAU Symposium 56, *Chromospheric Fine Structure*, ed by R.G. Athay (Reidel, Dordrecht 1974) p 161
350. H. Zirin: *Astrophysics of the Sun* (Cambridge University Press, Cambridge, UK 1988)
351. H. Zirin, R. Cameron: *Proceedings of an International Meeting Held in Guadeloupe, DOM, France* (1998)
352. H. Zirin, H. Wang: Solar Phys. **119**, 245 (1989)
353. H. Zirin, H. Wang: Space Rev. **11**, 147, 225 (1991)
354. J.B. Zirker, S. Koutchmy: Solar Phys. **131**, 107 (1991)

Index

The symbol ∼ in an entry replaces the entry's headword or head-expression

ASTRONOMY AND ASTROPHYSICS LIBRARY

A&A LIBRARY

The Early Universe Facts and Fiction
4th Edition By G. Börner

The Design and Construction of Large Optical Telescopes By P. Y. Bely

The Solar System 4th Edition
By T. Encrenaz, J.-P. Bibring, M. Blanc, M. A. Barucci, F. Roques, Ph. Zarka

General Relativity, Astrophysics, and Cosmology By A. K. Raychaudhuri, S. Banerji, and A. Banerjee

Stellar Interiors Physical Principles, Structure, and Evolution 2nd Edition
By C. J. Hansen, S. D. Kawaler, and V. Trimble

Asymptotic Giant Branch Stars
By H. J. Habing and H. Olofsson

The Interstellar Medium
By J. Lequeux

Methods of Celestial Mechanics (2 volumes)
Volume I: Physical, Mathematical, and Numerical Principles
Volume II: Application to Planetary System, Geodynamics and Satellite Geodesy
By G. Beutler

Solar-Type Activity in Main-Sequence Stars
By R. E. Gershberg

Relativistic Astrophysics and Cosmology
A Primer By P. Hoyng

Physical Constants

Elementary charge	$e = 1.602176462 \times 10^{-19}$ coulomb
Electron rest mass	$m = 9.10938188 \times 10^{-31}$ kilogram
Proton rest mass	$m_p = 1.67262158 \times 10^{-27}$ kilogram
Speed of light in vacuum	$c = 2.99792458 \times 10^{8}$ meters/second
Permittivity of vacuum	$\epsilon_0 = 8.854187817 \times 10^{-12}$ farad/meter
Permeability of vacuum	$\mu_0 \equiv 4\pi \times 10^{-7}$ henry/meter
Avogadro constant	$N_A = 6.02214199 \times 10^{23}$/mole
Boltzmann constant	$k = 1.3806503 \times 10^{-23}$ joule/kelvin
Planck constant	$h = 6.62606876 \times 10^{-34}$ joule second
Gravitational constant	$G = 6.673 \times 10^{-11}$ newton meter^2/kilogram^2

The Sun

Mass of the Sun	1.98596×10^{30} kilograms
Radius of the Sun	6.965×10^{8} meters

The Sun's photosphere

Acceleration of gravity	273 meters/second^2
Temperature	6420 kelvins
Free electron number density	6.44×10^{19}/meter^3
Proton number density	4.85×10^{19}/meter^3
H atom number density	1.16×10^{23}/meter^3
Mass density	2.72×10^{-4} kilogram/meter^3
Pressure	1.17×10^{4} pascals

The Earth

Mass	5.973×10^{24} kilograms
Equatorial radius	6.378139×10^{6} meters
Polar radius	6.35675×10^{6} meters
Inner core radius	1.3×10^{6} meters
Outer core radius	3.5×10^{6} meters
Outer core conductivity	3×10^{5} siemens/meter
Mantle radius	5.75×10^{6} meters
Mantle conductivity	≈ 0.1 siemens/meter
Equatorial gravity	9.780317 meters/second^2
Mean geothermal flux	$(6.15 \pm 0.34) \times 10^{-2}$ watt/meter^2
Mean orbital speed	2.977×10^{4} meters/second
Mean Sun–Earth distance	1.495×10^{11} meters
Solar constant	1.360 kilowatts/meter^2

References: Cohen and Taylor (2001), Stacey (1992), Vernazza et al. (1981)